Water Governance - Concepts, Methods, and Practice

Series Editors

Claudia Pahl-Wostl, University of Osnabrück, Osnabrück, Germany

Joyeeta Gupta, Faculty of Earth and Life Sciences, University of Amsterdam, Amsterdam, The Netherlands

This book series aims at providing a platform for developing an integrated perspective on major advances in the field of water governance. Contributions will build bridges across established fields of expertise, across disciplinary perspectives, across levels from global to local and across geographical regions. Topics to be covered include conceptual and methodological advances capturing the complexity of water governance systems, institutional settings, actor constellations, diagnostic approaches, and comparative studies of governance systems. The book series encompasses monographs, textbooks and edited, coordinated volumes addressing one theme from different perspectives. The book series will address mainly a wider scientific audience, but will also provide valuable knowledge to interested practitioners. The sustainable management of fresh water resources and the ecosystem services that they provide is one of the key challenges of the 21st century. Many water related problems can be attributed to governance failure at multiple levels of governance rather than to the resource base itself. At the same time our knowledge on water governance systems and conditions for success of water governance reform is still quite limited. The notion of water governance aims at capturing the complexity of processes that determine the delivery of water related services for societal needs and that provide the context within which water management operates. Water governance is a fast growing field of scholarly expertise which has largely developed over the past decade. The number of publications in peer reviewed journals has increased from less than 20 in the year 2000 to nearly 400 in the year 2013. The increasing popularity of the term in science and policy has not lead to conceptual convergence but rather to an increasing vagueness and competing interpretations of how the concept should be understood, studied, and analyzed; which disciplines are involved; and what methodological approaches are most suitable for the study and analysis of water governance. The time seems to be ripe for comprehensive synthesis and integration.

Francine van den Brandeler

Scalar Mismatches in Metropolitan Water Governance

A Comparative Study of São Paulo and Mexico City

Springer

Francine van den Brandeler
Swiss Federal Institute of Aquatic Science
and Technology
Dübendorf, Switzerland

ISSN 2365-4961 ISSN 2365-497X (electronic)
Water Governance - Concepts, Methods, and Practice
ISBN 978-3-031-08060-9 ISBN 978-3-031-08061-6 (eBook)
https://doi.org/10.1007/978-3-031-08061-6

This Springer imprint is published by the registered company Springer Nature Switzerland AG
The registered company address is: Gewerbestrasse 11, 6330 Cham, Switzerland

Acknowledgements

Concluding this journey and book would not have been possible without the support and guidance that I received along the way. There are no proper words to convey my deep gratitude and respect for my promoter, Prof. Joyeeta Gupta, and co-supervisor, Dr. Michaela Hordijk. Joyeeta, your dedication, great knowledge and hard work have been an inspiration. Thank you for sharing your knowledge and pushing me through guidance and critique. Michaela, thank you for believing in me from the beginning, for encouraging me to embark on this journey and for your invaluable input along the way.

I am deeply grateful to the Centre for Latin American Research and Documentation (CEDLA) and the Latin American Studies Programme (LASP) for funding this research and fostering a welcoming and dedicated community of Latin America scholars.

I am thankful for the institutional support extended by the University of Amsterdam, the Netherlands. I would like to thank the members of the Governance and Inclusive Development (GID) Group for their support, I am especially thankful for my Ph.D. colleagues.

Thank you to Jillian for your kindness and moral support and for joining me in therapeutic writing sessions, in Amsterdam and virtually.

My gratitude also goes to all the kind and generous individuals in Mexico and Brazil who accepted to participate in this research. I learned a lot from their unique experiences, extensive knowledge and valuable insights into the topics of this thesis. They also helped me see certain issues from a different standpoint and reconsider my own biases. My friends in Mexico City and São Paulo helped me disconnect and enjoy my fieldwork experience.

I was able to spend months and months living with my aunt Francien and uncle Christian, and I am eternally grateful for how they always welcomed me in their home whenever I needed it. My fieldwork in Brazil was also made easier and more enjoyable thanks to my wonderful parents-in-law, who gave me a home in São Paulo, treated me like their own daughter and taught me so much about Brazil's culture and way of life.

I am grateful to my parents for their love, trust and for always having supported my education and the pursuit of my dreams, no matter how impractical or where they would take me. Without you, none of this would have been possible. Thank you to my Willem, Kiren, Pietro, Francesca and to my friends who are family for your support these past few years and for needed distractions. Lastly, I have no words to express my gratitude to Filippo, my husband. Without him, I would not have had the courage to start this journey nor the confidence that I could finish it. I had your unwavering support throughout, even though it meant spending many months apart. Thank you for reading and re-reading this manuscript, for sharing your thoughts on it and making sure I used a reference manager, for holding me up when I was down and for celebrating every small (and big!) milestone along the way. I could not be luckier to share my life with you.

Contents

Abbreviations

ANA	Agência Nacional de Água (National Water Agency)
APRM	Áreas de Proteção e Recuperação dos Mananciais (Areas of Protection and Recovery of Springs)
ARSESP	Agência Reguladora de Saneamento e Energia do Estado de São Paulo (Regulating Agency for Sanitation and Energy)
ATB	Alto-Tietê Basin
BESS	Biodiversity and ecosystem services
CBH-AT	Comitê de Bacia Hidrografica do Alto-Tietê (Alto-Tietê River Basin Committee)
CETESB	Companhia Ambiental do Estado de São Paulo (São Paulo State Environment Agency)
CHR	Conselho Estadual de Recursos Hídricos (State Council on Hydrological Resources)
CNHR	Conselho Nacional de Recursos Hídricos (National Council on Hydrological Resources)
CODEGRAN	Conselho Deliberativo da Grande São Paulo (Deliberative Council of Greater São Paulo)
COFEHIDRO	Conselho do Fundo Estadual de Recursos Hidricos (Council for the State Fund for Water Resources)
CONAFOR	Comisíon Nacional Florestal (National Forestry Commission)
CONAGUA	Comisíon Nacional De Água (National Water Commission)
CORENA	Mexico City's Natural Resources Commission
CSO	Civil society organization
DAEE	Departamento de Águas e Energia Elétrica (Department of Water and Electricity)
EMAE	Empresa Metropolitana de Águas e Energia (Metropolitan Company of Water and Electricity)
EMPLASA	Empresa Paulista de Planejamento Metropolitano S.A. (São Paulo Company of Metropolitan Planning)

FABHAT	Fundação da Agência da Bacia Hidrografica do Alto-Tietê (Foundation for the Agency of the Alto-Tietê Basin Committee)
FEHIDRO	Fundo Estadual de Recursos Hídricos (State Fund for Water Resources)
FIESP	Federação das Indústrias do Estado de São Paulo (Federation of Industries of the State of São Paulo)
IBGE	Instituto Brasileiro de Geografia e Estatística (Brazilian Institute of Geography and Statistics)
IDB	Inter-American Development Bank
INECC	Instituto Nacional de Ecología y Cambio Climático (National Institute for Ecology and Climate Change)
IPT	Instituto de Pesquisa Tecnológica (Institute for Technological Research)
ITB	Instituto Trata Brasil
IWRM	Integrated Water Resources Management
MASL	Metres above sea level
MMP-ATB	Macro-drainage Master Plan of the Alto-Tietê Basin
MRSP	Metropolitan Region of São Paulo
MVMC	Metropolitan Valley of Mexico City
PES	Payment for ecosystem services
PLANASA	Plano Nacional de Saneamento (National Basic Sanitation Plan)
SAAE	Serviço Autonomo de Agua e Esgoto (Autonomous Water and Sanitation Service)
SABESP	Companhia de Saneamento Básico do Estado de São Paulo (Company of Basic Sanitation of São Paulo State)
SACMEX	Sistema de Aguas de la Ciudad de Mexico (Water Systems of Mexico City)
SDG	Sustainable Development Goal
SEMARNAT	Ministry for the Environment and Natural Resources
SIGRH	Sistema de Informações sobre o Gerenciamento de Recursos Hídricos (System of Information on the Management of Hydrological Resources)
SIMA	Secretaria de Infraestrutura e Meio Ambiente do Estado de São Paulo (São Paulo State Environmental Secretariat)
SINGREH	Sistema Nacional De Gerenciamento De Recursos Hídricos (National Hydrological Resources Management System)
SNIS	Sistema Nacional de Informações sobre Saneamento (National Sanitation Information System)
SNSA	Secretaria Nacional de Saneamento Ambiental (National Secretariat for Environmental Sanitation)
SSRH	Secretaria de Saneamento e Recursos Hídricos do Estado de São Paulo (São Paulo State Sanitation and Water Resources Secretariat)
UGRHI	Unidades de Gerenciamento de Recursos Hídricos (Units of Management of Hydrological Resources)
USP	Universidade de São Paulo (University of São Paulo)

VMB	Valley of Mexico Basin
Wat&San	Water and sanitation
WRM	Water resources management

List of Figures

List of Tables

List of Boxes

List of Maps

List of Maps

Chapter 1
Introduction

1.1 Introduction

Large cities have heavy impacts on their rural hinterlands, while also depending on these and the natural resources they provide. Between 2013 and 2015, a major drought brought the metropolitan region of São Paulo to the edge. Reservoirs outside the city were running so low that bars and restaurants all over the city shut doors or started using plastic cups because they had no water to rinse dishes. Household taps ran dry, especially in the outskirts, and protests ensued. Rains eventually brought water to parched reservoirs but, by late 2021, much of the South and Southeast of Brazil—including São Paulo—faced the worst drought since these extreme weather events began to be recorded in 1910. In 2018, the city of Cape Town made international news headlines for months as it was battling a severe water shortage and heading closer to 'Day Zero', the day that municipal water supplies would be shut off (Alexander 2019). The reallocation of water earmarked for agriculture to urban residents helped mitigate the looming disaster. In 2003, a dam that supplies water to Mexico City through an inter-basin water transfer flooded 300 ha of fields cultivated by the Mazahua indigenous community (Marcos and Fernández 2016). The Federal government did not respond adequately to their claims, leading to peaceful but long and highly mediatized protests by Mazahua women for compensation and access to drinking water.[1]

Such examples illustrate the rising challenges in terms of water quantity, quality and climate change adaptation for cities and their rural hinterlands. These tensions are triggered by a combination of population growth, urbanization, economic growth, consumption patterns, anthropogenic climate change, land use and other driving forces at multiple levels (Vörösmarty et al. 2000; Elmqvist et al. 2013; Nobre and Marengo 2016; UN-HABITAT 2016). Water challenges are particularly severe in

[1] Despite living near the large dam, local Mazahua communities were not connected to the public water supply network.

F. van den Brandeler, *Scalar Mismatches in Metropolitan Water Governance*,
Water Governance - Concepts, Methods, and Practice,
https://doi.org/10.1007/978-3-031-08061-6_1

megacities of the Global South, marked by stark inequalities within the urban agglomeration and between the city and its rural hinterlands, and where the urbanization process is unfolding at an accelerated pace (Elmqvist et al. 2013; Azzam et al. 2014). Although the world's 100 largest cities occupy less than 1% of the planet's land area, the basins that provide their water resources cover more than 12% of it (ARUP 2018). Estimates indicate that cities with populations larger than 750,000 people draw water from almost half of the global land surface and transport it over a cumulative distance of 27,000 km (McDonald et al. 2014).

This book examines the tensions between cities and their river basins through interactions between metropolitan governance regimes with integrated basin management regimes. More specifically, it explores the role that institutions play in urban water challenges, how effective existing policy instruments are in addressing these challenges within metropolitan regions and how more sustainable and inclusive institutions could be designed for this purpose. It does so by focusing on the cases of São Paulo in Brazil and Mexico City, in Mexico—two major megacities facing a wide range of water-related challenges (see Sect. 2.2.2).

This introductory chapter presents the growing worldwide tensions between water use at urban and river basin scales, their theoretical underpinnings, the gap in scholarly knowledge, and their policy implications (see Sect. 1.2). It then introduces the ensuing research questions that this book aims to answer, as well as its focus and limits (see Sect. 1.3), provides a background on the nature of water (see Sect. 1.4), discusses the position of the researcher (see Sect. 1.5) and, finally, the overall structure of the book (see Sect. 1.6).

1.2 Water Challenges and Metropolitan Regions

1.2.1 Rising Water Challenges in an Urbanizing World

Today, around 56.2% of the global population is urban (UN-DESA 2018).[2] As this number rises in the coming decades, the burden cities impose on their river basins is likely to intensify, even if in an uneven manner (see Map 1.1). The Global South will lead this urban growth. Asia and Africa will add 2.5 billion urban residents by 2050 (see Fig. 1.1) (UN-DESA 2018). Latin America, already one of the world's most urbanized regions, should see urban populations increase from 81% in 2015 to 88% by 2050 (see Fig. 1.2) (UN-DESA 2018). By comparison, 82% of Northern America's population lived in urban areas in 2018 (UN-DESA 2018).

As the global urban population rises, so does the number of cities of various sizes. More and more people are living in metropolitan regions, characterized by a contiguous urban area often governed by multiple political jurisdictions (see Box 1.1). In 1950, 177 cities had more than 500,000 inhabitants. This increased to 1067 by

[2] Recent research suggests that urbanization levels are much higher, due to the fact that countries self-report their demographic statistics and use very different standards (Scruggs 2018).

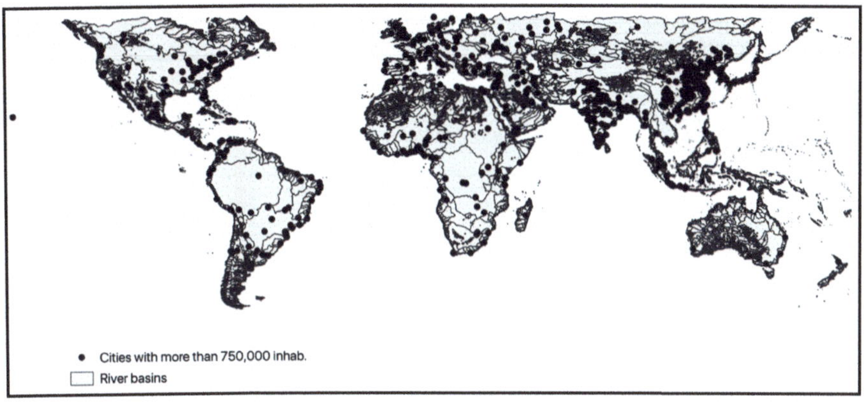

Map 1.1 Large cities and basins around the world in 2015. *Source* Author

Fig. 1.1 Global expansion of urban population, 1950–2050 (thousands)

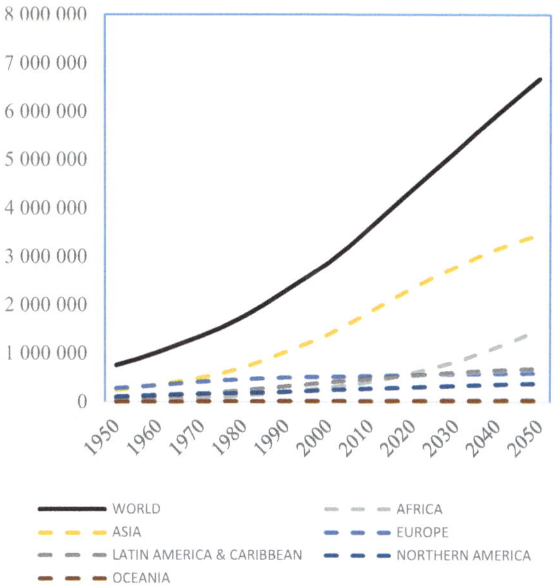

2015 and it is expected to further rise to 1416 by 2030 (see Table 1.1) (UN-DESA 2018). Megacities, the focus of this book, are defined as cities of 10 million inhabitants or more; they have increased from 2 in 1950 to 29 in 2015 and possibly 43 in 2030 (UN-DESA 2018). Urban settlements with less than 300,000 inhabitants will remain the largest in number and in total population. Nevertheless, the population of larger cities is increasingly predominant both in relative and in absolute terms. In 1950, less than 10% of the global population lived in cities larger than 500,000 inhabitants; by 2015 this population had increased to more than 27%, and it is projected to reach more than 33% by 2030 (see Fig. 1.3) (UN-DESA 2018). This recent, yet

Fig. 1.2 Urban population per region (percentage)

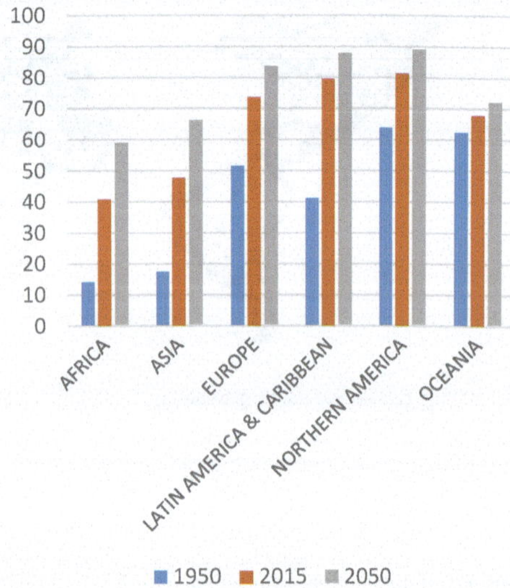

Table 1.1 Evolution of the number of urban settlements per population size

Population size	Number of urban settlements		
	1950	2015	2030
10 million or more	2	29	43
5–10 million	5	45	66
1–5 million	69	439	597
500,000–1 million	101	554	710
300,000–500,000	129	707	827

Source Based on raw data from UN-DESA (2018)

Fig. 1.3 Evolution of population per size classes of urban settlements

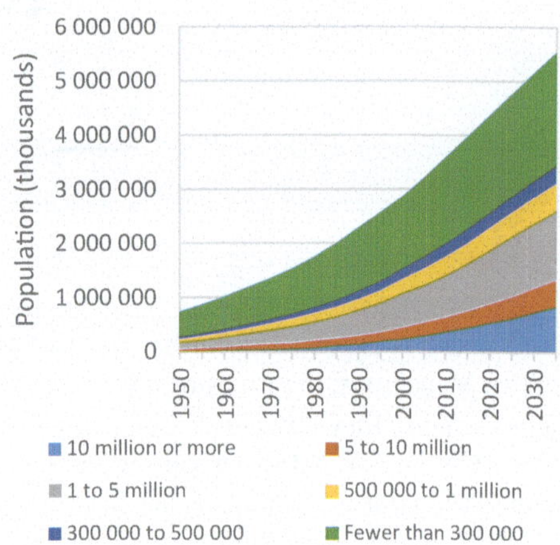

accelerated, worldwide transformation of the natural environment through urbanization is a characteristic of the 'Anthropocene'.[3]

While agriculture represents 70% of global water use, large urban agglomerations can have disproportionate impacts on their river basins. As cities around the world grow, so does their demand for goods and services, including resources such as water, food and energy, which come largely from surrounding areas (Jenerette and Larsen 2006). Besides population growth, the economic development that often follows urbanization further increases per capita water use in cities (McDonald et al. 2014). Urban water demand is expected to increase by 80% by 2050 while total available freshwater remains more or less constant (Flörke et al. 2018). This demand is unevenly distributed across the world's river basins. Between 1.6 and 2.4 billion people live in river basins that experience water scarcity (Gosling and Arnell 2016). In quantitative terms, 'chronic water shortages' take place when an area's annual water supply drops below 1000 m^3 per person, and 'absolute water scarcity' takes place below 500 m^3 per person (FAO 2012).[4] Managing water resources across large cities and their river basins has led to increasing competition, tensions and conflicts (Varis et al. 2006; Tortajada 2008). Despite the far-reaching impacts of these cities, and their potential to influence basin management, cities invest very little in their basins (ARUP 2018).

Box 1.1 Definitions of Large Cities

There are multiple terms to refer to cities but no internationally-recognized definitions with standardized criteria for determining the boundaries of any given urban area (Slack 2007; Knieling 2014; United Nations 2016b). 'City proper' is used to define a city according to an administrative boundary (United Nations 2016b). Terms associated with large, multi-jurisdictional urban areas include metropolis, metropolitan area, metropolitan region, megacity, urban agglomeration, and more. Definitions generally refer to a large urban core with adjacent urban and rural areas that are socially and economically integrated with the core (Slack 2007). The key terms are:

Urban Agglomeration: This definition is based on physical characteristics as it considers the extent of the contiguous urban area, or built-up area, as the limits of the city's boundaries (United Nations 2016b).

[3] The term 'anthropocene' is a geologic term for an epoch that starts when human activities began to have a significant global impact on the Earth's ecosystem (Crutzen and Stoermer 2000).

[4] Population growth and climate variability and change may lead to as many as 3.1 billion people (37% of the global population) living in water scarce river basins by 2050 (Gosling and Arnell 2016).

Functional Urban Areas: Urban area defined by a method that relies on settlement patterns and commuting flows rather than administrative borders (OECD 2012).

Metropolitan Regions: The term 'metropolitan region' is used by international institutions (OECD, World Bank, etc.) and European authors (Herrschel and Newman 2002; Salet et al. 2003; Sellers et al. 2013) to describe highly urbanized, city-regional areas characterized by high population densities and the concentration and interconnectedness of economic, political and cultural activities (Knieling 2014; United Nations 2016b). These cities are typically composed of multiple jurisdictions with independent political authorities. Minimum population thresholds for the city core are not necessarily very high (i.e. 50,000 or 100,000 in some cases), but adjacent areas of lower density are connected to the core and under its influence (United Nations 2012; Knieling 2014).

Megacities: The term 'megacity' has been defined by the United Nations as an urban agglomeration of at least 10 million inhabitants (United Nations 2012).

Metacities: UN-Habitat introduced the term 'metacity' to describe "massive conurbations of more than 20 million people" (UN-Habitat (United Nations Human Settlements Programme) 2006).

Megalopolis: This term refers to a clustered network of cities. There is no consensus on population size, with definitions ranging between 10 million (Doxiadis 1970) and 25 million (Gottmann and Harper 1990).

The definition of a city's boundaries has implications for population assessments (United Nations 2016b). Although the two case studies in this study—São Paulo and Mexico City are 'megacities', this study favours the terms metropolitan region or area as these are the terms used by the relevant authorities of each jurisdiction.

Rapid urbanization and land use changes have also caused water quality deterioration through drastic interferences in ecosystems and the hydrological cycle (Azzam et al. 2014). Deteriorating water quality poses significant risks to human and environmental health (OECD 2015b). Estimates indicate that around one third of all rivers in Latin America, Africa and Asia are affected by severe pathogen pollution, although it is not clear how many people are at risk of coming into contact with polluted waters as current estimates only account for rural populations (UNEP 2018). Water quality in urban rivers is often heavily impacted by point source pollution, such as untreated wastewater discharge, and this is worsened by high population density and the concentration of polluting activities (Vlachos and Braga 2001; Elmqvist et al. 2013). Diffuse pollution from agriculture (e.g. fertilizers and pesticides) and urban sources (e.g. runoff from sealed surfaces and roads) also affects urban areas and is particularly challenging to regulate (Martinez-Santos et al. 2014). Water contamination by large and megacities aggravates issues such as regional water stress and unequal access to water resources (Varis et al. 2006). In addition, treating water to meet adequate drinking water standards can represent a considerable cost for some countries (OECD 2015b). However, inaction is also costly, as contaminated water bodies can lead to outbreaks of waterborne diseases and negatively impact both urban residents and communities and the environment far downstream (Vlachos and Braga 2001; OECD 2015b).

Extreme weather events can cause floods, landslides and droughts with devastating effects on urban and rural settlements. Many large urban agglomerations are located in the Global South and have limited coping capacities (Kraas et al. 2014). As cities grow, they tend to expand into risk-prone areas as available land becomes scarcer and more expensive (UCLG 2016). In 2000, about 30% of global urban land was in high-frequency flood zones. By 2030, this will rise to 40% (Güneralp et al. 2015). These hazards can be part of seasonal variations (e.g. monsoons) and climate variability, but climate change is expected to aggravate their frequency and intensity by causing changes in hydrological patterns, with more evaporation and melting through warming, and more frequent and intensive extreme weather events (Engel et al. 2011). Large cities are particularly vulnerable to climate change, as they are often located in coastal areas, flood-prone areas or areas suffering from water scarcity and droughts (Biswas 2004; Varis et al. 2006; Hansjürgens and Heinrichs 2014). In addition, water-related risks are compounded by human factors such as population density, socio-economic inequality, poor urban planning and the environmental impact of land use changes (e.g. erosion from deforestation, rapid urbanization) (Rietveld et al. 2016).

1.2.2 The Policy Challenge: Implementing IWRM Is Key

There have been many discussions within global policy circles on water-related challenges since the UN Conference on the Human Environment in 1997, including special attention paid to Agenda 21 adopted in 1992 (see Conti 2017; Obani 2018 for details). In 2015, within the context of Agenda 2030, the UN General Assembly adopted water-related goals within its Sustainable Development Goals (UNGA 2015). These Goals highlight areas of priority for the global community to work on. Goals 6 (Ensure availability and sustainable management of water and sanitation for all), 11 (Make cities and human settlements inclusive, safe, resilient and sustainable) and 13 (Take urgent action to combat climate change and its impacts), and their associated targets and indicators, are relevant for this research. These goals are linked to water quantity, water quality and climate change adaptation in multiple ways (see Table 1.2).

SDG target 6.5 promotes the implementation of IWRM (Integrated Water Resources Management) at all levels, implicitly recognizing it as the most appropriate management approach to the world's diverse water-related challenges and as necessary to attain all other SDG 6 targets (UNEP 2018). The suggestion that this implementation should take place "at all levels" highlights the multi-scalar nature of these challenges. IWRM is deemed critical for the 2030 SDG agenda as a way of allocating water resources efficiently, equitably and sustainably and coordinating sustainable development in the global context of increasing water scarcity and pollution. Progress on SDG 6.5 is measured by two indicators: a score of 0–100 on the degree of IWRM implementation and the proportion of transboundary basins with cooperation agreements.[5] Nonetheless a 2018 self-assessment survey answered by 172 countries as part of a UN Progress Report on SDG 6, indicated that around 60% were unlikely to implement IWRM by 2030 (UNEP 2018). A 2021 UNEP report on progress on SDG 6 indicators showed that some countries made progress but that overall the world was not on track to achieve target 6.5 (UNEP 2021). Survey from the 2018 results further revealed that sub-national levels lag even further behind and emphasized the need for coordination across levels to ensure the flow of resources to where they are most needed and effective. The survey results mention links with

Table 1.2 The SDGs and targets and their links to water quantity, water quality and climate change adaptation

<table>
<tr><th>Water quantity</th><th>Water quality</th><th>Climate change adaptation</th></tr>
<tr><td colspan="2">6.1 Access to drinking water
6.2 Access to sanitation</td><td>11.5 Reduce effects of water-related disasters</td></tr>
<tr><td colspan="3">11.1 Access to housing, basic services and slum upgrading</td></tr>
<tr><td rowspan="2"></td><td rowspan="2">6.3 Reduce water contamination
11.6 Reduce cities' environmental impact, including through waste management</td><td>11.B Increase the number of cities with integrated policies and plans for inclusion, resource efficiency, climate change adaptation and disaster resilience. Develop holistic multilevel disaster risk management</td></tr>
<tr><td>13.1 Strengthen resilience and adaptive capacity to climate-related hazards and natural disasters
13.2 Integrate climate change measures into national policies, strategies and planning
13.3 Improve education, awareness-raising and human/institutional capacity on climate change adaptation, impact reduction and early warning</td></tr>
</table>

(continued)

[5] Indicator 6.5.2 on transboundary agreements concerns basins and aquifers shared by at least two countries.

Table 1.2 (continued)

Water quantity	Water quality	Climate change adaptation
6.4 Increase water use efficiency **11.B** Increase the number of cities with integrated policies and plans for inclusion, resource efficiency, climate change adaptation and disaster resilience. Develop holistic disaster risk management at all levels		
6.5 Implement IWRM at all levels **6.6** Protect and restore water-related ecosystems **11.4** Protect the world's natural heritage		
6.A Expand international cooperation and capacity-building support to developing countries in water- and sanitation-related activities and programmes		
11.C Support least developed countries, including through financial and technical assistance, in building sustainable and resilient buildings using local materials		
6.B Support and strengthen local communities' participation in improving water and sanitation management **11.3** Enhance inclusive and sustainable urbanization and participatory planning		

Source Compiled from https://sustainabledevelopment.un.org/?menu=1300

SDG 11 on 'Sustainable Cities and Communities' and SDG 13 on 'Climate Action', but they do not provide clarifications on how to develop synergies between them. There seem to be clear interlinkages between the goals: SDG target 6.5 (Implement IWRM) relates to SDG 11.5 (Reduce effects of water-related disasters) and to SDG 13.1 (Strengthen resilience and adaptive capacity to climate-related hazards and natural disasters). In fact, the first indicators for the SDGs 11.5 and 13.1 are identical: "Number of deaths, missing persons and persons affected by disaster per 100,000 people". In addition, targets 6.1 and 6.2 (access to drinking water and sanitation) are closely intertwined with target 11.1 (ensure access for all to adequate, safe and affordable housing and basic services and upgrade slums). Target 6.3 (reduce pollution, including from untreated wastewater) could be linked to target 11.6 (reduce the adverse per capita environmental impact of cities), but the latter is only focused on pollution from solid waste and air pollution. Moreover, IWRM planning has potential synergies with SDG 11.B and SDG 13.1, which both promote the adoption of local and national disaster risk reduction strategies.

This brief review demonstrates that there are clear interlinkages between these three Goals. Nonetheless, the abovementioned reports do not clarify how countries can explore these links, allowing for coordinated responses and promoting win–win strategies. This research aims to help fill this gap, contributing to the SDGs by focusing on how these inter-linkages positively and negatively impact urban water challenges and how these synergies can be harnessed (Fig. 1.4).

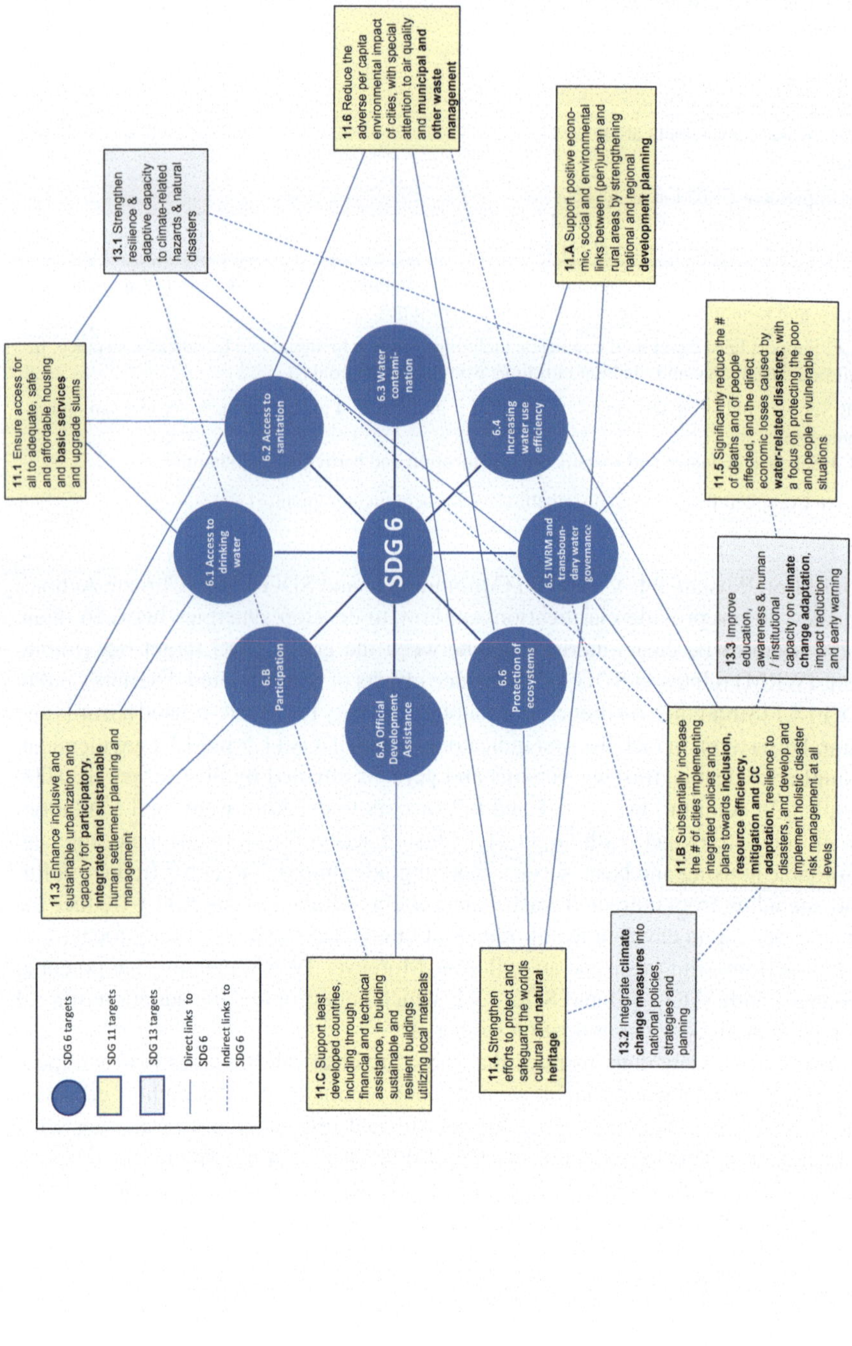

Fig. 1.4 Links between SDGs 6, 11 and 13

1.2.3 The Gap in Scholarly Knowledge: The Disconnect Between Cities and River Basins

The abovementioned social and policy challenge has led to a growing literature on River Basin Management (see Chap. 3) and Urban Water Management (see Chap. 4). While extensive, three main gaps emerge on how the existing literature addresses water-related challenges in large metropolitan regions.

First, the dominant scholarly bodies of literature that analyse water management in river basins are the Integrated Water Resources Management (IWRM) and Integrated River Basin Management (IRBM) literatures (see Chaps. 2 and 3). IWRM and IRBM adopt the river basin as the ideal unit for water management, although IWRM also has a national spatial focus (Watson 2004; Jones et al. 2006; Bahri 2012). They address water quantity concerns such as managing water resources for multiple users and avoiding conflicts, water supply through large-scale infrastructure, and the need for more water demand-focused measures and soft approaches (Abdullah and Christensen 2004; Watson 2004; Barrios et al. 2009). Concerns about water quality centre on contamination risks within the river basin and the adoption of ecosystem-oriented approaches (Renner et al. 2017). Adaptation to climate change risks is not a focus of the literature, being only laterally addressed through specific topics such as droughts (Hurlbert 2016) and irrigation practices, minimum environmental flows, water allocation, and others (Barrios et al. 2009; Brandeler et al. 2019). Both IWRM and IRBM take a holistic approach to water management, emphasizing the need to integrate upstream and downstream issues, surface and groundwater, land and water systems, humans and ecosystems, multiple sectors, and viewpoints (Jønch-Clausen and Fugl 2001; Watson 2004; Grigg 2008; Medema et al. 2008; Savenije and Van der Zaag 2008; Barrios et al. 2009; Molle 2009a; Agyenim 2011; Closas et al. 2012; Foster and Ait-Kadi 2012; Anokye 2013). However, it is not clear how water services and planning at local levels should be addressed. Although IWRM/IRBM aim to integrate urban water concerns, including water services, this is challenging in practice, as natural hydrological borders rarely coincide with political-administrative borders (Bahri 2012; OECD 2016). A review of the literature on IWRM/IRBM in titles, abstracts and key words in ScienceDirect indicated that, despite an exponential increase in the number of publications in the last 15 years, few contained terms referring to the urban scale (urban, city/ies, megacity/ies and metropolitan) (see Fig. 1.5).

Second, the (social sciences) literature on urban water has largely focused on the provision of water services and infrastructure, as well as issues of risk and adaptation (Engel et al. 2011; GTT 2014). However, adaptation and 'resilience thinking' are often limited to the field of disaster management (Baud and Hordijk 2009). The urban water literature has evolved from a technocratic and sectoral approach with a focus on public health and risk control, towards including considerations for sustainability, integration and participatory decision-making (Brown and Farrelly 2009; Rauch and Morgenroth 2013). This has led to new paradigms, such as Sustainable Urban Water Management (SUWM) and Integrated Urban Water Management (IUWM). Yet, the

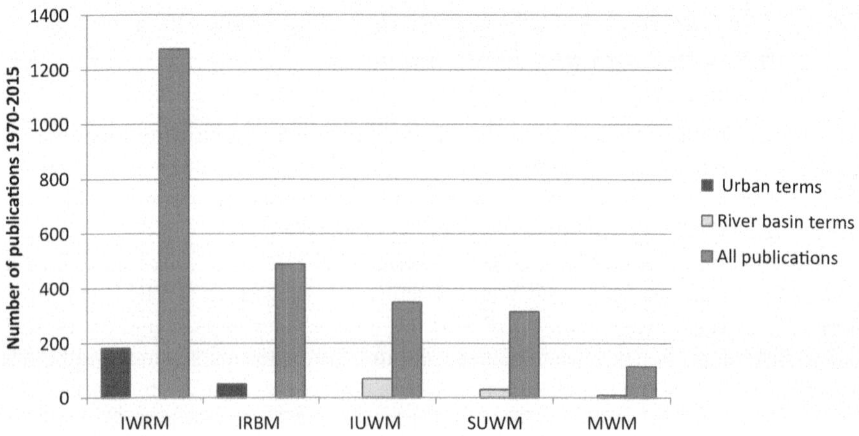

Fig. 1.5 The occurrence of terms linked to the urban and river basin within the main concepts. *Source* Brandeler et al. (2019)

literature often leaves out the broader context of the river basin, and the relationship between the city and its rural hinterlands, including that of their respective water systems (Pearson et al. 2010; Brandeler et al. 2019). Only a fraction of the literature reviewed with 'SUWM' or 'IUWM' in titles, abstracts and keywords included terms related to the river basin scale (see Fig. 1.5). The pressure on local officials to provide certain services within their jurisdiction and the relatively short time span and scope of their mandates mean that they often lack strong incentives to consider the long-term impacts on the wider basin. The SUWM literature emphasizes local level, infrastructure-oriented initiatives (Marlow et al. 2013; Rietveld et al. 2016). However, as the river basin is typically considered as the ideal unit for water resources analysis and management, addressing water problems at the local level could lead to issues of administrative/institutional mismatch between the basin and city scales, with various social, environmental, economic and relational implications.

Third, neither the RBM nor the UWM literature clarifies how their principles apply to the context of large metropolitan regions, despite the multiplication of the world's large urban agglomerations (Brandeler et al. 2019). A review of publications with the term 'metropolitan water management' revealed only 113 results between 1970 and 2015, compared with 314 for 'SUWM' and 349 for 'IUWM' (see Sect. 2.3). They made virtually no mention of the river basin. Megacities sometimes expand across large parts of river basins and even beyond their boundaries, and impacts are felt even further. They are often composed of a multitude of local governments with autonomous decision-making powers, and the coordinated management of urban water and water resources is particularly challenging in these contexts. A relevant question is thus whether institutions dealing with the metropolitanization of cities should be coordinated with those for water management (Roche et al. 2001). Despite their unique characteristics and challenges, the existing scholarship fails to address whether metropolises should receive special consideration in urban water or river basin management, or whether they require a completely different approach.

1.3 Research Questions, Approach and Limits

1.3.1 Research Questions

Hence, this book investigates river basin and urban water governance in relation to metropolitan water challenges through the following overarching question: **How do interactions between drivers and institutions at different spatial and institutional scale levels shape metropolitan water challenges, and how can policy instruments from river basin and urban water governance frameworks be (re)designed to foster more sustainable and inclusive metropolitan water governance?**

This book is grounded within the water governance literature. More specifically, in the absence of a theory of metropolitan water governance, it explores what the literature on river basin governance and on urban water governance can tell us about how to understand and address metropolitan water challenges. The ensuing theoretical chapters combine a more normative perspective to evaluate how the literature frames urban and river basin governance in relation to inclusive and sustainable development and are critical to examine the spatial scalar dimensions of water governance. Chaps. 3 and 4 engage with the literature on IWRM and IRBM and UWM respectively. River basin governance and urban water governance are focused on different spatial scales and have a different scope, which determines the actors that are involved and the institutional arrangements that shape water-related decision-making and policy. This book adopts and integrates elements of IWRM/IRBM and UWM to analyse the relevant institutions, instruments, actors and drivers that shape water-related challenges in metropolitan areas. Based on this theoretical framework, the empirical analysis follows three steps, inspired by the three research foci of causality, performance and design of the Science Plan of the International Human Dimensions Programme's Institutional Dimensions of Global Environmental Change (IHDP/IDGEC 2005).

As a first step, this study examines causality, which means understanding the role that institutions in general, and more specifically river basin and urban water governance regimes, play in causing and confronting metropolitan water challenges. Institutions are often treated as intervening variables that affect the impact of underlying forces (or drivers) but are not such forces themselves (Young et al. 2008: 9). However, institutions can affect the behaviour of a variety of individual actors, something critical in shaping outcomes of human/environmental relations (IHDP/IDGEC 2005). In addition, institutions at multiple levels interact with other forces, such as biophysical and cultural systems. Studying institutions thus requires understanding 'joint effects', as institutional mechanisms interact with other mechanisms to coproduce outcomes (Underdal 2008: 66). Within environmental and resource regimes there are so many underlying factors (biophysical and socioeconomic forces) that contribute to collective outcomes that it is almost impossible to separate the causal linkages from different elements and evaluate the salience of each for explaining collective outcomes (Young et al. 2008). This research then relies on a case study method and comparative analyses as a methodology that enables the assessment of the profound complexities of the interactions between human and biophysical systems, although it has limitations in terms of providing generalizations (see Sect. 2.2).

The second step is to evaluate the performance of existing institutions on actors' behaviour towards urban water challenges, given the drivers and contextual factors that affect their behaviour. According to the environmental governance literature, specific regimes are established to address well-defined problems, and performance analysis is used to evaluate how successfully it solves, or at least alleviates, these problems (Young 2003a: 100; Mitchell 2008). As a more normative lens for looking at institutions, performance analysis can be used to evaluate various dimensions (or criteria), such as sustainability or equity (Young et al. 2008: 21). It corresponds to an 'actual-versus-aspiration' comparison that establishes how much an institution has contributed to a specified goal (Mitchell 2008: 79). This thus requires using a performance scale or measurement system for each dimension being evaluated, as well as a reference point to which outcomes can be compared, leading to a numeric or non-numeric performance score (Mitchell 2008: 80). Thus, this research selected and evaluated instruments of each governance regime in relation to four dimensions of inclusiveness and sustainability, defined below (see Sect. 1.5.1).

The final step considers how to improve institutional design so as to better address urban water challenges in metropolitan regions. It consists in the (re)design of policy instruments for both Mexico City and São Paulo in order to enhance the sustainability and inclusiveness of policy responses to metropolitan water challenges. This is built on the greater understanding of the interactions between the urban water and river basin governance regimes and the effectiveness of existing policy instruments. The specific format and purpose of these instruments therefore depends on the findings of the first two steps (causality and performance).

Consequently, to answer the main research question, this study explores the following sub-questions:

1. What does the literature on urban water governance and on river basin governance tell us about how to understand and address metropolitan water challenges?
2. How do multiple drivers as well as river basin and urban water governance institutions shape current metropolitan water challenges? [**causality**]
    (a) What are the current drivers of urban water challenges in the metropolitan regions of São Paulo and Mexico City and in their respective river basins?
    (b) What IWRM/IRBM actors and institutions exist at multiple levels to address urban water challenges in São Paulo and Mexico City?
    (c) What UWM actors and institutions exist at multiple levels to address urban water challenges in São Paulo and Mexico City?

3. Which policy instruments are effective,[6] and which are not, in dealing with these water challenges and their drivers? [**performance**]
    (a) In relation to water sharing (Quantity)
    (b) In relation to water preservation (Quality)
    (c) In relation to unpredictable and extreme water-related weather events (Climate Change)
    (d) How are the policy instruments at the urban scale linked to those at the river basin scale, and are these coherent within the metropolitan region and across multiple levels of governance?
4. Based on this analysis, how can more appropriate instruments be designed to address metropolitan water challenges, with the aim to guide metropolitan regions towards inclusive and sustainable development? [**(re)design**]

1.3.2 Focus of the Research

This research focuses on large cities within Latin America and their unique characteristics and water-related challenges. The urban population produces about 80% of global wealth, counts for over half of the global population, accounts for three quarters of global demand for natural resources, and produces 50% of its waste (IRP 2018). Latin America is the most urbanized region of the Global South and is still urbanizing. As most of the world's future urbanization will happen in the Global South, the experiences of Latin America's cities may provide lessons or insights that are more relatable for other growing cities than the experiences of cities in the Global North, which underwent very different development paths.

1.3.3 Limits of the Research

A first limitation of this research is in terms of the spatial scope, as it focuses on large metropolitan areas. Smaller cities and rural areas far from large cities also face important water-related challenges. However, the book argues that the size and complexity of large cities, and political and economic factors, lead to challenges of a different nature.

For both case studies, historical developments of the past centuries shaped their water-related problems and their water governance regimes. However, the research mainly focused on developments dating back to the early 1990s, when IWRM received international recognition and began to be implemented in Brazil and Mexico. Moreover, the research limits its forward outlook to 2030, as this is the timeframe for attaining the Sustainable Development Goals. This corresponds to 15 years from

---

[6] "Effective" here should be understood to mean: enhance sustainability and inclusiveness.

the start of the research, which is enough to estimate medium-term challenges and implement significant reforms. The timeframe of 2050 is occasionally considered for longer-term dynamics, such as climate change effects. This study favours scope over depth by considering a wide range of water-related challenges and responses to these within the metropolitan regions of Mexico City and São Paulo. Analysing these challenges and the effectiveness of existing responses to attain sustainable and inclusive metropolitan water governance regimes requires considering the metropolitan water cycle as a whole, including both the ecosystem functions it depends on and land use considerations, and the efforts of actors across different sectors and at multiple levels. This limits a deeper understanding of large metropolitan regions' specific water-related challenges, in order to shed light on the complexity, transdisciplinarity and multi-dimensionality of metropolitan water governance. Moreover, the research mostly addresses the governance of 'blue water' (water resources in surface and groundwater bodies), as well as 'grey' and 'black' waters (wastewater), as current water governance regimes and policies are mainly limited to these types of water. Further research is first needed on 'green water' (soil moisture) and 'rainbow water' (atmospheric moisture), before policies can be designed to better manage these within and around large cities.

1.4 The Nature of Water

Before delving deeper, it is important to briefly overview the elementary features of water. Water has one simple chemical formula composed of two hydrogen atoms and one oxygen atom, but there are multiple types or 'colours' of water, notably: rainbow water, green water, blue water, grey water, black water (see Table 1.3) (Hayat and Gupta 2016). Most freshwater is not (easily) accessible as it is stored in saltwater in oceans and seas or locked away in the polar caps.

Blue water can be captured for human uses and is discharged as grey or black water. Besides taking different forms, water also flows through the hydrological cycle, and does not follow socially constructed boundaries such as national or subnational borders. Managing these flows is necessary to sustain and protect human settlements, in particular in large cities.[7] The delimitation of river basins and aquifers as governance scales developed as societies were particularly concerned with blue water and led to the emergence of IWRM/IRBM as dominant paradigms for water resources management. Given that water flows can be transformed through engineering for the benefit of certain groups (Swyngedouw 2009), choosing the river basin as a management unit is also a political process as it transforms these spaces into territories of governance and raises questions about who will take decisions and how (Warner et al. 2008). Consequently, the flow of water can then be understood as a hybridization of physical and social processes.

---

[7] The circulation of water and other material inflows and outflows in and out of cities is sometimes referred to as "urban metabolism" (Heynen et al. 2005; Carlo Delgado-Ramos 2015).

Table 1.3 Colours of water and their characteristics

Colours of water	Characteristics
Blue water	Accessible clean freshwater resources (surface and groundwater). Blue water availability is the volume of water that can be consumed without causing adverse ecological impacts
Rainbow water	Waters in their vaporized stage (atmospheric moisture). It can travel long distances as clouds, is depleted by precipitation and replenished through evaporation
Grey water	Volume of water required to dilute pollutants so that water quality remains above agreed water quality standards. This includes water polluted by pesticides from agriculture and household wastewater (excluding inputs from toilets)
Black water	Water polluted with human faeces (sewage). Ecological buildings nowadays separate blackwater and greywater
Green water	Rainwater that is stored in the soil, absorbed by plant roots and evaporates through plant transpiration. It corresponds to 60% of freshwater flow

Source Hayat and Gupta (2016: 1233–1237)

Furthermore, water contributes in multiple ways, directly or indirectly, to humans and human well-being (Hayat and Gupta 2016). Green water, largely ignored by research and policy until recently, is the largest freshwater water resource, the basis of rain-fed agriculture and all life on land and performs crucial functions by sustaining water-dependent ecosystems such as forests and wetlands and their biodiversity (Ringersma et al. 2003). These ecosystems are essential for sustaining blue water as they often hold the springs of rivers and streams. Therefore, the degradation of ecosystems in the urban–rural interface impacts sustainable water resources management and water supply for urban users (Avissar and Werth 2005). The disruption of water-related ecosystem services within the hydrological cycle leads to challenges in terms of water quantity and water quality, and risks related to climate change.

1.5 Positionality

This study combines a sustainable and inclusive development perspective with a multi-level governance approach to examine the social, economic, environmental and relational impacts of metropolitan water governance regimes.

1.5.1 Sustainable and Inclusive Development Perspective

Sustainable development is a set of well-established principles that was institutionalized within the U.N. Sustainable Development Goals and is defined by the balance of three pillars: Economic, social and ecological. Inclusive development is a pull away from the neoliberal bias towards economic growth, through its focus on marginalized people, environmental sustainability and empowerment (Gupta et al. 2015). The SDG framework does not mention 'inclusive development' but includes the term 'inclusive' 41 times (Gupta and Vegelin 2016). Combining both perspectives allows for integrating social, ecological, economic and relational dimensions (see Fig. 1.6). This book will consider these four dimensions to analyse the impact of policy instruments on the multiple water challenges experienced by large cities and their river basins and in the process clearly focuses on the more marginalized people in society.

These four dimensions have the following characteristics:

- Social inclusiveness implies accounting for the multiple dimensions of poverty and striving for the social inclusion of specific individuals and groups. It reduces their exposure to various risks and enhances their well-being by investing in human capital and increasing opportunities for participation, meeting basic needs, reducing inequality and promoting equal rights (Gupta and Vegelin 2016; Pouw and Gupta 2017).
- Ecological inclusiveness requires recognizing that the growing demand for limited natural resources due to population growth and economic development may lead to 'ecospace' grabbing (i.e. large-scale transfer of these resources from local communities to governments, corporation and the private sector) (Gupta et al. 2015; Gupta and Vegelin 2016). It aims for local to global sharing of water resources, protecting equitable access to and ownership of resources and preserving ecosystems and their biodiversity (Gupta et al. 2015; Gupta and Vegelin 2016). People's well-being, particularly the poorest, is closely related to investments in preserving ecosystem services (Chopra et al. 2005; Gupta et al. 2015).
- The economic pillar refers to cities' ability to manage their resources and develop (Gupta et al. 2015). Water is crucial for economic activities that sustain lives and livelihoods within and outside urban areas. Large cities are embedded in local to global political economic structures (Hordijk et al. 2015), but are also important economic actors that can spearhead changes in economic development models and push for economic efficiency (e.g. rational water use, cost-efficient technology and infrastructure) to ensure the viability of various policy instruments. The focus of

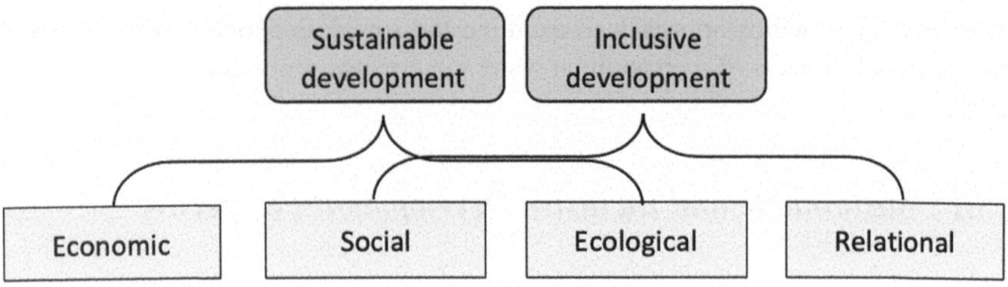

Fig. 1.6 Dimensions of inclusive and sustainable development

the economic pillar has often been on profit-maximization, reproducing inequalities and allowing for environmental degradation to persist through neo-liberal thinking and market-driven approaches (Pahl-Wostl 2018). This book aims to rectify this imbalance and considers that sustainable and inclusive economic development should occur within planetary boundaries.

- Relational inclusiveness recognizes the role of politics and local and global drivers of inequality in shaping poverty and ecological degradation (Gupta et al. 2015; Gupta and Vegelin 2016). These relations operate at multiple levels, and addressing these requires dealing with all actors (Gupta and Vegelin 2016). This requires addressing inequality and aims for the redistribution of power and resources, and ensuring that the most vulnerable and marginalized populations have access to affordable water services and receive protection from water-related risk (Swyngedouw 2009; Pouw and Gupta 2017).

This framework is used to examine the impact of institutional responses to water-related challenges. Institutions and instruments can empower actors to respond to water quantity, water quality and risks related to climate change through measures that stimulate certain behaviours and discourage others (Gupta et al. 2010; Majoor and Schwartz 2015). Their effectiveness determines whether cities and their river basins can develop sustainably and inclusively. This research focuses on institutions and instruments derived from IWRM/IRBM and UWM, as these were identified as the dominant paradigms. IWRM and IRBM emphasise the sustainable management of all water and related ecosystems (linking to SDGs 6.4, 6.5 and 6.6), with an emphasis on decentralized and participatory decision-making (SDG 6.B) (Watson 2004; Medema et al. 2008; Molle 2009a; Huitema and Meijerink 2014). UWM addresses the provision of water and sanitation services (SDGs 6.1 and 6.2), including the treatment of wastewater and its recycling (SDG 6.3), as well as mitigation and adaptation to risks from extreme weather events (SDGs 13.1, 13.2 and 13.3) and control of waterborne diseases (SDGs 6.1 and 6.2) (Barraqué 2011; Engel et al. 2011; GTT 2014). The linkages between the three SDGs are useful to consider the policy implications of metropolitan water governance.

1.5.2 Multi-level Governance Through a Political Science and Geography Perspective

Metropolitan water challenges are shaped by biophysical and social factors at multiple levels. Yet, many analyses of urban sustainability have divorced the local from other governance levels (Marvin and Guy 1997). While environmental impacts, such as water scarcity, are often experienced locally, the complex web of factors behind them cannot realistically be addressed by local governments alone. When externalities appear in different jurisdictions than those that caused them, this requires multilevel responses (Hooghe and Marks 2003; Newig and Fritsch 2009; Termeer et al. 2010). A multilevel governance perspective allows for examining how local problems are constructed and contested at different scales of governance and through multiple political spaces (Hooghe and Marks 2003; Bulkeley and Betsill 2005).[8] This is also the case for addressing water-related challenges in metropolitan regions, which are nested within social, economic and political contexts at local, national and global levels.

Multilevel governance is broadly understood as the participation in decision-making processes of governmental actors and non-state actors at different administrative levels and across administrative jurisdictions (Pierre and Peters 2000; Hooghe and Marks 2003; Bache and Flinders 2005; Papadopoulos 2007; Kluvánková-Oravská et al. 2009; Newig and Fritsch 2009). This involves vertical interactions across administrative levels and horizontal interactions between agencies at central level, and different actors at sub-national level (Peters and Pierre 2004; OECD 2011, 2015b). These interactions may involve continuous negotiation among nested governments at multiple territorial levels (Pierre and Peters 2000; Hooghe and Marks 2003). Multilevel governance can also be an alternative to hierarchical government, with more complex and contextually defined relationships. In fact, the concept of multilevel governance emerged in order to make sense of the "unravelling of central state control" and the diffusion of authority more generally (Hooghe and Marks 2003). Nevertheless, the traditional state-centric and constitutional perspective has not lost relevance (Peters and Pierre 2004: 75). Rather, the "'shift' towards multilevel governance" is occurring in an incremental manner, where relations between institutions at different government levels are "fluid, negotiated and contextually defined".

In addition, multilevel governance allows for a more flexible approach to issues of spatial fit between natural and institutional (governance) scales, thereby striving for better institutional performance in terms of sustainability (Young 2003b; Newig and Fritsch 2009). For instance, the provision of public services and policy responses can be designed at the most appropriate scale in each context (Pahl-Wostl 2015). This flexibility is necessary in order to face rapid and unpredictable environmental changes and to foster sustainable development (Bavinck and Gupta 2014: 78). However, designing institutions to fit natural scales is not a panacea. Such institutions are more likely to face coordination challenges with other actors, particularly in terms of financing due to the difficulty of sustaining joint funding schemes (Ingram 2008; Newig and Fritsch 2009). It requires collaboration between actors who may otherwise compete for political and financial support. Political leaders whose constituencies have little relationship to the territory defined by river basin institutions also tend to feel estranged from the latter. Indeed, they are not directly held accountable to the population within the institutionalized river basin territory (Ingram 2008).

---

[8] 'Governance' is defined as a process, and more specifically as "the shaping and sustaining of the arrangements of authority and power within which actors make decisions and frame policies that are binding on individual and collective actors within different territorial bounds" (Hanf and Jansen 1998: 3). I take this perspective to examine (urban water and river basin) management approaches and their associated policy instruments.

Furthermore, dynamics at multiple levels not only shape metropolitan water challenges, and the institutions that address these, but also the power of actors at different levels. This research recognizes the fundamentally political nature of these regimes in both shaping and responding to water-related challenges and the responses to these, from the dominant discourses to the chosen policy instruments. However, it is beyond the scope of this study to formulate recommendations that require important shifts in the political, economic, social and cultural fabric of the two case studies and of other large metropolitan regions. The objectives of this study are limited to (re)designing policy instruments that can lead to changes within existing metropolitan water governance regimes. Nevertheless, such changes can have broader repercussions when there is sufficient political will and public support.

1.6 Data Collection and Methods

1.6.1 Structure of Thesis

Following this introductory chapter, Chap. 2 grounds the research methodology and methods. The theoretical framework unpacks the literature on RBM and UWM from a multilevel governance perspective in Chaps. 3 and 4 respectively. The four following chapters discuss the findings from the empirical research on the two case studies. Chaps. 5 and 7 analyse the driving forces causing metropolitan water challenges and the institutions and instruments of IWRM/IRBM in place to address them in São Paulo and Mexico City respectively. In Chaps. 6 and 8 a similar analysis focuses on UWM. Chapter 9 draws comparisons between the two cases, with a particular focus on the interactions between IWRM/IRBM and UWM, and reflects on the scalar mismatches that arise in each case's metropolitan water governance regime. It answers the research questions in relation to São Paulo and Mexico City, discusses the policy implications and proposes a redesign of their policy instruments. Finally, Chap. 10 concludes this book by answering the research questions in relation to the literature review, reflecting on the implications for international development studies and providing recommendations, not least in the context of the Sustainable Development Goals.

References

Abdullah D, Christensen B (2004) Integrated river basin management. Bullet Ingenieur 22:21–23

Agyenim JB (2011) Investigating institutional arrangements for integrated water resource management in developing countries: the case of white volta basin. Vrije Universiteit Amsterdam, Ghana

Alexander C (2019) Cape town's "Day Zero" water crisis, one year later. CityLab, 12 Apr 2019

Anokye NA (2013) Stakeholder participation in water resources management: the case of Densu Basin in Ghana. Vrije Universiteit Amsterdam

ARUP (2018) Cities alive: water for people. Leeds, UK

Avissar R, Werth D (2005) Global hydroclimatological teleconnections resulting from tropical deforestation. J Hydrometeorol 6L134–145. Available at: https://doi.org/10.1175/JHM406.1. Accessed 1 Dec 2018

Azzam R et al (2014) Water quality and socio-ecological vulnerability regarding urban development in selected case studies of megacity Guangzhou, China. In: Megacities: our global urban future. Internatio. Springer

Bache I, Flinders M (2005) Multi-level governance: conclusion and implications. In: Bache I, Flinders M (eds) Multi-level governance. Oxford University Press, New York, pp 195–206

Bahri A (2012) Integrated urban water management: briefing note. Technical Committee Background Paper 16. Stockholm

Barraqué B (2011) Urban water conflicts. CRC Press. Available at: https://www.ingentaconnect.com/content/rout/c4z92/2011/00000001/00000001/art00002. Accessed 11 Dec 2018

Barrios JE, Rodríguez-Pineda JA, de la Maza Benignos M (2009) Integrated river basin management in the conchos river basin, Mexico: a case study of freshwater climate change adaptation. Clim Dev 1(3):249–260

Baud ISA, Hordijk MA (2009) Dealing with risks in urban governance: what can we learn from "resilience thinking". In: The 4th International conference of the international forum on urbanism (IFoU)—the new urban question—urbanism beyond neo-liberalism, Amsterdam/Delft. Available at: http://www.ifou.org/conferences/2009delft/proceedings/8NewApproachesofUrbanGovernance/fullpapers/F001BaudandHordijkTheNewUrbanQuestion.pdf. Accessed 30 Sep 2019

Bavinck M, Gupta J (2014) Legal pluralism in aquatic regimes: a challenge for governance. Curr Opin Environ Sustain 11:78–85. https://doi.org/10.1016/j.cosust.2014.10.003

Biswas AK (2004) Integrated water resources management: is it working? Wat Int 29(2):248–256. https://doi.org/10.1080/07900620701871718

Brown RR, Farrelly MA (2009) Delivering sustainable urban water management: a review of the hurdles we face. Water Sci Technol. https://doi.org/10.2166/wst.2009.028

Bulkeley H, Betsill M (2005) Rethinking sustainable cities: multilevel governance and the "urban" politics of climate change. Environ Polit 14(1):42–63. https://doi.org/10.1080/0964401042000310178

Carlo Delgado-Ramos G (2015) Water and the political ecology of urban metabolism: the case of Mexico City. J Polit Ecol 22:98–114. https://doi.org/10.2458/v22i1.21080

Chopra K et al (2005) Ecosystems and human well-being: policy responses. The millen. Island Press, Washington, DC

Closas A, Schuring M, Rodriguez D (2012) Integrated urban water management—lessons and recommendations from regional experiences in Latin America, Central Asia, and Africa. Washington, DC

Conti KI (2017) Norms in multilevel groundwater governance & sustainable development. University of Amsterdam

Crutzen PJ, Stoermer EF (2000) The "Anthropocene." Global Change Newslett 41:17–18

Doxiadis C (1970) The great lakes megalopolis. Doxiadis Assoc

Elmqvist T et al (eds) (2013) Urbanization, biodiversity and ecosystem services: challenges and opportunities. A global assessment. Springer

Engel K et al (2011) Big cities. Big water. Big Challenges. Water in an urbanizing world. Berlin. Available at: https://www.wwf.se/source.php/1390895/BigCities_BigWater_BigChallenges_2011.pdf. Accessed 27 Dec 2018

FAO (2012) Coping with water scarcity: an action framework for agriculture and food security. Rome. Available at: http://www.fao.org. Accessed 11 Dec 2018

Flörke M, Schneider C, McDonald RI (2018) Water competition between cities and agriculture driven by climate change and urban growth. Nat Sustain 1(1):51–58. https://doi.org/10.1038/s41893-017-0006-8

Foster S, Ait-Kadi M (2012) Integrated water resources management (iwrm): how does groundwater fit in? Hydrogeol J 20:415–418. https://doi.org/10.1007/s10040-012-0831-9

Gosling SN, Arnell NW (2016) A global assessment of the impact of climate change on water scarcity. Climat Change 134(3):371–385. https://doi.org/10.1007/s10584-013-0853-x

Gottmann J, Harper RA (1990) Since megalopolis: the urban writings of Jean Gottmann. Available at: https://www.worldcat.org/title/since-megalopolis-the-urban-writings-of-jean-gottmann/oclc/19724007. Accessed 26 Jun 2019

Grigg NS (2008) Integrated water resources management: balancing views and improving practice. Water Int 33(3):1–20. https://doi.org/10.1080/02508060802272820

GTT (2014) Integrated urban water resources management, in humanity. In: Gulbenkian think tank on water and the future humanity- revisiting water security. Springer International Publishing, pp 109–132

Güneralp B, Güneralp İ, Liu Y (2015) Changing global patterns of urban exposure to flood and drought hazards. Glob Environ Chang 31:217–225. https://doi.org/10.1016/J.GLOENVCHA.2015.01.002

Gupta J, Vegelin C (2016) Sustainable development goals and inclusive development. Int Environ Agreements Polit Law Econ 16:433–448. https://doi.org/10.1007/s10784-016-9323-z

Gupta J et al (2010) The adaptive capacity wheel: a method to assess the inherent characteristics of institutions to enable the adaptive capacity of society. Environ Sci Policy 13(6):459–471. https://doi.org/10.1016/J.ENVSCI.2010.05.006

Gupta J, Pouw NRM, Ros-Tonen MAF (2015) Towards an elaborated theory of inclusive development. Europ J Dev Res 27(4):541–559. https://doi.org/10.1057/ejdr.2015.30

Hanf K, Jansen A-I (1998) Governance and environment in Western Europe: politics, policy and administration. Longman

Hansjürgens B, Heinrichs D (2014) Megacities and climate change: early adapters, mainstream adapters and capacities. In: Megacities—our global common future. Springer, Dordrecht, pp 9–24

Hayat S, Gupta J (2016) Kinds of freshwater and their relation to ecosystem services and human well-being. Water Policy 18(5):1229–1246. https://doi.org/10.2166/wp.2016.182

Herrschel T, Newman P (2002) Governance of Europe's city regions: planning, policy, and politics. Routledge, London

Heynen N, Kaika M, Swyngedouw E (2005) Urban political ecology: politicizing the production of urban natures. In: The nature of cities: urban political ecology and the politics of urban metabolism. Routledge, pp 16–35. https://doi.org/10.4324/9780203027523-8

Hooghe L, Marks G (2003) Unraveling the central state, but how? Types of multi-level governance. Am Polit Sci Rev 97(2). Available at: https://www.unc.edu/~gwmarks/assets/doc/hooghe.marks.unravelingcentralstate.apsr.2003.pdf. Accessed 7 May 2018

Hordijk M et al (2015) Participatory instruments and practices in urban governance. In: Gupta J et al (eds) Geographies of urban governance: advanced theories, methods and practices. Springer International Publishing, Amsterdam, pp 127–146

Huitema D, Meijerink SV (2014) The politics of river basin organisations: institutional design choices, coalitions and consequences. Edward Elgar Publishing Limited, Cheltenham. Available at: https://www.e-elgar.com/shop/eep/preview/book/isbn/9781782549222/. Accessed 11 Jun 2018

Hurlbert M (2016) Adaptive governance of disaster: drought and flood in rural areas. Springer, Amsterdam

IHDP/IDGEC (2005) Science plan: institutional dimensions of global environmental change. Bonn, Germany

Ingram H (2008) Beyond universal remedies for good water governance: a political and contextual approach. Paper presented at the Rosenberg Forum for Water Policy, Zaragoza, Spain. Available at: https://www.zaragoza.es/contenidos/medioambiente/cajaAzul/IngramACC.pdf. Accessed 27 Jun 2019

IRP (2018) The weight of cities: resource requirements of future urbanization. In: Swilling M et al (ed) A report b. United Nations Environment Programme, Nairobi

Jenerette GD, Larsen L (2006) A global perspective on changing sustainable urban water supplies. Global Planet Change 50:202–211. https://doi.org/10.1016/j.gloplacha.2006.01.004

Jønch-Clausen, T., & Fugl, J. (2001). Firming up the conceptual basis of integrated water resources management. Int J Water Resour Dev 174:501–510. https://doi.org/10.1080/07900620120094055

Jones T, Newborne P, Phillips B (2006) Applying the principles of integrated water resource and river basin management—an introduction

Kluvánková-Oravská T et al (2009) From government to governance for biodiversity: the perspective of central and Eastern European transition countries. Environ Policy Govern 19(3):186–196. https://doi.org/10.1002/eet.508

Knieling J (2014) Metropolitan regions definitions, typologies and recommendations for development cooperation introduction to the publication series sustainable development of metropolitan regions. Eschborn, Germany. Available at: www.giz.dewww.bmz.de. Accessed 18 Oct 2018

Kraas F et al (2014) In: Kraas F et al (eds) Megacities: our global urban future. Springer, Dordrecht

Majoor S, Schwartz K (2015) Instruments of urban governance. In: Gupta J et al (eds) Geographies of urban governance: advanced theories, methods and practices. Springer International

Marcos J, Fernández A (2016) Un ejército de mujeres para defender el agua. Pueblos—Revista de Información y Debate (68). Available at: http://www.revistapueblos.org/blog/2016/01/28/un-ejercito-de-mujeres-para-defender-el-agua/. Accessed 26 Jun 2019

Marlow DR et al (2013) Towards sustainable urban water management: a critical reassessment. Water Res 47:7150–7161. https://doi.org/10.1016/j.watres.2013.07.046

Martinez-Santos P, Aldaya MM, Llamas MR (2014) Integrated water resources management in the 21st century: revisiting the paradigm. CRC Press/Balkema, Leiden

Marvin S, Guy S (1997) Creating myths rather than sustainability: the transition fallacies of the new localism. Local Environ 2(3):311–318. https://doi.org/10.1080/13549839708725536

McDonald RI et al (2014) Water on an urban planet: urbanization and the reach of urban water infrastructure. Global Environ Change 27(1). https://doi.org/10.1016/j.gloenvcha.2014.04.022

Medema W, Mcintosh BS, Jeffrey PJ (2008) From premise to practice: a critical assessment of integrated water resources management and adaptive management approaches in the water sector. Ecol Soc 13(2) 29. Available at: http://www.ecologyandsociety.org/vol13/iss2/art29/

Mitchell RB (2008) Evaluating the performance of environmental institutions: what to evaluate and how to evaluate it? In: Institutions and environmental change. The MIT Press, pp 79–114

Molle F (2009) River-basin planning and management: the social life of a concept. Geoforum 40:484–494. https://doi.org/10.1016/j.geoforum.2009.03.004

Newig J, Fritsch O (2009) Environmental governance: participatory, multi-level—and effective? Environ Policy Govern 19(3):197–214. https://doi.org/10.1002/eet.509

Nobre CA, Marengo JA (2016) National Center for monitoring and early warning of natural disasters (CEMADEN), Brazil. Available at: http://www.globalwaterforum.org/2016/10/17/water-crises-and-megacities-in-brazil-meteorological-context-of-the-sao-paulo-drought-of-2014-2015/?pdf=13381. Accessed 23 Sep 2017

Obani PC (2018) Strengthening the human right to sanitation as an instrument for inclusive development. University of Amstedam

OECD (2015) Water and cities: ensuring sustainable futures. OECD Publishing, Paris. https://doi.org/10.1787/9789264230149-en

OECD (2011) Water governance in OECD countries: a multi-level approach. OECD studies on Water, OECD Publishing

OECD (2012) Redefining "Urban": a new way to measure metropolitan areas. OECD Publishing. Available at: https://www.oecd-ilibrary.org/docserver/9789264174108-en.pdf?expires=1540210663&id=id&accname=ocid56024160&checksum=FA16A31438E88630AD218261DE2A8130. Accessed 22 Oct 2018

OECD (2016) Water governance in cities. Paris

Pahl-Wostl C (2018) The role of governance modes and meta-governance in the transformation towards sustainable water governance. Environ Sci Policy 91:6–16. https://doi.org/10.1016/j.envsci.2018.10.008

Pahl-Wostl C (2015) Water governance in the face of global change: from understanding to transformation. Springer. https://doi.org/10.1007/978-3-319-21855-7

Papadopoulos Y (2007) Problems of democratic accountability in network and multilevel governance. Eur Law J. 13(4):469–486. https://doi.org/10.1111/j.1468-0386.2007.00379.x

Pearson LJ et al (2010) A sustainable decision support framework for urban water management. Water Resour Manage 24:363–376. https://doi.org/10.1007/s11269-009-9450-1

Peters BG, Pierre J (2004) Multi-level governance and democracy: a faustian bargain? In: Multi-level governance. Oxford University Press, pp 75–90. https://doi.org/10.1093/0199259259.003.0005

Pierre J, Peters BG (2000) Governance, politics and the state. Macmillan, Basingstoke

Pouw N, Gupta J (2017) Inclusive development: a multi-disciplinary approach. Curr Opinion Environ Sustain 24:104–108. https://doi.org/10.1016/j.cosust.2016.11.013

Rauch W, Morgenroth E (2013) Urban water management to increase sustainability of cities. Water Res 47:7149. https://doi.org/10.1016/j.watres.2013.10.042

Renner T, Meijerink S, Van Der Zaag P (2017) The evolution of regional cross border water regimes, the case of Deltarhine. J Environ Plan Manage 61(10):1701–1721

Rietveld LC et al (2016) Improving health in cities through systems approaches for urban water management. Environ Health Global Access Sci Sour 15(Suppl 1). https://doi.org/10.1186/s12940-016-0107-2

Ringersma J, Batjes N, Dent D (2003) Green water: definitions and data for assessment. Wageningen. Available at: https://pdfs.semanticscholar.org/7d8f/8b8ff81a0aa80758fdc2dbb4585a0323991e.pdf?_ga=2.95548825.44534484.1568230520-213102760.1568230520. Accessed 11 Sep 2019

Roche PA et al (2001) Infrastructure integration issues. In: Maksimović Č, Tejada-Guibert JA (eds) Frontiers in urban water management: deadlock or hope. IWA Publishing, Cornwall

Salet W, Thornley A, Kreukels A (2003) Metropolitan governance and spatial planning; comparative comparative case studies of European city-regions. Routledge and Spon, London

Savenije HHG, Van der Zaag P (2008) Integrated water resources management: concepts and issues. Phys Chem Earth 33:290–297. https://doi.org/10.1016/j.pce.2008.02.003

Scruggs G (2018) Everything we've heard about global urbanization turns out to be wrong—researchers. Place, 10 July. Available at: http://www.thisisplace.org/i/?id=0150beca-e3f5-47e0-bc74-9ccc5ef1db8a. Accessed 17 Oct 2018

Sellers JM et al (eds) (2013) The political ecology of the metropolis: metropolitan sources of electoral behaviour in eleven countries. ECPR Press, Colchester

Slack E (2007) Managing the coordination of service delivery in metropolitan cities: the role of metropolitan governance. The World Bank (Policy research working papers). https://doi.org/10.1596/1813-9450-4317

Swyngedouw E (2009) The political economy and political ecology of the hydro-social cycle. J Contemp Water Res Educ 142: 56–60. Available at: https://doi.org/10.1111/j.1936-704X.2009.00054.x. Accessed 6 Nov 2018

Termeer CJAM, Dewulf A, van Lieshout M (2010) Disentangling scale approaches in governance research: comparing monocentric, multilevel, and adaptive governance. Ecol Soc. Published at Carleton University by the Resilience Alliance. Available at: http://library.wur.nl/WebQuery/wurpubs/399683. Accessed 11 Dec 2018

Tortajada C (2008) Challenges and realities of water management of megacities: the case of Mexico City metropolitan area. J Int Aff 61(2):147–166

UCLG (2016) Co-creating the urban future: the agenda of metropolises, cities and territories. Barcelona

Underdal A (2008) Determining the causal significance of institutions: accomplishments and challenges. In: Young OR, Schroeder H, King LA (eds) Institutions and environmental change. The MIT Press, pp 49–78

UN-DESA (2018) World urbanization prospects—population division—United Nations. Available at: https://population.un.org/wup/Download/. Accessed 5 Dec 2018

UNEP (2018) Progress on integrated water resources management. Global baseline for SDG 6 Indicator 6.5.1: degree of IWRM implementation

UNEP (2021) Progress on integrated water resources management. Tracking SDG 6 series: global indicator 6.5.1 updates and acceleration needs

UNGA (United Nations General Assembly) (2015) 2015 time for global action for people and planet. New York. Available at: https://sustainabledevleopment.un.org/%0Apost2015

UN-Habitat (United Nations Human Settlements Programme) (2006) The state of the world's cities 2006/2007: 30 years of shaping the Habitat Agenda. London: Earthscan. Available at: https://unhabitat.org/books/state-of-the-worlds-cities-20062007/. Accessed 19 Oct 2018

UN-HABITAT (2016) Urbanization and development: emerging futures. Nairobi, Kenya. Available at: http://wcr.unhabitat.org/wp-content/uploads/2017/02/WCR-2016-Full-Report.pdf. Accessed 6 May 2018

United Nations (2012) 2012: World urbanization prospects The 2011 revision. New York. Available at: http://www.un.org/en/development/desa/population/publications/pdf/urbanization/WUP2011_Report.pdf. Accessed 19 Oct 2018

United Nations (2016b) The world's cities in 2016 data booklet. Available at: www.unpopulation.org. Accessed 19 Oct 2018

van den Brandeler F, Gupta J, Hordijk M (2019) Megacities and rivers: scalar mismatches between urban water management and river basin management. J Hydrol 573:1067–1074. https://doi.org/10.1016/j.jhydrol.2018.01.001

Varis O et al (2006) Megacities and water management. Int J Water Resour Dev 22(2):377–394. https://doi.org/10.1080/07900620600684550

Vlachos E, Braga B (2001) The challenges of urban water management. In: Maksimović Č, Tejada-Guibert JA (eds) Frontiers in urban water management: deadlock or hope. IWA Publishing, Cornwall

Vörösmarty CJ et al (2000) Global water resources: vulnerability from climate change and population growth. Science 289(5477): 284 LP–288. Available at: http://science.sciencemag.org/content/289/5477/284.abstract

Warner J, Wester P, Bolding A (2008) Going with the flow: river basins as the natural units for water management? Water Policy 10(2):121–138. https://doi.org/10.2166/wp.2008.210

Watson N (2004) Integrated river basin management: a case for collaboration. Int J River Basin Manag 2(4):243–257. https://doi.org/10.1080/15715124.2004.9635235

Young OR (2003b) Environmental governance: the role of institutions in causing and confronting environmental problems. Int Environ Agreements Polit 3(4):377–393. https://doi.org/10.1023/B:INEA.0000005802.86439.39

Young OR, Schroeder H, King LA (2008) Institutions and environmental change. The MIT Press

Young OR (2003a) Determining regime effectiveness: a commentary on the oslo-potsdam solution. Global environmental politics. MIT Press, Cambridge, MA, USA 3(3):97–104. https://doi.org/10.1162/152638003322469295

Chapter 2
Methodology

2.1 Introduction

This chapter develops the qualitative methodology employed to answer the research questions. The methodology has several components: the comparative case study method (see Sect. 2.2), a literature review (see Sect. 2.3), an analytical framework (see Sect. 2.4), the units of analysis (see Sect. 2.5), the qualitative content analysis of the policies (see Sect. 2.6), the fieldwork and ethical considerations (see Sect. 2.7) and the integration of these different elements (see Sect. 2.8).

2.2 The Comparative Case Study Method

2.2.1 Justification for the Comparative Case Study Method

This research adopts an inductive comparative case study approach examining São Paulo in Brazil, and Mexico City in Mexico. This means investigating the metropolitan water challenges in Mexico City and São Paulo in the context of the interrelations between water governance at urban scale and at river basin scale. Many driving forces and institutional factors shape such challenges and a case study may help elucidate their links (Gehring and Oberthur 2008).

Comparative analyses allow for building upon case studies of institutions in different settings and gaining insights that could not be attained by observing a single case (IDGEC 2005). Considering similar problems in different locations and contexts is important to gain valuable insights on policy processes at multiple levels and the performance of similar regimes (e.g. the inclusiveness and sustainability of water governance regimes in metropolitan regions) (Young et al. 2008; Kuzdas and Wiek 2014).

F. van den Brandeler, *Scalar Mismatches in Metropolitan Water Governance*,
Water Governance - Concepts, Methods, and Practice,
https://doi.org/10.1007/978-3-031-08061-6_2

2.2.2 The Choice of São Paulo and Mexico City

Brazil and Mexico are the most populated countries of Latin America and are both highly urbanized.[1] People and water resources are unevenly distributed, with populations more densely concentrated in relatively water scarce regions. Besides experiencing seasonal climate variation, both countries are likely to suffer increases in extreme weather events due to climate change (see Sect. 5.2 and 7.2).

Brazil and Mexico have relatively similar GDP per capita (i.e. USD 9812 in Brazil and 8910 in Mexico in 2017) and both are upper middle-income economies with stark socio-economic inequalities (high Gini coefficients of 53.7 in Brazil and 43.4 in Mexico in 2016) (World Bank 2019a, b). In Mexico, economic growth has been concentrated in the drier parts of the country, while the humid Southeast has only 16% of economic output (Hearne 2004). In Brazil, the South and Southeast are significantly wealthier than the rest of the country. Both experienced rapid population growth, have inadequate urban planning, and serious water-related challenges for urban dwellers (Tortajada 2008; Kelman 2015).

Both countries are federal regimes, with responsibilities shared between three levels of government (federal, state and municipal), which has important implications for water resources management, water services and water-related risks.[2] The implementation of IWRM is estimated to be relatively advanced in comparison to other Latin American nations, and slightly above average for their HDI (Human Development Index) score (UNEP 2018). They were also the first two countries in Latin America to have legally mandated river basin organizations (Tortajada 2001). These are significant developments for the relatively young democracies: Brazil began transitioning to democracy in 1985, and Mexico initiated democratic reforms in the early 1990s (Martínez-Lara 1996; Vegelin 2016; Ginsburg 2017).

São Paulo and Mexico City are two of the world's largest urban agglomerations, and the largest in Latin America. Throughout the twentieth century, they experienced explosive demographic growth and rapid industrialization prompted by the Import Substitution Industrialization policies of the 1930s (see Fig. 2.1) (Blouet and Blouet 2015). As of the 1970s and 1980s, population growth concentrated in their peripheries (Escamilla and Santos 2012). They have similar population and surface area sizes and are composed of a large number of municipalities (see Table 2.1). Mexico City is the capital of Mexico, whereas São Paulo is a state capital. However, both are their country's financial and economic centres. The former spreads over three different federal entities, whereas São Paulo is contained within one state.

[1] According to 2021 estimates, Brazil had a population of just under 214 million inhabitants (IBGE 2021). 87% of the population was urban as of the latest census (IBGE 2010). In Mexico, 79% of the population was urban out of approximately 126 million inhabitants in 2020. Mexico's population quadrupled since 1950 and went from being predominantly rural (57.4%) to mainly urban (76.5%) (INEGI 2022).

[2] Mexico has 31 states and the federal entity of Mexico City. Brazil has 26 states and the Federal District.

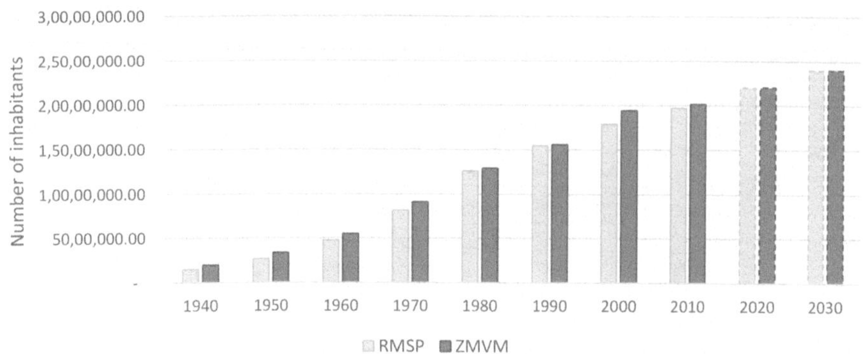

Fig. 2.1 Demographic growth in the metropolitan regions of São Paulo and Mexico City. *Source* Based on raw data from São Paulo Pref. (2018); UNDESA (2018); Pradilla Cobos (2016)

Table 2.1 Characteristics of Mexico City and São Paulo

	National capital	National GDP per capita	City proper pop.	Metro pop.	# of munici-palities in metro area	# of federative entities in metro area	Surface area of metro area
São Paulo	No	9821.4 USD in 2017	12 million	21.5 million	39	1 state	7946 km^2
Mexico City	Yes	8902.8 USD in 2017	9 million	21.6 million	60 + 16 districts of Mexico City	3 states	7866 km^2

Sources EMPLASA, no date; World Bank, no date; World Population Review, no date; State (1994); INEGI (2014), Brazil (2015)

The two megacities are both located inland and upstream in their river basins (see Maps 2.1 and 2.2). They have experienced extreme transformations of their natural landscape to accommodate growing populations and their demand for water, and to protect themselves from flood risks. However, they both continue to struggle with water-related challenges in terms of quantity, quality and the effects of extreme weather events. The Valley of Mexico Basin endured a water crisis in the Spring of 2009, after facing floods months earlier. Similarly, in 2010 and 2011 the Upper-Tietê Basin, where São Paulo is located, experienced heavy precipitations and floods. In 2013–2015 the region suffered a historical drought. These extreme contradictions in such short timeframes could be linked to climate change, but also indicate a failure in the water management model.

In sum, the two cases share many common features in terms of size, economic development and historical background. Their experiences addressing water-related challenges—both successful and unsuccessful—can bring valuable insights to many

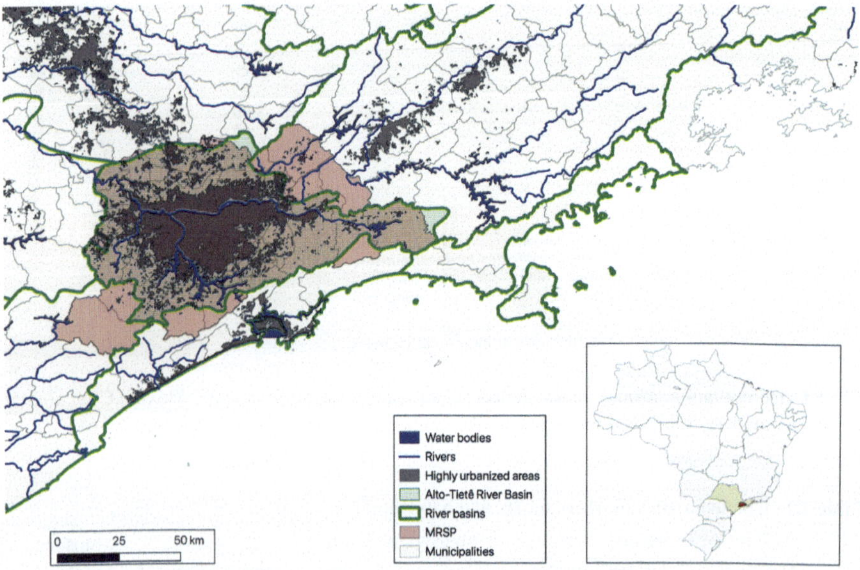

Map 2.1 Spatial characteristics of São Paulo. *Source* Author

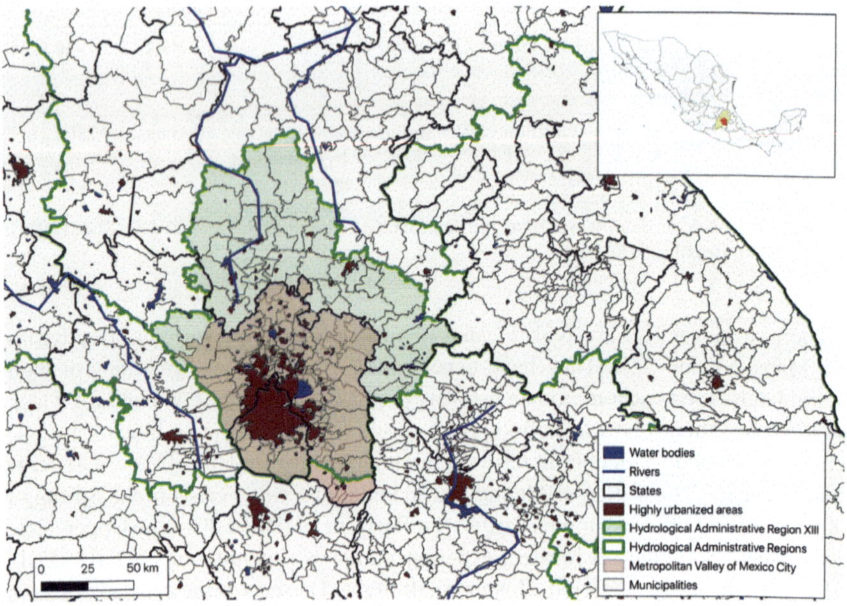

Map 2.2 Spatial characteristics of Mexico City. *Source* Author

cities in the Global South that have similar biophysical, political, economic and demographic characteristics, and that are likely to face growing water-related challenges.

2.3 Literature Review

2.3.1 Literature Review on Key Concepts

An extensive literature review on UWM, IWRM, IRBM and MWM, Inclusive and Sustainable Development, Multilevel Governance and Institutional Analysis identified relevant principles and instruments and included scholarship from both the Global North and Global South. For UWM/IWRM/IRBM/MWM, the literature survey focused on articles published between 1990 and 2018, as earlier literature on these topics was virtually non-existent.

The literature review on IWRM/IRBM and Integrated Urban Water Management (IUWM)/ Sustainable Urban Water Management (SUWM)/ Metropolitan Water Management (MWM) allowed for an assessment of the evolution of publications over time. This search was conducted in ScienceDirect and was limited to the occurrence of these terms in titles, abstracts and key words between 1970 and 2015. The resulting graph (see Fig. 2.2) reveals the quasi non-existence of all five terms prior to 1990. IWRM shows the steepest rise, increasing from 19 publications in 1995, to 54 in 2005 and 183 in 2015. IRBM is marginally ahead of IUWM and SUWM, with 62, 38 and 50 publications respectively in 2015. Finally, the graph shows that the publications on MWM are negligible, with results only entering the double digits in 2013. In the

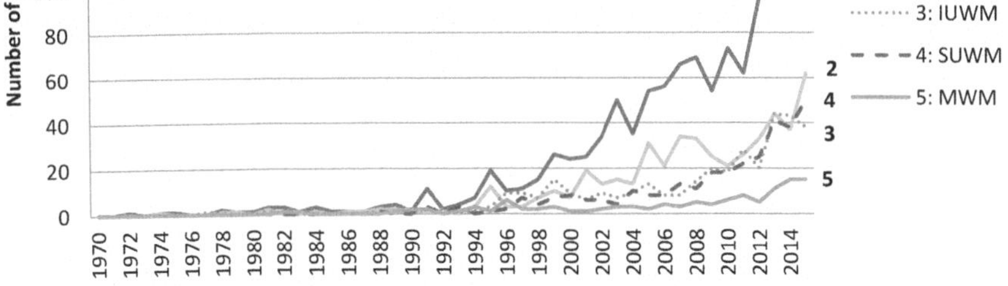

Fig. 2.2 Evolution of concepts between 1970 and 2015. *Source* Author

case of MWM, there were no results between 1970 and 2015. This indicates that there is no clearly defined metropolitan water management approach.

In a second step, I analysed the urban-river basin linkages by examining the incidence of terms in titles, abstracts and key words between 1970 and 2015 associated with the urban (i.e. urban, city/ies, megacity/ies and metropolitan) in the IWRM/IRBM literature, and to the river basin (i.e. river basin, watershed and catchment) in the IUWM/SUWM/MWM literature. The results did not coincide or overlap between the two sets of literature (see Fig. 1.5). Of the articles that included the term *Metropolitan Water Management* in their titles, abstracts or keywords, only 7 contained the terms *river basin, catchment* or *watershed* in the body of the article. The literature therefore does not differentiate water management in metropolitan regions from other urban areas, nor does it link to the scales relevant for IWRM/IRBM.

2.3.2 Literature Review on Case Studies

Prior to conducting fieldwork, I conducted literature reviews for each case study on the existing scholarship on water governance, urban governance, multilevel governance and environmental issues in Mexico City and São Paulo. In addition, I reviewed approximately 50 policy documents and a dozen legal documents for each case study. This involved legal and policy documents at local, state and federal levels on water resources management and urban water management, but also environmental policy, climate change and urban planning.

2.4 Analytical Framework

To address my research questions, I use the institutional analysis model of the Institutional Dimensions of Global Environmental Change (IDGEC) as the foundation for my methodological framework (IHDP/IDGEC 2005). This framework is part of a stream of analysis known as 'new institutionalism', which focuses on how institutions affect society. More specifically, it examines environmental and resource regimes, which are types of institutions that address situations where actions can degrade ecosystems through overuse of natural resources or due to unintended side effects (Young et al. 2008). This led to the following six-step framework, which allows for examining the causal mechanisms of metropolitan water challenges, the effectiveness of instruments that address these and the (re)design of such instruments for more sustainable and inclusive water governance:

(1) *Define the major drivers of water challenges in the metropolitan region*

First, I analysed the driving forces of the water-related tensions between the metropolises and the river basins (see Sect. 5.2 for São Paulo, and Sect. 7.2 for Mexico City). This evaluation was initially done through document analysis and the results were then triangulated through semi-structured interviews with a variety of respondents. This answered Research Question 2a.

(2) *Actor and institutional analysis*

The next step was to identify the key actors addressing water challenges in the metropolitan regions of São Paulo and Mexico City at multiple levels (Research questions 2b and 2c). A crucial element when examining environmental regimes are the spatial boundaries of specific institutions and their environments. To examine metropolitan water challenges, this research considers both the institutions addressing water-related issues at the urban scale (UWM regime) and at the hydrological scale (IRBM regime). This further required awareness of the linkages between cities and the environment that sustains them (i.e. assessing urban water demand and appraising investment needs beyond traditional city boundaries) (OECD 2015a).

I defined actors as entities that represent specific interests and have mandates relevant for addressing metropolitan water challenges. Who the actors are and how they frame the policy problems is crucial as this process is highly political and reflects power imbalances in society (Majoor and Schwartz 2015: 121). Their behaviour is shaped by driving forces but also by the institutions in place. More specifically, the institutions analysed were chosen because they aim to influence challenges linked to water quantity, water quality and climate variability and change in terms of sustainability and inclusiveness (see Sects. 5.3, 6.3, 7.3 and 8.3).

(3) *Instrument design analysis*

The institutional framework determines the policy instruments that help or hinder key actors address metropolitan water challenges. The instrument analysis serves to identify the regulatory, economic, suasive and infrastructural instruments within the urban water and river basin governance regimes that target challenges linked to water quantity, water quality and climate change adaptation (Research questions 3a, 3b and 3c). A comparable mix of instruments for both case studies was selected based on their estimated salience for water-related challenges (see Sects. 5.4 and 6.4 for São Paulo and Sects. 7.4 and 8.4 for Mexico City). Each instrument's design is examined in terms of objectives, scale of implementation and main characteristics. Design is assessed in terms of sustainability and inclusiveness criteria.

(4) *Effect on actors given drivers*

Driving forces and institutional factors can both influence actors' behaviour. This book assesses the effectiveness of policy instruments in terms of key actors' behavioural changes given contextual drivers (Research question 3). To determine the effectiveness of instruments, this research examines whether UWM and IWRM/IRBM actors fulfil their mandates in relation to the instruments' specific objectives (see Sects. 5.4 and 6.4 for São Paulo and Sects. 7.4 and 8.4 for Mexico City). It measures this effectiveness based on the instruments' stated goals (identified through content analysis) and the perception of respondents on the extent to which these goals are met (based on semi-structured interviews and other sources[3]).

[3] Newspaper articles, evaluation reports and technical assessments.

Beyond the focus on actors' compliance, this step also involves evaluating additional changes in behaviour (Underdal 2008). In addition, policy instruments are usually created and used by those in power (Gupta et al. 2015: 218). This is important to consider for a comprehensive understanding of instruments' effects. The analysis may also highlight whether the current instruments in place fail to address certain drivers.

(5) *Impacts*

Ultimately, the goal of designing institutions is to effectively address the challenges deriving from human/environment interactions. Changes in actors' behaviour do not guarantee impacts on the challenges at hand, in this case metropolitan water challenges, and as such they can prove unsatisfying if they remain the only dimension for evaluating institutional performance (Mitchell 2008). The impact of instruments is therefore considered (based on semi-structured interviews, technical and scientific assessments, newspapers) in the context of existing driving forces and assessed in terms of the four dimensions of sustainable and inclusive development, defined in Sect. 1.5.1.

The instruments' design, their effect on actors given drivers and their impacts in terms of sustainability and inclusiveness have been measured through an ordinal scale and inductive 'calculations' of institutional performance. This calculation is an estimation of each instrument's aggregated effects, symbolized by "–" (very negative), "–" (negative), "0" (neutral effect), " + " (positive) and "++" (very positive). The choices of scores reflect estimates based on available quantitative and qualitative data and perceptions from key actors. They should be considered in relation to each other to assess which instruments have a comparatively more positive or more negative effect. A score of "–" or " ++" therefore does not mean that an instrument could not perform any worse or better. A neutral effect means either that there is no discernible effect of the instrument in question, or that the positive effects are balanced out by the negative effects. It is important to distinguish between each instrument's potential positive and negative effects when considering redesign (Research question 3).

(6) *Redesign (of instruments)*

Finally, the proposals to redesign policy instruments of the urban water and river basin governance regimes are based on the instruments' effects on actors and impacts on sustainability and inclusiveness. The book proposes to maintain instruments that met these goals, to discard or alter ineffective instruments, and to (re)design existing or new instruments that could address gaps (see Sects. 5.5, 6.5, 7.5 and 8.5). It considers overall policy coherence across multiple levels and between urban and river basin scales (Research question 4).

The analytical scheme below illustrates the analytical framework (see Fig. 2.3).

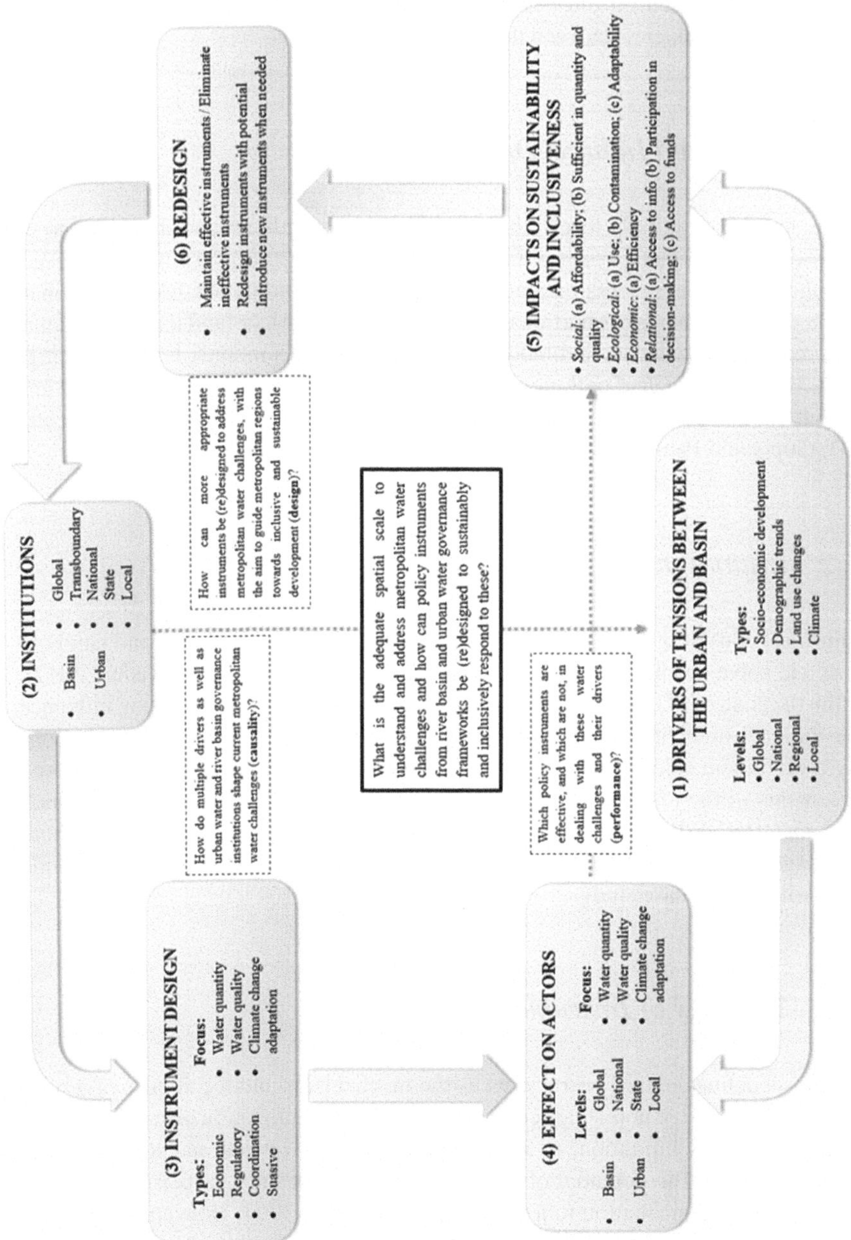

Fig. 2.3 Analytical framework for an institutional approach to urban water challenges

2.5 Units of Analysis

This research is based on three units of analysis—drivers, institutions and instruments—and the linkages between them.

2.5.1 Direct and Indirect Drivers

Drivers unfold at multiple levels, they can be direct or indirect and they need to be addressed by governance systems (Gupta and Pahl-Wostl 2013). Examples of drivers are demographic trends, socio-economic development patterns (i.e. concentration of wealth, water use that is proportional to population growth), or land use changes that affect regional ecosystems and biodiversity. Direct drivers include land use change such as deforestation, and pollution, and indirect drivers concern climate change and variability, demography, economy, infrastructure and technology (Postel and Richter 2003; Gupta and Pahl-Wostl 2013).

2.5.2 Institutions

Institutions consist of formal and informal norms, laws, and policies, and can both cause and solve problems with environmental governance (Vatn and Vedeld 2012). Within the context of the *Anthropocene*, where human action has increasing influence over the environment, creating effective institutions becomes crucial. Institutions play a prominent role both "as sources of large-scale environmental problems and as elements in the responses humans make to actual or anticipated environmental problems" (IHDP/IDGEC 2005: 35). This research examines institutions of both the urban water governance regimes and river basin governance regimes at multiple levels within each case study.

2.5.3 Typology of Instruments

This thesis defines policy instruments as the mechanisms and techniques of governance used by state or non-state actors and involving the utilization of state resources or their conscious limitation, to achieve policy goals (Howlett and Rayner 2007; Anderson 2010). The potential of a policy instrument lies not in its isolated application but in its contribution to a policy mix (Chapman 2003; Howlett and Rayner 2007). The choice of instruments reflects the mode of governance, and changes in instruments tend to signify a shift in governance modes (Majoor and Schwartz 2015: 114). This choice is not neutral, but rather reflects the balance of power between actors (Kassim and Le Galès 2010; Majoor and Schwartz 2015). Policy instruments may also shape and shift power relations. This research distinguishes four types of instruments:

– **Regulatory instruments**

These are regulatory measures enforced by government or institutions with legal mandates (Hurlbert 2016), and which prohibit (i.e. bans), empower (i.e. property rights) or compel behaviour (i.e. standards) (Majoor and Schwartz 2015). They tend to be effective as they have a direct impact on their goals and are relatively predictable (Ibid). However, they are top-down and often inflexible, they do not provide incentives for actors to achieve more than the minimum standards and their coercive nature may lead to resistance (Ibid). These instruments may enhance sustainability and inclusiveness by applying legally binding principles (e.g. equity, human rights), social norms (e.g. good governance), standards and mandatory inclusion targets, inclusive spatial planning, the provision of civic amenities and public infrastructure (i.e. water and sanitation), safety net schemes and target subsidies (Gupta et al. 2015).

– **Economic instruments**

These instruments follow market rather than government directives and aim to encourage or discourage certain behaviours (rather than enforcing or prohibiting them, as with regulatory instruments) through financial (dis)incentives (Stavins 2003; Majoor and Schwartz 2015). They are often more politically feasible than regulatory instruments as they do not directly intervene in actors' affairs (Bähr 2010; Majoor and Schwartz 2015). However, their costs vary considerably according to the instrument and context, and they are less self-enforcing than regulatory instruments. They often have to be supported by regulatory mechanisms.

– **Suasive instruments**

Suasive instruments aim to internalize behaviour into individual decision-making through persuasion (Majoor and Schwartz 2015: 112). This requires the provision of information, such as general education programmes, guidelines and codes of practice, training programmes, and research and development (Aramyan et al. 2016). They are most effective when applied in combination with other types of policy instruments (Majoor and Schwartz 2015). While relatively cheap and less intrusive than regulatory instruments, as they rely on voluntary compliance, their impact is uncertain and dependent on the quality of information available (Majoor and Schwartz 2015).

– **Infrastructural instruments**

Infrastructural instruments can directly shape actors' behaviour and impact the goals they are designed to attain by physically generating certain behaviour (Majoor and Schwartz 2015). To be effective, they require considerable knowledge of social and biophysical processes (Ibid).

In practice, instruments can frequently be classified in more than one category. For instance, Payment for Ecosystem Services (PES) programmes are economic

Table 2.2 Criteria for instrument selection

Criteria	Options
Scope	UWM and IWRM/IRBM
Focus	Water quantity; Water quality; Climate change adaptation
Instrument goal	Inclusiveness and sustainability, in terms of environmental, social, economic and relational dimensions
Type of instrument	Regulatory; Economic; Infrastructure; Suasive
Levels of implementation	National; State; Basin; Metropolitan; Municipal

Source Author

instruments as they involve a financial incentive for a certain behaviour, but they may also be considered a suasive instrument as they are usually voluntary. Nonetheless, using the labels above to reflect their primary goals is useful in terms of classifying and evaluating the effectiveness of these instruments (Table 2.2).

2.5.4 Instrument Selection and Evaluation Criteria

To be selected for this study, instruments had to fulfil multiple criteria to ensure relevance in relation to the research questions and comparability between the two cases (see Table 2.3). First, selected instruments had to, directly or indirectly, address metropolitan water challenges through the urban water or river basin governance regimes, in order to determine how they affect these (positively or negatively). The focus of each instrument must be on water quantity, water quality or climate change adaptation, or a combination of these. They should be designed to impact water-related challenges in terms of inclusiveness and sustainability. Moreover, they must be either a regulatory, economic, infrastructural or suasive instrument. The instruments may be implemented at national, state, basin, metropolitan or municipal levels, as long as they aim to address the water-related tensions between cities and their basins.

This led to a comprehensive list of instruments to potentially analyse in both case studies (see Annex B: Comprehensive List of Policy Instruments). This was further refined to a selection of several instruments by considering available data and feasibility, and these were analysed in Chaps. 6 and 8 (for an overview of selected instruments see Annex C: Policy Instruments Selected for Analysis). The major focus is on regulatory and economic instruments, the most common and relevant in both cases. The same or similar instruments are often implemented at different administrative levels in each case.

Generally, policy instruments are not implemented in isolation, but rather different types of instruments are used simultaneously and interact with each other (Howlett and Rayner 2007). This mix of instruments is embedded within a particular socio-political context and reflects a particular ideological foundation (Majoor and

Table 2.3 Examples of potential instruments

Instruments	Examples
Regulatory	*Water quality standards*: Regulatory limits for the amounts of certain contaminants in water provided by public water systems (Helmer and Hespanhol 1997) *Environmental Impact Assessments*: Evaluation of the likely environmental impacts of a proposed project or development, and development and assessment of measures to avoid or minimize these (Komínková 2008) *Water use permits*: Rights to withdraw water from rivers and aquifers, typically allocated by national or state governments for different water uses (i.e. domestic, industrial and agricultural uses) (OECD 2011). These are meant for quantitative and qualitative control of water resources, and to guarantee the right to access water resources *Environmental licensing*: Legally binding requirements to protect human health and the environment applied by a public authority. It should be carried out as part of the planning process, prior to the approval of projects (World Bank 2012)
Economic	*Payment for ecosystem services (PES)*: Incentives offered to people in exchange for preserving ecosystems and their services. Some argue that protecting certain ecosystems is only viable if their economic value is considered. Others claim that this valuation should be a strategy to create a new 'rural–urban compact', where cities reward rural dwellers for their provision of private and public goods (Corbera et al. 2009) *Polluter-pays principle*: Those who pollute must internalize its costs into their production costs (Porto and Costa 2004) Other instruments include subsidies and bounties, tax concessions, special purpose grants, performance bonds and guarantees, tradable quotas, resource rents and royalties, and sliding charges for utilities (Buckley 1991)
Suasive	Campaigns to reduce water consumption, environmental education programmes, demonstration projects, labelling schemes (i.e. trademarks and brand names that assure buyers of the authenticity of a seller's product(s) or service(s)), the provision of information to the public, flood warnings, and more
Infrastructure	From small-scale technology to large-scale infrastructure, this includes water metres, water-saving technology, public toilets, rainwater harvesting systems and inter-basin water transfers. It can involve hard, grey infrastructure as well as green infrastructure

Source Author

Schwartz 2015). While a large range of instruments may be available, government authorities tend to use specific combinations of instruments, based on cost, effectiveness, feasibility and an estimation of their joint effect (Lascoumes and Le Gales 2007; Perevochtchikova and Torruco Colorado 2014). Instrument mixes can be more effective than single instruments. However, their design and use also reflects the power balance between actors and are therefore political (Kassim and Le Galès 2010; Majoor and Schwartz 2015). Consequently, it is essential to consider the context in which instruments are selected and the possible motivations and constraints behind this. Instruments are not a panacea and have often have mixed results in practice, and their potential to induce behavioural change has to be put in perspective, especially for complex systems such as large cities (Majoor and Schwartz 2015).

2.6 Content Analysis

This thesis makes use of a qualitative analysis of content using coding categories derived from the problem definition and the literature review. As a first step, it uses content analysis to assess the SDGs and linkages between their specific targets and urban/metropolitan water challenges in a context of anthropogenic climate change. It identified SDG 6 (Clean water and sanitation), 11 (Sustainable cities and communities) and 13 (Climate action) as the main SDGs aiming to address these issues (see Sect. 1.2.2). This was based on the SDGs' targets and indicators, as well as progress reports.

Following this, this thesis uses content analysis to identify and analyse the drivers, institutions and instruments at multiple levels of governance in both case studies that related to urban water and river basin challenges. This was done through a systematic examination of the relevant laws, policies, regulations and other information such as planning documents (master plans, plans for the metropolitan regions, river basin plans), newspaper articles and local government publications. The purpose is to describe and make inferences about the characteristics of drivers, institutions and instruments that relate to metropolitan water challenges. Materials were categorized in terms of policy problem (i.e. water quantity, water quality, climate change), different geographic levels (i.e. international, transboundary, national, provincial and local) and focus on the urban or river basin scales. Instruments were also categorized according to their typology (i.e. regulatory, economic, infrastructural and suasive). The literature identified UWM and IWRM/IRBM as the main paradigms from which institutions and instruments are derived to address such challenges. Documents were analysed for language relating to the main characteristics of these paradigms, the effectiveness of instruments and impacts in terms of sustainability and inclusiveness. This was further supplemented by a review of previous studies on Mexico City and São Paulo. The resulting themes were then complemented and triangulated with the transcripts from semi-structured interviews. This forms the core of Chaps. 5, 6, 7 and 8.

2.7 Fieldwork

2.7.1 *Fieldwork Research Approach*

I conducted fieldwork to collect primary and secondary data on metropolitan water challenges and how these are addressed by the urban water and river basin governance regimes in the metropolitan regions of São Paulo and Mexico City. Semi-structured interviews were used to discover perceptions of key informants on the drivers, institutions and instruments related to metropolitan water challenges. These were particularly useful to analyse the effects of instruments on actors and their impacts in terms of sustainability and inclusiveness. The difficulty to establish causal linkages within environmental and resources regimes means that it is difficult to use statistical procedures to explain the causal significance of institutions (Young et al. 2008: 20). The combination of the results from the content analysis and a wide range of perspectives from key informants allowed for a more balanced assessment.

The fieldwork took place in two stages: From February to August 2016 in Mexico City, and from September 2017 to January 2018 in São Paulo.[4] Sixty respondents were interviewed in Mexico City and 38 in São Paulo (see Annex D: Interview List). Interviews were conducted with state and non-state actors at multiple levels of governance with influence on the metropolitan and river basin spatial scales.

2.8 Integration

This chapter has explained the value of conducting a comparative case study to gain in-depth knowledge of specific cases and identify causal mechanisms, and then identify similarities and differences that support theory-building. As this research investigates only two cases, it aims to provide insights on the causal mechanisms of metropolitan challenges and the effectiveness of institutions and instruments that address these and, if possible, to identify lessons learnt for other metropolises. This is supported by an analytical framework based on the IDGEC's institutional analysis approach and that focuses on water related challenges and their drivers, and the institutions and instruments that address these. Such an approach allows for exploring the roles of both urban water governance regimes (see Chap. 3) and river basin governance (see Chap. 4) in shaping water challenges in the metropolitan regions of São Paulo and Mexico City.

References

Anderson JE (2010) Public policymaking: an introduction. Cengage Learning, Hampshire

Aramyan L et al (2016) Market-based instruments and other socio-economic incentives enhancing food waste prevention and reduction. Wageningen, The Netherlands

Bähr H (2010) The politics of means and ends: policy instruments in the European Union. Ashgate. Available at: https://www.routledge.com/The-Politics-of-Means-and-Ends-Policy-Instruments-in-the-European-Union/Bahr/p/book/9781409410706. Accessed 27 May 2019

Blouet BW, Blouet OM (2015) Latin America and the Caribbean: a systematic and regional survey. Available at: https://www.wiley.com/en-us/Latin+America+and+the+Caribbean%3A+A+Systematic+and+Regional+Survey%2C+7th+Edition-p-9781118729847. Accessed 20 Jul 2018

Brazil (2015) Lei n. 13.089, Estatuto da Metrópole. Available at: http://www.planalto.gov.br/ccivil_03/_ato2015-2018/2015/lei/l13089.htm. Accessed 19 Jul 2018

[4] The second fieldwork was shorter as I was more familiar with the context thanks to previous research.

Buckley R (1991) Economic instruments of environmental policy. In: Perspectives in environmental management. Springer, Berlin, pp 80–92. https://doi.org/10.1007/978-3-642-76502-5_5

Chapman R (2003) A policy mix for environmentally sustainable development—learning from the Dutch experience. New Zealand J Environ Law 7(1):29–51

Corbera E, Soberanis CG, Brown K (2009) Institutional dimensions of payments for ecosystem services: an analysis of Mexico's carbon forestry programme. Ecol Econ 68(3):743–761. https://doi.org/10.1016/J.ECOLECON.2008.06.008

Escamilla IH, Santos CC (2012) La Zona Metropolitana del Valle de México: Transformación Urbano-Rural en la Región Centro de México. In: *XII Coloquio Internacional de Geocrítica.* Bogotá, p 15

Gehring T, Oberthur S (2008) Interplay: exploring Institutional Interaction. In: Young OR, Schroeder H, King LA (eds) Institutions and environmental change. The MIT Press, pp 187–224

Ginsburg T (2017) Introduction. In: Castagnola A, López Noriega S (eds) Judicial politics in Mexico: the supreme court and the transition to democracy. Routledge, New York. Available at: https://books.google.nl/books?hl=en&lr=&id=ijFuDQAAQBAJ&oi=fnd&pg=PP1&dq=mexico+democracy&ots=OM8dpZz0JD&sig=MUGJKQY6urvVRCchUmmLoCD-Siw#v=onepage&q=mexicodemocracy&f=false. Accessed 16 Dec 2018

Gupta J, Pahl-Wostl C (2013) Global water governance in the context of global and multilevel governance: its need, form, and challenges. Ecol Soc Resilience Alliance 18(4):art53. https://doi.org/10.5751/ES-05952-180453

Gupta J, Pfeffer K et al (eds) (2015) Geographies of urban governance: advanced theories, methods and practices. Springer International Publishing, Amsterdam

Gupta J, Pouw NRM, Ros-Tonen MAF (2015) Towards an elaborated theory of inclusive development. Eur J Dev Res 27(4):541–559. https://doi.org/10.1057/ejdr.2015.30

Hearne RR (2004) Evolving water management institutions in Mexico. Water Resour Res 40(12). https://doi.org/10.1029/2003WR002745

Helmer R, Hespanhol I (eds) (1997) Water pollution control—a guide to the use of water quality management principles. E. & F. Spon. Available at: https://www.who.int/water_sanitation_health/resourcesquality/wpcchap5.pdf. Accessed 27 Jun 2019

Howlett M, Rayner J (2007) Design principles for policy mixes: cohesion and coherence in "new governance arrangements." Policy Soc 6(4):1–18

Hurlbert M (2016) Adaptive governance of disaster: drought and flood in rural areas. Springer, Amsterdam

IBGE (2010) Taxa de Urbanização. Censo demográfico 1940–2010.

IBGE (2021) Projeção da população do Brasil e das Unidades da Federação. Available at: https://www.ibge.gov.br/apps/populacao/projecao/index.html

IHDP/IDGEC (2005) Science plan: institutional dimensions of global environmental change. Bonn, Germany

INEGI (2014) Cuaderno estadístico y geográfico de la zona metropolitana del Valle de México 2014. México, D.F. Available at: http://internet.contenidos.inegi.org.mx/contenidos/productos/prod_serv/contenidos/espanol/bvinegi/productos/nueva_estruc/valle_mex/702825068318.pdf. Accessed 10 May 2018

INEGI (2022) Población. Instituto Nacional de Estadística y Geografía. INEGI, Mexico City. Available at: https://www.inegi.org.mx/temas/estructura/

Kassim H, Le Galès P (2010) Exploring governance in a multi-level polity: a policy instruments approach. West Eur Polit 33(1):1–21. https://doi.org/10.1080/01402380903354031

Kelman J (2015) Water supply to the two largest Brazilian metropolitan regions. Aquat Proc 5:13–21. https://doi.org/10.1016/j.aqpro.2015.10.004

Komínková D (2008) Environmental impact assessment and application—part 2. In: Encyclopedia of ecology. Academic Press, pp 1329–1339. https://doi.org/10.1016/B978-008045405-4.00053-7

Kuzdas C, Wiek A (2014) Governance scenarios for addressing water conflicts and climate change impacts. Environ Sci Policy 42:181–196. https://doi.org/10.1016/j.envsci.2014.06.007

Lascoumes P, Le Gales P (2007) introduction: understanding public policy through its instruments? From the nature of instruments to the sociology of public policy instrumentation. Governance 20(1):1–21. https://doi.org/10.1111/j.1468-0491.2007.00342.x

Majoor S, Schwartz K (2015) Instruments of urban governance. In: Gupta J et al (eds) Geographies of urban governance: advanced theories, methods and practices. Springer International

Martínez-Lara J (1996) Building democracy in Brazil: the politics of constitutional change, 1985–95. Palgrave Macmillan, St Anthon

Mitchell RB (2008) Evaluating the performance of environmental institutions: what to evaluate and how to evaluate it? In: Institutions and environmental change. The MIT Press, pp 79–114

OECD (2011) Water governance in OECD countries: a multi-level approach. OECD studies on Water, OECD Publishing

OECD (2015) OECD territorial reviews: Valle de México, Mexico. OECD Publishing, Paris. Available at: https://books.google.com/books?id=cEO8CgAAQBAJ&printsec=frontcover#v=onepage&q&f=false. Accessed 20 Aug 2019

Perevochtchikova M, Torruco Colorado VM (2014) Análisis comparativo de dos instrumentos de conservación ambiental aplicados en el Suelo de Conservación del Distrito Federal Comparative Analysis of Two Environmental Conservation Instruments Used on Conservation Land in the Federal District. Sociedad y Ambiente 1(3):3–25

Porto MFDA, Da Costa FJL (2004) Mechanisms of water management: command & control and social mechanisms. REGA—Revista de Gestão de Água da América Latina 1(2). Available at: https://www.abrh.org.br/SGCv3/index.php?PUB=2&ID=64&SUMARIO=805. Accessed 26 May 2019

Postel S, Richter BD (2003) Rivers for life: managing water for people and nature. Island Press, Washington, DC

Pradilla CE (2016) Zona Metropolitana del Valle de México: neoliberalismo y contradicciones urbanas. Sociologias, Porto Alegre, ano 18, no 42, mai/ago 2016, p. 54–89. Accessible at: 5. Zona Metropolitana del Valle de México: (scielo.br)

São PP (2018) Cidade de São Paulo. Relações Internacionais. Accessible at: Cidade de São Paulo | Secretaria de Relações Internacionais | Prefeitura da Cidade de São Paulo

State SP (1994) *Lei Complementar nº 760, de 01 de agosto de 1994 - Assembleia Legislativa do Estado de São Paulo*. Assembleia Legislativa do Estado de São Paulo. Available at: https://www.al.sp.gov.br/repositorio/legislacao/lei.complementar/1994/lei.complementar-760-01.08.1994.html. (Accessed 6 Feb 2019)

Stavins RN (2003) Experience with market-based environmental policy instruments. In: Mäler KG, Vincent J (eds) Handbook of environmental economics. I. Elsevier Science, Amsterdam, pp 355–435. Available at: https://scholar.harvard.edu/stavins/publications/experience-market-based-environmental-policy-instruments. Accessed 27 May 2019

Tortajada C (2001) Institutions for integrated river basin management in Latin America. Int J Water Resour Dev 17(3):289–301. https://doi.org/10.1080/07900620120065084

Tortajada C (2008) Challenges and realities of water management of megacities: the case of Mexico city metropolitan area. J Int Aff 61(2):147–166

Underdal A (2008) Determining the causal significance of institutions: accomplishments and challenges. In: Young OR, Schroeder H, King LA (eds) Institutions and environmental change. The MIT Press, pp 49–78

UNDESA (2018) 2018 revision of world urbanization prospects. Population division of the united nations department of economic and social affairs (UNDESA). Accessible at: World Urbanization Prospects—Population Division—United Nations

UNEP (2018) Progress on integrated water resources management. Global baseline for SDG 6 Indicator 6.5.1: degree of IWRM implementation

Vatn A, Vedeld P (2012) Fit, interplay, and scale: a diagnosis. Ecol Soc 17(4):12

Vegelin CL (2016) Global changes, national development and urban poverty: political engagement among the poor in Mexico City. University of Amsterdam. Available at: http://dare.uva.nl. Accessed 26 May 2019

World Bank (2012) Integrated urban water management: case study Sao Paulo. Washington, DC. Available at: www.worldbank.org

World Bank (2019a) GDP per capita (current US$). Available at: https://data.worldbank.org/indicator/NY.GDP.PCAP.CD?locations=MX. Accessed 27 Jun 2019

World Bank (2019b) GINI index (World Bank estimate). Available at: https://data.worldbank.org/indicator/SI.POV.GINI?locations=MX. Accessed 27 Jun 2019

Young OR, Schroeder H, King LA (2008) Institutions and environmental change. The MIT Press

Chapter 3
Urban Water Governance

3.1 Introduction

This chapter explores the literature on urban water governance[1] to answer the following question: What does the literature on urban water governance tell us about how to understand and address metropolitan water challenges? I address this question through a literature review and content analysis. First, I examine the evolution of urban water governance over time, in particular the paradigmatic shift towards more integrated and sustainable approaches to water management in cities (see Sect. 3.2.1). I then describe the main characteristics of Urban Water Management (UWM) (see Sect. 3.2.2), as well as the main drivers, institutions and instruments influencing this approach (see Sect. 3.2.3). Following this, I debate the strengths and weaknesses of UWM using as reference the reviewed literature (see Sect. 3.3). Finally, the chapter links back to the knowledge gaps identified in Sect. 1.2.3, and provides insights from the literature on how UWM can address metropolitan water challenges are discussed (see Sect. 3.4).

3.2 Overview

3.2.1 Evolution Towards More Sustainable and Integrated UWM

Urban water governance paradigms have undergone significant changes over the past decades, starting with Urban Water Management (UWM) and subsequently

[1] This chapter draws heavily on Brandeler et al. (2019).

F. van den Brandeler, *Scalar Mismatches in Metropolitan Water Governance*, Water Governance - Concepts, Methods, and Practice,
https://doi.org/10.1007/978-3-031-08061-6_3

Integrated Urban Water Management (IUWM), Sustainable Urban water Management (SUWM), and a number of other approaches with overlapping goals.[2] During the twentieth century, modernist visions of city planning rose to prominence around the world. Urban water governance was shaped by UWM and reflected a technocratic approach, aiming for controlling water flows, promoting public health, safety, and property protection, while ignoring environmental considerations (Kaika 2005; Swyngedouw 2006; Brown and Farrelly 2009; Farrelly and Brown 2011; Rauch and Morgenroth 2013; Winz et al. 2014). Actions focused on bringing water in and wastewater out of the city in a centralized, hierarchical and linear manner, with less attention to upstream and downstream linkages or longer-term environmental and social impacts (Engel et al. 2011; Donoso 2014; Winz et al. 2014; Jacobi et al. 2015). Scholars and policymakers considered that water should circulate constantly in the city, to wash it off waste and then leave it as sewage (Swyngedouw 2006)—a process enabled through standardized and large-scale infrastructure and technological solutions (Farrelly and Brown 2011; GTT 2014; Winz et al. 2014).

The modernist paradigm behind UWM reinforces dichotomies or boundaries that separate the city and nature (i.e. human/non-human, urban/rural, culture/nature, centre/periphery) (Zimmer 2010; Follmann 2016). Land use planning institutionalizes boundary-making as it "formalizes the separation between nature and abstract space through the written codes of legal statute and professional conduct which impose a site-based, rather than system-based, narrative structure on its treatment of the environment" (Whatmore and Boucher 1993: 169). Land use planning therefore addresses the physical space, but often leaves out ecological processes and the fluidity of multiple relations between different spaces and places (Murdoch 2006: 127).

The Political Ecology scholarship is critical of conventional UWM and argues that the separation between the 'natural' and the 'social' will be undermined by ecological relations, as nature will eventually overcome human society and disrupt economic and social relations (Murdoch 2006). It also illustrates how urban works and UWM create mechanisms of exclusion to access adequate quantities of potable water (Swyngedouw 2006). This remains stark in cities of the Global South, where the urban elites often live in permanently irrigated tropical gardens, while the poor live in urban deserts.

Since the 1990s, the urban water literature has shifted towards a greater emphasis on social, cultural, environmental and economic aspects (Winz et al. 2014). This shift was influenced by Agenda 21—a comprehensive, multi-level plan of action to address human impacts on the environment adopted at the United Nations Earth Summit in 1992—and the broader realization that convention UWM could not address rising issues of water scarcity and future uncertainty. In recent years, the concept of Integrated Urban Water Management (IUWM) has emerged to help policymakers and local governments think about urban water. IUWM is nested within the broader

[2] Other concepts include Water-Sensitive Cities and Water-Sensitive Urban Design, for instance. The literature review found SUWM and IUWM to be the dominant concepts.

framework of IWRM (Bahri 2012), and is aligned with the 'sustainable development' paradigm formalized by Agenda 21 (Gabe et al. 2009). IUWM is defined as a "participatory planning and implementation process, based on sound science, which brings together stakeholders to determine how to meet society's long-term needs for water and coastal resources while maintaining essential ecological services and economic benefits" (GDRC 2015: 1). IUWM promotes a holistic view of the urban water cycle, with a coordinated and flexible mode of strategic planning, and decision-making processes that must involve broad stakeholder participation (Brown 2005; Varis et al. 2006; Closas et al. 2012). The participation of marginalized communities and the inclusion of gender issues aim to enhance equitable access to water (GDRC 2015).

SUWM (Sustainable Urban Water Management) is another approach that departs from traditional UWM through a focus on the urban water cycle, emphasizing adaptation, decentralization, participation and integration (Brown and Farrelly 2009; Daigger 2011; Marlow et al. 2013). Water is considered a central element of sustainable urban areas and a potential starting point for urban planning (Daigger 2011). This requires that water professionals and urban planners become strategic partners. This view promotes a circular rather than linear approach to managing the urban water cycle, one where water is collected to respond to urban demands, treated to meet quality requirements, distributed to end users, collected again as wastewater and treated for reuse (Daigger 2011). This combines large, centralized infrastructure with alternative, local and distributed technologies and participation (Van de Meene et al. 2011; Younos 2011; Closas et al. 2012). However, the emphasis is particularly on local level, infrastructure-oriented initiatives (Marlow et al. 2013; Rietveld et al. 2016). Integrating infrastructure and biophysical systems (e.g. stormwater treatment and rainwater harvesting systems) requires considering social, economic, environmental and political contexts (Vlachos and Braga 2001; Mitchell 2006; Brown and Keath 2008). In addition, it requires changing the existing (traditional) management structure, which will involve institutional reform (Daigger 2011). The barriers to transitioning to SUWM are more often socio-institutional than technological (Brown and Farrelly 2009).

Along with the shift to sustainable and integrated approaches to UWM, the end of the twentieth century also experienced an evolution from state led UWM towards private management, inspired by neoliberal principles. Neoliberalism promotes a shift from "state responsibility for providing services to provision of services by private economic actors in which the role of the state is limited to enabling and regulating" (Ahlers et al. 2014: 3). It is also associated with liberal notions of state reform, where decentralization reduces the burden of inefficient central bureaucracies and market incentives lead firms to incorporate the economic value of water into production costs (Abers and Keck 2013: 50–51). Under this approach, UWM embraces full cost pricing, competition and privatisation (Bakker 2002; Van de Meene et al. 2011). Although there are claims that market governance aims to allocate resources efficiently and empower citizens (Pierre and Peters 2000), this market-driven approach has also been criticized for leading to private monopolies, restrictive contracts limiting citizen access to previously accessible water resources and institutional fragmentation (Van de Meene et al. 2011). In recent years, municipalities and regions have been re-municipalising water services in areas where these had been privatized (Parker and Sewell 1988; Bakker 2002, 2003; McDonald 2018).[3] This trend is far from uniform however, with forms of re-municipalisation that are autocratic (i.e. aimed at enhancing control by a ruling elite), highly marketized (i.e. focused on market-based performance indicators), or observing social democratic principles (i.e. committed to promoting social and economic justice), among others (McDonald 2018).

3.2.2 *Main Characteristics of UWM and IUWM/SUWM*

UWM and IUWM/SUWM have similarities and differences in their goals, their scale of implementation, their approach to urban water, key actors and their mandates, water users, and what constitutes 'inputs' and 'outputs' of water. These are summarized in Table 3.1.

Goals: UWM's main goals are to address the need for drinking water, sewage collection, treatment and disposal, stormwater management, flood protection and pollution control for diverse urban users (i.e. domestic, commercial, industrial). Historically, UWM has focused on affordability (sometimes recognizing diverging abilities to pay), universalization and public health. The shift towards alternative approaches such as SUWM/IUWM reflects a concern for broader social, economic and environmental outcomes than improving water quality and quantity (Gabe et al. 2009; Maheepala 2010; Daigger 2011; Porse 2013). This means recognizing the ecological functions of water, diversifying water sources, and promoting adaptation, health and human welfare, social and economic vitality, overall community improvement and stakeholder satisfaction (Maheepala 2010; GDRC 2015; Neto 2016). Examples include improving urban biodiversity, exposing and rehabilitating urban water flows to create more 'liveable' cities, extending the life of urban water infrastructure, and developing blue-green infrastructure that manages stormwater, removes pollutants and recharges aquifers (Gabe et al. 2009). The IUWM framework also strives for optimizing the interface between urban water and other urban sectors (e.g. housing, transportation), as well as with relevant activities beyond urban boundaries (e.g. agriculture, downstream water use) (Brown 2005; Varis et al. 2006; Closas et al. 2012; GDRC 2015).

[3] Since the turn of the twenty-first century, France has shifted towards the re-municipalisation of water services, defined as the transfer of water services from private companies to municipal authorities, with Paris in 2010 as the most notable case (Pigeon et al. 2012; Hall et al. 2013). This was spurred by corruption between large private water companies and local politicians, and a report by France's public audit body highlighting widespread lack of transparency and overcharging in the water sector (Hall et al. 2013). In Paris, re-municipalisation led to economic savings and a reduced water tariff (Pigeon et al. 2012; Hall et al. 2013). The focus is the shift from private to public management, rather than the territorial scale of this management.

Table 3.1 Main characteristics of UWM/SUWM/IUWM

Characteristics	UWM	SUWM/IUWM
Goals	Affordability, universalization and public health. Flood control	Also concerned with sustainability, adaptation, human welfare
Principle scale of policy implementation	Urban	Urban Sub-basin (within municipal boundaries)
Approach	Technocratic, centralized, sectoral, linear	Integrating grey and green infrastructure. Combining centralized and participatory, traditional and alternative technology, hard and soft measures
Actors	City departments, water utility/ies	City departments (including environmental department, transport), water utility/ies, civil society organizations, private sector
Mandates	Drinking water, sewage collection, treatment and disposal, stormwater management, flood protection and pollution control for diverse users (i.e. domestic, commercial, industrial) with divergent abilities to pay	In addition to conventional mandates, they strive for synergies with environmental management/conservation and with other urban actors (e.g. planning, housing)
Users	All users within the urban area (mainly public/urban users)	All users within the urban area (mainly public/urban users)
Water 'inputs'	Blue water (i.e. groundwater from within urban area and surface water often imported from beyond urban area)	Blue water, including local water sources after rehabilitation and rainwater Greywater (i.e. treated wastewater)
Wastewater/stormwater 'outputs'	Expelled out of the urban area Focus on hygiene and controlling urban flood/mudslides risks	Recycled and reused or infiltrated Climate change adaptation increasingly considered

Source Author

Spatial scale: Scholars (Niemczynowicz 1996; Lee 2000; Daigger 2011) and practitioners (Bahri 2012; OECD 2015) emphasise the local level as the most appropriate for managing urban water. This involves municipal, district and lower levels, such as the neighbourhood and household. The form of urbanization, the type of urban planning and the demand for water in urban areas and its use and discharge, requires policies at the urban scale (Jha et al. 2012). The shift to IUWM/SUWM leads to an emphasis on the use of small-scale infrastructure and technology (i.e. rainwater harvesting devices, smart water systems) and community infrastructure (Gabe et al. 2009; OECD 2015).

Water-related risks may be addressed by local actors (e.g. municipal civil defence) but certain risks require interventions by higher levels of government or multi-level responses, when drivers of risk originate beyond the local level. Similarly, water distribution in cities can be entirely under the authority of local governments (e.g. groundwater extraction within municipal borders) but often involve higher levels of government as water must be transported from parts of the basin beyond the city. Moreover, cities lack authority over the ownership of water resources and their allocation as higher levels of government generally decide *who* owns water and *how* it can be allocated to different users.

Approach: UWM has historically been implemented through large-scale infrastructure and technocratic, centralized management (Farrelly and Brown 2011; Rauch and Morgenroth 2013). UWM promotes managing different steps of the urban water cycle separately, through sectoral entities that do not coordinate the needs for water supply, environmental quality and flood management (Niemczynowicz 1999; Abdullah and Christensen 2004; Brown 2008; Daigger 2011; GTT 2014). The result is a linear approach of taking, using and rapidly discharging water (Kayaga et al. 2007; Brown and Farrelly 2009; Barraqué and Zandaryaa 2011; Gregory and Hall 2011; GTT 2014). This approach reflects the human desire to control nature and translates into flood protection infrastructure (e.g. dikes, dams), long-distance transfers of water, elevation of water with pumps and a networked water distribution system (Barraqué and Zandaryaa 2011; Nastar 2014).

SUWM attempts to integrate infrastructure and biophysical systems (e.g. stormwater treatment and rainwater harvesting systems), thereby considering social, economic, environmental and political contexts (Vlachos and Braga 2001; Mitchell 2006; Brown and Keath 2008; Brown et al. 2008). It combines large, centralized infrastructure with alternative and distributed technologies and participation (Van de Meene et al. 2011; Younos 2011; Closas et al. 2012). Similarly, IUWM uses structural and non-structural measures (e.g. new knowledge and information technologies, education programmes, water pricing and regulations) (Maheepala 2010; Iacob 2013). This shift in approach also emphasises coordination between government levels and between different types of actors (i.e. private sector, civil society, government, educational and research institutions) (Jha et al. 2012).

Actors: UWM actors are generally centralized (at local or regional level) and hierarchical (Elzen and Wieczorek 2005; Saleth and Dinar 2005; Farrelly and Brown 2011; Porse 2013), although in the Global South, informal, decentralized water providers also play important roles (Ahlers et al. 2014), as do local and international NGOs (Brocklehurst 2004). Higher levels of government generally design legal and regulatory frameworks and provide financing. SUWM/ IUWM promote decentralization, the devolution of administrative functions, inter-sectoral coordination, innovative funding solutions, co-management with communities and the private sector, and flexible institutional frameworks (i.e. public–private partnerships) (Bahri 2012; Closas et al. 2012; Whitler and Warner 2014). Within the IUWM framework, local governments are seen as well-placed to oversee urban water programmes, in combination with top-down regulatory responsibility and bottom-up user needs and obligations (GDRC 2015).

Mandates: The core UWM functions are drinking water supply, sewage collection and treatment, and urban drainage (Barraqué and Zandaryaa 2011; Engel et al. 2011), supplemented by flood mitigation and control of waterborne diseases (GTT 2014). These are often the responsibility of local governments, although they may delegate certain functions to higher levels of government or private actors or call in help from outside (e.g. to deal with disasters) (Baud and Hordijk 2009). However, local governments are rarely in the forefront of developing policies that address these challenges (Ibid). More recent approaches to UWM promote coordination between a wider range of stakeholders, such as urban planning, and in the Global South sometimes the development community (Whitler and Warner 2014). The IUWM/SUWM literature further advocates broadening mandates to incorporate ecosystem health, basin management, biodiversity conservation, conflicts and competing water uses, wastewater treatment and disposal, risk prevention, and integrating all urban activities (Coccossis and Nijkamp 2002; GTT 2014).

Since 1990, urban rivers have gained prominence in urban planning (Levin-Keitel 2014) and urban river rehabilitation projects have multiplied (Deason et al. 2010; de Haan et al. 2015). Green/blue-green infrastructure and Low Impact Development (LID) Best Management Practices (BMPs), such as infiltration basins, grass swales and green roofs manage stormwater runoff as close as possible to the source and reduce impacts on downstream receiving rivers (Jia et al. 2013). However, such projects are usually municipal projects and rarely extend beyond city boundaries. While IUWM emphasises urban–rural relationships (Bahri 2012), mandates are restricted to areas under the jurisdiction of the local authorities and/or water utility (Engel et al. 2011). Overall, responsibilities often remain unclear, fragmented and overlapping, leading to tensions between professionals and politicians with different values and views (Brown 2008; OECD 2016).

Users: Within UWM, water users are generally those who receive water from utilities for public/urban use, commercial and industrial. This may include informal users, such as residents of informal settlements, particularly where legal frameworks for water management have incorporated principles such as the human right to water (e.g. South Africa). In other cases, providers only serve registered consumers, as is the case of Hyderabad in India (Nastar 2014), despite the growing recognition of the human right to water (Obani and Gupta 2015). UWM generally prioritizes public/urban uses above other uses, but this is not always accompanied by incentives at local levels for rational water use, even as urban water use has increased (Lee 2000). Cities also attract industries and activities that increase water demand for energy generation, recreation and irrigation in their surrounding rural hinterlands (Lee 2000). This leads to an increase in the frequency of multiple and successive water uses, which affects river regimes and water quality (Lee 2000).

SUWM/IUWM highlight the need to consider upstream and downstream users, as well as non-urban users (Gabe et al. 2009; Closas et al. 2012; GTT 2014). Additionally, they increasingly recognize ecosystems as water users, for instance through efforts to maintain the minimum ecological water requirement (EWR) (Jia et al. 2011). However, it does not address equitable or sustainable water allocation among all users within a river basin.

Inputs and outputs: UWM relies on a combination of surface and groundwater from within or beyond the basin, treating these as never-ending sources of water (GTT 2014; Porse et al. 2015; Conti 2017), exhausting them until they dry up or are too contaminated for use due to inadequate pollution-control systems, and then turning to dams and inter-basin water transfers (Daigger 2011; Richter et al. 2013; OECD 2015). Large cities, due to their high water use, frequently import water from sources beyond their watersheds and struggle to maintain a reliable flow (Barrios et al. 2009; Daigger 2011; OECD 2015). Meanwhile, IUWM/SUWM promote the use of through decentralized infrastructure and technologies, such as rainwater harvesting systems and (blue-)green infrastructure, which can support aquifers, waterways and vegetation (Marlow et al. 2013; GTT 2014), and the reuse of greywater (Daigger 2011). However, such initiatives remain scattered and limited in practice. Maintaining sustainable flows for surface and groundwater sources requires land use regulation and preserving catchment areas, but this is less emphasized than in the IWRM/IRBM literature (see Chap. 4) (Bahri 2012; Closas et al. 2012). In part, this may be because urban water managers and city officials do not have mandates over these water producing areas.

Furthermore, grey and black waters have conventionally been seen as externalities that have to be rapidly expelled beyond the city to protect urban health, ignoring environmental sustainability, population growth, urbanization, industrialization and climate change (Kayaga et al. 2007; Makropoulos et al. 2008; Daigger 2011). IUWM/SUWM approaches wastewater as an opportunity and a potential resource by including it into the urban water cycle through infrastructural and institutional integration and reusing it in industrial activities, urban irrigation and groundwater recharge (Jia et al. 2005; Closas et al. 2012; GTT 2014). Wastewater can be used to extract heat and nutrients, and produce energy and soil-conditioning products (Daigger 2011). This shift requires an overhaul of urban water and wastewater infrastructure and coordination between water resources management and urban planning (Daigger 2011). However, combined sewers continue to affect human health and ecosystems (Porse 2013), especially in megacities which concentrate humans and polluting activities.

3.2.3 *Drivers, Institutions and Instruments*

A number of drivers shape urban water challenges, such as population growth and concentration, leading to rising demand for water for residential, agricultural and industrial use within and around the city (Lee 2000; Darrel Jenerette and Larsen 2006; Maheepala 2010; Gregory and Hall 2011; Van de Meene et al. 2011). Whereas countries in the Global North face aging populations and migration from abroad, countries in the Global South have rapidly growing populations and rural–urban migration (Neto 2016). Urbanization without proper sanitary infrastructure leads to environmental degradation and public health risks (Lee 2000; Van de Meene et al. 2011; Neto 2016). Moreover, agricultural, urban and industrial activities that pollute supply catchments also threaten urban water (Maheepala 2010). Meanwhile, droughts and heat waves caused by climate variability or climate change, are likely to reduce water supplies and affect water quality (Maheepala 2010; Van de Meene et al. 2011).

UWM institutions at multiple levels address the provision of urban water services and promote measures against water-related risks. Legislation, regulatory frameworks and norms are often set at higher levels of government. However, these have not always adjusted to the rapid changes and complex challenges facing urban water management (Lee 2000), or to new developments in innovation, technology, science, and changes in mandates and policy (Farrelly and Brown 2011). Local governments and citizens play a key role in implementing national and international legislation that shape and regulate urban water governance (OECD 2015).

Part of the UWM scholarship argues that institutions can only effectively address urban water challenges if users and polluters pay "adequately and justly for the services they enjoy from the water resource" (Lee 2000: 77). Supporters of market-based approaches to the provision of urban water services argue that removing decision-making from political arenas will allow for more competent water and sanitation (Wat&San) utilities (Lee 2000).

Shifting towards IUWM/SUWM requires that institutions can evolve in contexts of rapid changes and uncertainty, and function in an integrated fashion (Porse 2013; GTT 2014; Braga 2016). This involves flexible institutions and institutional arrangements that enable multiple actors to collaborate and meet a variety of objectives (Niemczynowicz 1999; Brown and Farrelly 2009; Floyd et al. 2014; GTT 2014). Flexibility and collaboration are crucial for facing water-related challenges that are multi-dimensional, multi-sectoral and multi-regional (Tortajada 2008). Resulting redundancies in institutional structures sometimes increases complexity, but may also lead to enhanced resilience (Lebel 2005).

Furthermore, other institutions relevant for UWM involve laws and regulations on spatial planning and land use, such as construction, zoning, land use parcelling, sanitary control and environmental conservation (GTT 2014). Rules regarding the delimitation and amalgamation of administrative regions also have important implications for the allocation of roles and responsibilities and the spatial scale at which urban water is managed (OECD 2014). In addition, informal institutions may play an important role at household and community levels, especially in cities of the Global South where formal institutions are absent in certain areas or highly bureaucratic (Jha et al. 2012). This can involve flood adaptation measures such as building structures around houses to block flood waters, raising the house's floor, moving

Table 3.2 Types of instruments of urban water management

Types of instruments	UWM
Economic	Water tariffs (user-pays principle); Water metering (linked to tariffs); Subsidies for connecting to the piped network; Sewage tariffs (polluter-pays principle); Financial incentives for water/wastewater reuse; Subsidized retrofits; Fines for improper solid waste disposal
Regulatory	Quality standards (minimum standards for drinking water quality); Access to sanitation with minimum standards; Zoning restrictions in flood prone areas; Stormwater ordinances; Ordinances for source water protection
Suasive	Water saving campaigns; Education programmes for promoting culture of civil protection; Flood risk warnings
Coordination	Flood control plans; Climate change adaptation plans

Source Author

furniture to higher floors, helping more vulnerable neighbours, or unclogging drains in surrounding streets (Simarmata 2015; Hordijk et al. 2016).

The institutional setting determines the instruments of UWM. Table 3.2 provides an inventory of commonly used instruments in UWM/IUWM/SUWM. Economic and suasive instruments are often targeted at the level of the individual, household, or municipality, and aim to change behaviour through incentives or persuasion. Many of the UWM instruments are biased towards the economic and social dimensions of sustainability and inclusiveness (e.g. financial incentives or educational measures to reduce water consumption), and to some extent towards relational dimensions (e.g. subsidies, differentiated tariff systems that address urban inequality). More recently, environmental concerns have led to an increase in instruments that promote sustainability (i.e. wastewater reuse, protected areas).

3.3 UWM for Sustainable and Inclusive Development

3.3.1 Steps Forward

Urban water demand has increased in absolute terms (as cities grow) and per capita, as urban dwellers in the Global South become more affluent. Increasing competition between users adds stress, pollution reduces water availability and extreme weather makes future availability less predictable (GTT 2014; OECD 2014). In the Global North, population ageing means water demand could decrease, with implications for infrastructure investments and potential risks of bacterial after-growth in drinking water (Hummel and Lux 2007). While conventional UWM is supply-oriented, adjusting infrastructure to meet water demand, IUWM/SUWM aims to better match water demand to water availability (Maheepala 2010; Gregory and

Hall 2011; GTT 2014). Flexible approaches can help reorient infrastructure investments according to demand (Neto 2016). IUWM/SUWM also seeks to diversify water sources (i.e. greywater reuse, rainwater harvesting), enhancing water security (Maheepala 2010).

IUWM/SUWM can reduce negative environmental impacts by considering upstream and downstream linkages and the basin as a whole (Closas et al. 2012), for example, by protecting green areas, including native flora and fauna in urban waterways and estuaries and around urban areas (Maheepala 2010). The decentralized, collaborative and participatory forms of water management promoted by IUWM/SUWM are also more inclusive of different views, interests and environmental values (Brown 2008; Maheepala 2010). Deliberation and decision-making processes are iterative, long-term and consider the total water cycle rather than isolated, localized processes (Closas et al. 2012).

3.3.2 *Lingering Obstacles*

Despite the paradigm shifts towards greater sustainability and integration, the water sector's path-dependent nature means that implementation of IUWM/SUWM is slow. Retrofitting existing infrastructure and technology involves high transitioning costs and can lead to stranded assets (Brown and Farrelly 2007; Brown 2008; Closas et al. 2012; Marlow et al. 2013; OECD 2015; Bos and Gupta 2018). Investment and technological 'lock in' delays the uptake of alternatives (Marlow et al. 2013). Local governments often lack the necessary funds or capacity to leverage capital investments, and they remain dependent on higher levels of government (Closas et al. 2012). Water management institutions tend to be resistant to change and have low adaptive capacity (Van de Meene et al. 2011; Marlow et al. 2013). Transitioning to IUWM/SUWM also relies on a change in values throughout society (Gabe et al. 2009). It is mainly pushed for by social groups with more inclusive views regarding UWM (Gabe et al. 2009; Winz et al. 2014). Moreover, switching to new infrastructure and processes may involve trade-offs that require careful consideration. For instance, SUWM encourages innovative solutions that may be energy-intensive (e.g. desalination, pumps for rainwater tanks) or risky (e.g. wastewater reuse) (Marlow et al. 2013).

The IUWM/SUWM literature remains largely prescriptive and studies available reveal a failure to go beyond ad hoc projects (Harding 2006; Mitchell 2006; Brown and Farrelly 2009). Learning experiments and innovation at local levels are not easily transferred to urban decision-making processes and policy (Pearson et al. 2010; Farrelly and Brown 2011). Such knowledge transfers require adequate social learning and coordination mechanisms, long-term strategic planning and the inclusion of research and development partners in key projects (Farrelly and Brown 2011; Van de Meene et al. 2011; Colenbrander 2018). Limited human and financial resources, a lack of available information, a lack of industry-wide experience and knowledge, a lack of monitoring and poor communication processes further hinder SUWM/IUWM implementation (Brown and Farrelly 2007; Adank et al. 2011).

Institutional inertia further represents a barrier to the implementation of SUWM/IUWM, in particular for proactive responses to increasingly complex urban water challenges (Lee 2000; Brown and Farrelly 2007). This includes over-centralisation, bureaucratic inefficiencies, inconsistent regulatory approvals processes, conflicting or unclear mandates amongst organisations, institutional and sectoral fragmentation (despite aims of integrating the urban water cycle), unproductive intergovernmental relations and poor collaboration, limited data sharing, unclear property rights and lack of authority of operational organizations (Lee 2000; Brown and Farrelly 2007; Adank et al. 2011). The result is often inadequate, fragmented planning and a lack of enforcement of existing plans (Adank et al. 2011). While it is increasingly recognized that institutions contribute to the slow pace of change, there is little understanding on how to overcome this (Brown and Farrelly 2007, 2009).

Moreover, IUWM/SUWM does not address power relations, nor implementation challenges resulting from a lack of political will, political leadership or political incentives (Brown 2008; Pearson et al. 2010; Van de Meene et al. 2011). While IUWM promotes local government participation in basin-wide planning spaces, these usually lack decision-making powers (Vlachos and Braga 2001). Community participation is often lacking and expert knowledge tends to dominate decision-making to the detriment of local, lay knowledge (Brown and Farrelly 2007; Van de Meene et al. 2011; Brandeler et al. 2014). IUWM/SUWM also mainly applies to formal settlements, and its references to informal settlements focuses on the need for land use management and land tenure (Porse 2013).

3.4 Scalar Limitations of Urban Water Governance Paradigms

3.4.1 Metropolitan Areas Have Unique Challenges and Opportunities

Regarding urban water challenges, a limitation of UWM and SUWM/IUWM paradigms is that they do not consider the spatial scale of the metropolis. The limited MWM literature covers principles of water management for metropolitan contexts and the expansion of decision-making structures and urban water networks to suburban constituencies or peripheral municipalities (Kallis and Coccossis 2002; Keil and Boudreau 2006)—topics left unaddressed by the IUWM/SUWM literature. However, water challenges in metropolitan regions can be difficult for UWM actors to address. For instance, a lack of sanitation infrastructure in one metropolitan municipality affects the neighbouring municipality downstream, through the health risks of contaminated water. Large metropolitan regions can affect water resources in the broader basin. High urban water demand reduces water available to the basin's other users, such as farmers. Aquifers within metropolitan regions are affected by land use and inadequate wastewater management (Foster and Ait-Kadi 2012). Metropolitan regions are therefore not just large cities, but experience water-related problems of a different nature than smaller urban areas. Peripheral metropolitan municipalities often face greater water challenges and have fewer capacities to address these alone.

Megacities and large metropolitan regions also present a number of opportunities, including to address water challenges, due to their size (economies of scale), their greater capacity (financial, human) relative to smaller urban or rural areas and the markets (for jobs, products) that they provide for the broader region (GTT 2014; Kraas et al. 2014; UN-HABITAT 2016). The resulting "fragmentation of policy making among multiple governmental units diminishes problems of concentrated powers and can promote competition and innovation" (Feiock 2009: 356). Their unique position also makes them centres for research and innovation, including on environmental practices (GTT 2014; Kennedy et al. 2015).

However, responsibilities for water distribution, wastewater collection, storm and flood management often remain at local levels. Within megacities, this involves multiple local governments and utilities. Municipalities are frequently the level at which policy is implemented, but they are often arbitrary and reflect ancient patterns (OECD 2015). This can result in incoherent and even contradictory water management practices, as well as inefficiencies, as decisions by one municipality may impose positive or negative externalities on others (Richardson 1989; Feiock 2009; OECD 2011; Sorensen 2011; Li et al. 2015). It may also hamper cities' capacity to build the necessary coalitions of actors or structures of governance (Aguilar 2008) and to foster productive cooperation (Kim et al. 2015). Metropolitan regions often experience governance fragmentation through "overlapping or disconnected institutional structures, national government intervention overpowering local authorities, and disconnections between land and water sectors" (Li et al. 2015: 603) and environmental planning and resource management (Kim et al. 2015). Neighbouring municipalities may be led by rival political parties, or struggle to develop cohesive responses to shared challenges due to the inconsistency of available data, data dispersion across agencies and the lack of information-sharing mechanisms (OECD 2016).

The growth and multiplication of metropolitan regions puts pressure on service delivery and infrastructure development, as well as on adequate management (Vlachos and Braga 2001: 4). Metropolitan municipalities often share water resources, and their water-related activities affect each other (e.g. lack of sewage treatment upstream, deforestation around areas of springs), but political fragmentation hinders coherent policies and plans to address shared concerns. Exclusively local practices leads to water resource degradation, irrational investments and sector-oriented management (Toledo Silva and Amaral Porto 2003; Foster and Ait-Kadi 2012). This is even more complex in metropolitan regions with large numbers of jurisdictions within the urban area. Moreover, many of the world's largest cities are located in the Global South, in areas particular vulnerable to climate change, they hold pockets of extreme vulnerability and lack coping mechanisms (Kraas et al. 2014). They often experience rapid growth, especially on their margins, challenging an already fragmented institutional set up and further weakening regional planning and coordination (Adank et al. 2011).

3.4.2 UWM Ignores the Linkages to the River Basin

The mismatch between the spatial boundaries of institutions (with administrative functions) and of biophysical systems (e.g. river basins and aquifers) can be an obstacle for addressing urban water challenges (Abdullah and Christensen 2004; Cumming et al. 2006; Young et al. 2008; Salzman et al. 2014; OECD 2015).[4] UWM focuses on urban (often municipal-level) concerns and does not address those of the river basin and the links between the two spatial scales, which hinders sustainable and inclusive policies (Brown 2008; Van de Meene et al. 2011). The IUWM and SUWM scholarship does so to a greater extent, by highlighting the importance of urban–rural relationships (Pearson et al. 2010; Bahri 2012) and advocates for broadening mandates to incorporate ecosystem health, basin management, biodiversity conservation, conflicts and competing water uses, wastewater treatment and disposal, risk prevention, and integrating all urban activities (Coccossis and Nijkamp 2002; GTT 2014). This is evidenced by the increased prominence since the early 1990s of water and rivers in urban planning (Levin-Keitel 2014). As water resources become scarcer, a better understanding of urban–rural interlinkages are needed (Pearson et al. 2010). However, although more projects incorporate basin considerations, they usually are municipal projects and they rarely extend beyond city boundaries (Neto 2016).

IUWM/SUWM actors also do not interfere in water allocation, which is typically the mandate of higher levels of government. Local governments have limited (or no) mandates on water resources outside their borders, even if these resources are part of the same basin (e.g. in irrigation or hydropower), which may have enormous implications for water availability (Maheepala 2010). They also generally do not manage surface and groundwater systems within their borders, nor address evapotranspiration and the preservation of environmental flows (Pearson et al. 2010). In addition, responsibilities for urban water are not clearly coordinated with basin actors. Rather, mandates are often fragmented and overlapping, leading to tensions between professionals and politicians with different values and views (Brown 2008; OECD 2016; Brandeler et al. 2019). As decisions within the river basin are (mainly) made by different actors than those within UWM, learning is not easily transferred between the two scales (Pearson et al. 2010).

Urban areas depend on biodiversity and ecosystem services (BESS), which can be produced within their borders, but more often in their rural hinterlands or further away (e.g. virtual water through food imports). These services are essential for the sustainable and inclusive development of cities, and cities can have profound effects on them. Ecologists have only recently started to show interest in urban areas and how to address urban environmental problems (Grimm et al. 2008). Urban water scholars

[4] This issue is also referred to as "hydro-administrative mismatch" by the OECD (2013).

and urban planners also increasingly recognize the benefits of BESS within cities [e.g. urban forests can reduce stormwater runoff (Xiao and McPherson 2016) and prevent pollution in urban waterways and groundwater resources (Livesley et al. 2016)] and those from the wider watershed [e.g. protecting upstream areas for urban water supply and water purification services (Kenny 2006)]. However, despite convergence between scholarly disciplines, cities continue to create externalities related to the management of water resources and related ecosystems (i.e. sewage contamination from affecting communities downstream, high water demand requiring reallocation to urban uses from other uses, soil sealing affecting infiltration and aggravating risks within the city and downstream). Conflicts between environmental sustainability and housing right agendas are common at the urban/rural interface of large cities, as low-income populations are pushed to the margins through a real estate driven process of social-spatial exclusion and occupy watershed areas (Refinetti 2006; Klink 2009).

3.5 Inferences

This chapter reviewed the urban water management literature to gain insights on how it understands and addresses metropolitan water challenges. It explored this in relation to two knowledge gaps identified in Sect. 1.2.3, namely (a) how the literature relates to the river basin context and conceptualizes urban/rural relations, (b) how the literature relates to the metropolitan context as a multi-jurisdictional and often heterogenous urban environment. Multiple factors identified in the UWM literature may facilitate or hinder sustainable and inclusive development. In conjunction with these factors, the research examined whether political fragmentation within metropolises and the lack of mandates over rural areas of their basin also play a role in determining the effectiveness of UWM institutions and instruments in terms of sustainability and inclusiveness.

UWM elements for sustainable and inclusive development

The literature review highlighted a shift from conventional approaches towards paradigms such as IUWM and SUWM, which embrace sustainable and inclusive development principles. These principles are promoted through institutions and instruments that implement water demand management, diversify water sources, protect ecosystem services, foster decentralization and participation and facilitate social learning. Common obstacles that hinder effective implementation are path dependency, a focus on technological fixes, the difficulty to scale up local initiatives, rapid demographic changes (shrinking cities in the Global North and rapid, unplanned urban growth in the Global South), limited local budgets and a lack of knowledge sharing. The empirical chapters evaluate whether such measures are part of the responses to metropolitan water challenges, and if so, how effective they are.

Political fragmentation within metropolitan areas

Metropolitan regions are often composed of multiple local governments with varying levels of autonomy in decision-making. As UWM mandates are mainly local, metropolitan regions can concentrate a significant number of actors. IUWM and SUWM embrace decentralization and participation, which may ensure more context-relevant policies and social and relational inclusion, particularly in highly unequal metropolises of the Global South. However, a constant prioritization of local interests, policies and practices can lead to political fragmentation within metropolitan areas. Localism could hinder solutions to shared problems, as actors have no incentives or authority to spend resources beyond their jurisdiction's boundaries. This may be aggravated when there are strong political differences in metropolitan regions. Partisan politics can lead to tensions between metropolitan municipalities and a lack of political will to coordinate actions on UWM. Differences in human and financial capacity, and power relations, also shape different local governments' ability to coordinate on an equal footing. Based on this framework, the empirical chapters explore the coherence of UWM across metropolitan regions in terms of characteristics identified in Sect. 3.2.2 (i.e. goals, mandates, approaches, users), and how the framework's shortcomings increase shared water challenges.

Urban dependency on rural areas

Large cities often rely on watersheds and other ecosystems beyond their boundaries. There is growing recognition of the importance of preserving these areas, particularly as cities expand. However, UWM actors lack mandates to do so directly, in particular when these areas fall outside the borders of their jurisdiction. Urban water governance regimes therefore generally fail to recognize these dependencies and address drivers that affect water resources and ecosystem services at the adequate level. This research therefore explores the coherence of UWM across the urban/rural interface, in terms of characteristics identified in Sect. 3.2.2 (i.e. goals, mandates, approaches, users), and its implications for addressing shared water challenges.

References

Abdullah D, Christensen B (2004) Integrated river basin management. Bullet Ingenieur 22:21–23

Abers R, Keck ME (2013) Practical authority: agency and institutional change in brazilian water politics—excerpts. In: American Political Science Association 2013 Annual Meeting. Available at: https://papers.ssrn.com/sol3/papers.cfm?abstract_id=2302112. Accessed 14 Oct 2018

Adank M et al (2011) Towards integrated urban water management in the Greater Accra Metropolitan Area: current status and strategic directions for the future. Accra, Ghana. Available at: http://switchurbanwater.lboro.ac.uk/outputs. Accessed 4 Jan 2019

Aguilar AG (2008) Peri-urbanization, illegal settlements and environmental impact in Mexico City. Cities. https://doi.org/10.1016/j.cities.2008.02.003

Ahlers R et al (2014) Informal space in the urban waterscape: disaggregation and co-production of water services. Water Alternat 7(1):1–14. Available at: www.water-alternatives.org. Accessed 2 Oct 2018

Bahri A (2012) Integrated urban water management: briefing note. Technical Committee Background Paper 16. Stockholm

Bakker K (2002) From state to market? water mercantilización in Spain. Environ Plann A 34(5):767–790. https://doi.org/10.1068/a3425

Bakker K (2003) Archipelagos and networks: urbanization and water privatization in the South. Geogr J 169(4): 328–341. Available at: https://www.jstor.org/stable/pdf/3451572.pdf?refreqid=excelsior%3Ac8e3dfe47a3383a68806db39fa6b4cf7. Accessed 24 Dec 2018

Barraqué B, Zandaryaa S (2011) Urban water conflicts: background and conceptual framework. In: Barraqué B (ed) Urban water conflicts. Routledge and GSE Research. Available at: https://www.ingentaconnect.com/content/rout/c4z92/2011/00000001/00000001/art00002. Accessed 27 Dec 2018

Barrios JE, Rodríguez-Pineda JA, de la Maza Benignos M (2009) Integrated river basin management in the Conchos River basin, Mexico: a case study of freshwater climate change adaptation. Climate Dev 1(3):249–260

Baud ISA, Hordijk MA (2009) Dealing with risks in urban governance: what can we learn from "resilience thinking". In: The 4th International Conference of the International Forum on Urbanism (IFoU)—The New Urban Question—Urbanism beyond Neo-Liberalism, Amsterdam/Delft. Available at: http://www.ifou.org/conferences/2009delft/proceedings/8NewApproachesofUrbanGovernance/fullpapers/F001BaudandHordijkTheNewUrbanQuestion.pdf. Accessed 30 Sept 2019

Bos K, Gupta J (2018) Climate change: the risks of stranded fossil fuel assets and resources to the developing world. Third World Q 39(3):436–453. https://doi.org/10.1080/01436597.2017.1387477

Braga BPF (2016) Integrated urban water resources management: a challenge into the 21st century. Water Resour Dev 17(4):581–599. https://doi.org/10.1080/07900620120094127

Brocklehurst C (2004) Local management models for water supply and sanitation for the urban poor

Brown RR (2005) Impediments to integrated urban stormwater management: the need for institutional reform. Environ Manag 36(3):455–468. https://doi.org/10.1007/s00267-004-0217-4

Brown RR (2008) Local institutional development and organizational change for advancing sustainable urban water futures. Environ Manag 41(2):221–233. https://doi.org/10.1007/s00267-007-9046-6

Brown RR, Keath NA (2008) Drawing on social theory for transitioning to sustainable urban water management: turning the institutional super-tanker. Aust J Water Resourc 12(2):73–83. https://doi.org/10.1080/13241583.2008.11465336

Brown RR, Farrelly M (2007) barriers to advancing sustainable urban water management: a typology. Rainwater Urban Des 229–236. Available at: http://search.informit.com.au.ezproxy.lib.monash.edu.au/documentSummary;dn=888968667496813;res=IELENG

Brown RR, Keath N, Wong T (2008) Transitioning to water sensitive cities: historical, current and future transition states. In: Proceedings of the 11th International Conference on Urban Drainage. Edinburgh, Aug 31–Sept 5

Brown RR, Farrelly MA (2009) Delivering sustainable urban water management: a review of the hurdles we face. Water Sci Technol. https://doi.org/10.2166/wst.2009.028

Closas A, Schuring M, Rodriguez D (2012) Integrated urban water management—lessons and recommendations from regional experiences in Latin America, Central Asia, and Africa. Washington, DC

Coccossis H, Nijkamp P (2002) Scarce water in modern cities. Built Environ (1978-), 28(2):92–95. Available at: http://www.jstor.org

Colenbrander DR (2018) Governing coastal risk and vulnerability: new pathways within developing city-scale contexts, Cape Town. University of Amsterdam, South Africa

Conti KI (2017) Norms in multilevel groundwater governance & sustainable development. University of Amsterdam

Cumming GS, Cumming DHM, Redman CL (2006) Scale mismatches in social-ecological systems: causes, consequences, and solutions. Ecol Soc 11(1):14. https://doi.org/10.5751/ES-01569-110114

Daigger GT (2011) Sustainable urban water and resource management. Bridge 41(1):13–18

Darrel Jenerette G, Larsen L (2006) A global perspective on changing sustainable urban water supplies. Global Planet Change 50(3–4):202–211. https://doi.org/10.1016/J.GLOPLACHA.2006.01.004

de Haan FJ et al (2015) Transitions through a lens of urban water. Environ Innov Soc Trans 15:1–10. https://doi.org/10.1016/j.eist.2014.11.005

Deason JP et al (2010) Integrated planning framework for urban river rehabilitation. J Water Resour Plann Manag 136(6)

Donoso G (2014) Integrated water management in Chile. In: Martínez-Santos P (ed) Integrated water resources management in the 21st century: revisiting the paradigm. Taylor & Francis Group, London

Elzen B, Wieczorek A (2005) Transitions towards sustainability through system innovation. Technol Forecast Soc Change 72(6):651–661. https://doi.org/10.1016/J.TECHFORE.2005.04.002

Engel K et al (2011) Big cities. Big water. Big challenges. Water in an urbanizing world. Berlin. Available at: https://wwwwwfse.cdn.triggerfish.cloud/uploads/2019/01/big-cities_big-water_big-challenges_2011.pdf. Accessed 27 Dec 2018

Farrelly M, Brown RR (2011) Rethinking urban water management: experimentation as a way forward? Glob Environ Chang. https://doi.org/10.1016/j.gloenvcha.2011.01.007

Feiock RC (2009) Metropolitan governance and institutional collective action. Urban Affairs Rev 44(3):356–377. https://doi.org/10.1177/1078087408324000

Floyd J et al (2014) Managing complexity in Australian urban water governance: transitioning sydney to a water sensitive city. Futures 61:1–12. https://doi.org/10.1016/J.FUTURES.2014.04.002

Follmann A (2016) Governing riverscapes: urban environmental change along the River Yamuna in Delhi, India. Megacities. Franz Steiner Verlag, Stuttgart

Foster S, Ait-Kadi M (2012) Integrated water resources management (iwrm): how does groundwater fit in? Hydrogeol J 20:415–418. https://doi.org/10.1007/s10040-012-0831-9

Gabe J, Trowsdale S, Vale R (2009) Achieving integrated urban water management: planning top-down or bottom-up? Water Sci Technol. https://doi.org/10.2166/wst.2009.196

GDRC (2015) Integrated urban water resources management. Available at: https://www.gdrc.org/uem/water/iwrm/index.html. Accessed 13 Oct 2015

Gregory A, Hall M (2011) Urban water sustainability. In: Prosser I (ed) Water: science and solutions for Australia. CSIRO Publishing, p 192

Grimm NB et al (2008) Global change and the ecology of cities. Sour Sci New Ser 319(5864):756–760. Available at: http://www.jstor.org

GTT (2014) Integrated urban water resources management. In: Humanity, Gulbenkian think tank on water and the future of humanity- revisiting water security. Springer International Publishing, pp 109–132

Hall D, Lobina E, Terhorst P (2013) Re-municipalisation in the early twenty-first century: water in France and energy in Germany. Int Rev Appl Econ 27(2):193–214. https://doi.org/10.1080/02692171.2012.754844

Harding R (2006) Ecologically sustainable development: origins, implementation and challenges. Desalination 187(1–3):229–239. https://doi.org/10.1016/J.DESAL.2005.04.082

Hordijk M, van den Brandeler F, Filippi ME (2016) Facing the floods: community responses to increased rainfall in Guarulhos, Brazil and Arequipa, Peru. In: Roy M et al (eds) Urban poverty and climate change: life in the slums of Asia, Africa and Latin America. Routledge, London

Hummel D, Lux A (2007) Population decline and infrastructure: the case of the German water supply system, pp 167–191. https://doi.org/10.1553/populationyearbook2007s167

Iacob V-S (2013) The wastewater—a problem of integrated urban water management. Proc Econ Fin 6:436–443. https://doi.org/10.1016/S2212-5671(13)00160-3

Jacobi P, Fracalanza AP, Silva-Sánchez S (2015) Governança da água e inovação na política de recuperação de recursos hídricos na cidade de São Paulo. Cadernos Metropole 17(33):61–81. https://doi.org/10.1590/2236-9996.2015-3303

Jha AK, Bloch R, Lamond J (2012) Cities and flooding a guide to integrated urban flood risk management for the 21st century. Washington, DC

Jia H et al (2005) Research on wastewater reuse planning in Beijing central region. Water Sci Technol 51(10):195–202

Jia H, Ma H, Wei M (2011) Calculation of the minimum ecological water requirement of an urban river system and its deployment: a case study in Beijing central region. Ecol Model 222(17):3271–3276. https://doi.org/10.1016/J.ECOLMODEL.2011.05.026

Jia H, Yao H, Yu SL (2013) Advances in LID BMPs research and practice for urban runoff control in China. Front Environ Sci Eng 7(5):709–720. https://doi.org/10.1007/s11783-013-0557-5

Kaika M (2005) City of flows: modernity, nature, and the city. Routledge, London. Available at: https://books.google.com.br/books?hl=en&lr=&id=srhjfcA5x5sC&oi=fnd&pg=PP2&ots=d9NW1sH83R&sig=EJbOYk_kvayRKHx_IOiWTIR8WZg&redir_esc=y#v=onepage&q&f=false. Accessed 22 Dec 2018

Kallis G, Coccossis H (2002) Water for the city: lessons from tendencies and critical issues in five advanced metropolitan areas. Built Environ 28(2):96–110

Kayaga S, Smout I, Al-Maskati H (2007) Water demand management—shifting urban water management towards sustainability. Water Sci Technol Water Supply 7(4):49–56. Available at: http://creativecommons.org/licenses/by-nc-nd/2.5/

Keil R, Boudreau JA (2006) Metropolitics and metabolics: rolling out environmentalism in Toronto. In: In the nature of cities: urban political ecology and the politics of urban metabolism. Routledge, pp 56–77. doi: https://doi.org/10.4324/9780203027523-10

Kennedy CA et al (2015) Energy and material flows of megacities. In: Proceedings of the National Academy of Sciences of the United States of America. Natl Acad Sci 112(19):5985–5990. doi: https://doi.org/10.1073/pnas.1504315112

Kenny A (2006) Ecosystem services in the New York City watershed. Ecosyst Marketplace, 10 Feb. https://doi.org/10.1016/j.trsl.2007.06.005

Kim JH, Keane TD, Bernard EA (2015) Fragmented local governance and water resource management outcomes. J Environ Manag 150:378–386. https://doi.org/10.1016/j.jenvman.2014.12.002

Klink J (2009) New governance for metropolitan areas. The international scene and prospects for the Brazilian case. Cadernos Metrópole 11(22):415–433. Available at: http://www.redalyc.org/pdf/4028/402837806006.pdf. Accessed 15 Nov 2018

Kraas F et al. (2014) Megacities: our global urban future. In: Kraas F et al (eds). Springer, Dordrecht

Lebel L (2005) The politics of scale, position, and place in the governance of water resources in the Mekong Region. Ecol Soc 10(2). https://doi.org/10.5751/ES-01543-100218

Lee TR (2000) Urban water management for better urban life in Latin America. Urban Water 2(1):71–78. Available at: www.elsevier.com/locate/urbwat. Accessed 13 Dec 2018

Levin-Keitel M (2014) Managing urban riverscapes: towards a cultural perspective of land and water governance. Water Int 39(6):842–857. https://doi.org/10.1080/02508060.2014.957797

Li E, Endter-Wada J, Li S (2015) Characterizing and contextualizing the water challenges of megacities. J Am Water Resour Assoc 51(3):589–613. https://doi.org/10.1111/1752-1688.12310

Livesley SJ et al (2016) Soil carbon and carbon/nitrogen ratio change under tree canopy, tall grass, and turf grass areas of urban green space. J Environ Qual 45(1):215. https://doi.org/10.2134/jeq2015.03.0121

Maheepala S (2010) towards the adoption of integrated urban water management approach for planning. In: Swayne D et al. (eds) International congress on environmental modelling and software modelling for environment's sake. Ottawa, Canada, p 8. https://doi.org/10.2175/193864710798207017

Makropoulos CK et al (2008) Decision support for sustainable option selection in integrated urban water management. Environ Model Softw. https://doi.org/10.1016/j.envsoft.2008.04.010

Marlow DR et al (2013) Towards sustainable urban water management: a critical reassessment. Water Res 47:7150–7161. https://doi.org/10.1016/j.watres.2013.07.046
McDonald DA (2018) Remunicipalization: the future of water services? Geoforum 91:47–56. https://doi.org/10.1016/J.GEOFORUM.2018.02.027
Mitchell VG (2006) Applying integrated urban water management concepts: a review of Australian experience. Environ Manage 37(5):589–605. https://doi.org/10.1007/s00267-004-0252-1
Murdoch J (2006) Post-structuralist geography: a guide to relational space. Sage, London
Nastar M (2014) What drives the urban water regime? An analysis of water governance arrangements in Hyderabad, India. Ecol Soc. https://doi.org/10.5751/ES-06570-190257
Neto S (2016) Water governance in an urban age. Utilities Policy, 43. https://doi.org/10.1016/j.jup.2016.05.004
Niemczynowicz J (1996) Megacities from a water perspective. Water Int 21(4):198–205
Niemczynowicz J (1999) Urban hydrology and water management—present and future challenges. Urban Water 1(1):1–14. https://doi.org/10.1016/S1462-0758(99)00009-6
Obani P, Gupta J (2015) The evolution of the right to water and sanitation: differentiating the implications. Rev Eur Compar Int Environ Law 24(1):27–39. https://doi.org/10.1111/reel.12095
OECD (2011) Water governance in OECD countries: a multi-level approach. OECD studies on water. OECD Publishing
OECD (2013) Making water reform happen in Mexico. OECD studies on water, Paris. https://doi.org/10.1787/9789264187894-en
OECD (2014) Managing water for future cities. Paris
OECD (2015) Water and cities: ensuring sustainable futures. OECD Publishing, Paris. https://doi.org/10.1787/9789264230149-en
OECD (2016) Water governance in cities. Paris
Parker DJ, Sewell DR (1988) Evolving water institutions in England and Wales: an assessment of two decades of experience. Nat Resour J 28(4). Available at: http://digitalrepository.unm.edu/nrj/vol28/iss4/6. Accessed 24 Dec 2018
Pearson LJ et al (2010) A sustainable decision support framework for Urban water management. Water Resour Manage 24:363–376. https://doi.org/10.1007/s11269-009-9450-1
Pierre J, Peters BG (2000) Governance, politics and the state. Macmillan, Basingstoke
Pigeon M et al (2012) Remunicipalisation: putting water back into public hands. In: Pigeon M et al (eds) Transnational Institute, Amsterdam. Available at: https://idl-bnc-idrc.dspacedirect.org/bitstream/handle/10625/49061/IDL-49061.pdf?sequence=1. Accessed 30 Sept 2018
Porse EC (2013) Stormwater governance and future cities. Water 5:29–52. https://doi.org/10.3390/w5010029
Porse E et al (2015) Pumping for the masses: evolution of groundwater management in metropolitan Los Angeles. GeoJournal 81:793–809. https://doi.org/10.1007/s10708-015-9664-0
Rauch W, Morgenroth E (2013) Urban water management to increase sustainability of cities. Water Res 47:7149. https://doi.org/10.1016/j.watres.2013.10.042
Refinetti ML (2006) Moradia e Mananciais: tensão e diálogo na metrópole. FAUUSP/FAPESP, São Paulo
Richardson HW (1989) The big, bad city: mega-city myth? Third World Plann Rev 11(4):355–372. Available at: http://www.ncbi.nlm.nih.gov/pubmed/12343008. Accessed 29 May 2019
Richter BD et al (2013) Tapped out: how can cities secure their water future? Water Policy 15(3):335–363. https://doi.org/10.2166/wp.2013.105
Rietveld LC et al (2016) Improving health in cities through systems approaches for urban water management. Environ Health Global Access Sci Sour 15(Suppl 1). https://doi.org/10.1186/s12940-016-0107-2
Saleth RM, Dinar A (2005) Water institutional reforms: theory and practice. Water Policy 7(1):1–19. https://doi.org/10.2166/wp.2005.0001
Salzman J et al (2014) The most important current research questions in urban ecosystem services. Duke Environ Law Policy Forum 25(1)

Simarmata HA (2015) How informal institutions manage flood-risk at community level: an empirical study of Kampung Muara Baru, Jakarta. Available at: https://www.unisdr.org/campaign/resilientcities/assets/documents/privatepages/Howinformalinstitutionsmanageflood-riskatcommunitylevelAnempiricalstudyofKampungMuaraBaru,Jakarta.pdf. Accessed 30 Dec 2018

Sorensen A (2011) Megacity sustainability: urban form, development, and governance. In: Sorensen A, Okata J (eds) Megacities: urban form, governance, and sustainability. Springer, Tokyo, pp 397–418. https://doi.org/10.1007/978-4-431-99267-7_17.

Swyngedouw E (2006) Power, water and money: exploring the nexus power. Oxford, UK. Available at: http://hdr.undp.org/sites/default/files/swyngedouw.pdf. Accessed: 20 June 2018

Toledo Silva R, do Amaral Porto MF (2003) Caracterização da Bacia do Alto Tietê. Estudos Avançados 17(1747):129–145

Tortajada C (2008) Challenges and realities of water management of megacities: the case of Mexico city metropolitan area. J Int Aff 61(2):147–166

UN-HABITAT (2016) Urbanization and development: emerging futures. Nairobi, Kenya. Available at: http://wcr.unhabitat.org/wp-content/uploads/2017/02/WCR-2016-Full-Report.pdf. Accessed 6 May 2018

van de Meene SJ, Brown RR, Farrelly MA (2011) Towards understanding governance for sustainable urban water management. Glob Environ Chang 21(3):1117–1127. https://doi.org/10.1016/j.gloenvcha.2011.04.003

van den Brandeler F et al (2014) Decentralization, participation and deliberation in water governance: a case study of the implications for Guarulhos, Brazil. Environ Urban 26(2):489–504. https://doi.org/10.1177/0956247814544423

van den Brandeler F, Gupta J, Hordijk M (2019) Megacities and rivers: scalar mismatches between urban water management and river basin management. J Hydrol 573:1067–1074. https://doi.org/10.1016/j.jhydrol.2018.01.001

Varis O et al (2006) Megacities and water management. Int J Water Resour Dev 22(2):377–394. https://doi.org/10.1080/07900620600684550

Vlachos E, Braga B (2001) The challenges of urban water management. In: Maksimović Č, Tejada-Guibert JA (eds) Frontiers in urban water management: deadlock or hope. IWA Publishing

Whatmore S, Boucher S (1993) Bargaining with nature. The discourse and practice of environmental planning gain. Trans Inst British Geogr 17/2:66–88. Available at: https://research-information.bristol.ac.uk/en/publications/bargaining-with-nature-the-discourse-and-practice-of-environmental-planning-gain(3de44436-b383-49cf-b392-ba925d9c164c).html. Accessed 26 May 2019

Whitler J, Warner J (2014) Integrated urban water management for planners. PAS Memo. Available at: https://www.planning.org/publications/document/9137354/. Accessed 15 Oct 2018

Winz I, Trowsdale S, Brierley G (2014) Understanding barrier interactions to support the implementation of sustainable urban water management. Urban Water J 11(6):497–505. https://doi.org/10.1080/1573062X.2013.832777

Xiao Q, McPherson EG (2016) Surface water storage capacity of twenty tree species in Davis, California. J Environ 45(1):188. https://doi.org/10.2134/jeq2015.02.0092

Young OR, Schroeder H, King LA (2008) Institutions and environmental change. The MIT Press

Younos T (2011) Paradigm shift: holistic approach for water management in urban environments. Front Earth Sci. https://doi.org/10.1007/s11707-011-0209-7

Zimmer A (2010) Urban political ecology: theoretical concepts, challenges, and suggested future directions. Erdkunde 64(4):343–354. https://doi.org/10.2307/25822107

Chapter 4
River Basin Governance

4.1 Introduction

This chapter reviews the literature on river basin governance[1] to answer the following question: What does the literature on river basin governance tell us about how to understand and address metropolitan water challenges? First, it examines the evolution of river basin governance over time, and more specifically its shift towards Integrated Water Resources Management (IWRM) and Integrated River Basin Management (IRBM) (see Sect. 4.2.1). It then describes the main characteristics of IWRM and IRBM (see Sect. 4.2.2), as well as the main drivers, institutions and instruments shaping these concepts (see Sect. 4.2.3). Subsequently, the chapter debates the strengths and weaknesses of IWRM/IRBM based on the literature reviewed (see Sect. 4.3). Finally, the chapter links back to the knowledge gaps identified in Sect. 1.2.3 and provide three main insights relevant for analysing metropolitan water challenges from an IWRM/IRBM perspective (see Sect. 4.4).

4.2 Overview

4.2.1 Evolution of River Basin Governance

Water resources management (WRM) can be traced back to the first river basin civilizations, with water rules defined in the Indus Valley around 2500 BCE and flood management and irrigation techniques developed in Ancient Egypt (Dellapenna and Gupta 2009). For many early civilisations, controlling rivers was part of the formation of the centralized state (Wittfogel 1955). Already in prehistoric times, people learned to adapt to seasonal changes in water availability, and water management was crucial

[1] This chapter draws heavily on Brandeler et al. (2019).

F. van den Brandeler, *Scalar Mismatches in Metropolitan Water Governance*,
Water Governance - Concepts, Methods, and Practice,
https://doi.org/10.1007/978-3-031-08061-6_4

during the transition from hunting-gathering to farming (Mithen 2010). Rivers and coasts have historically attracted populations, as they provide resources, allow for navigation and serve as natural borders.

From the early twentieth century onwards, the hydraulic paradigm permeated water management worldwide with its focus on controlling the natural environment through human transformations at the service of society (Swyngedouw 1999; Lopez-Gunn 2009; Custódio 2012). Water resources management practices were developed mainly by technical experts working under the auspices of the state (Pahl-Wostl et al. 2007). The assumption was that water resources could be controlled, especially through large-scale infrastructural. These engineering works required large investments and centralized coordination and management, and became synonymous with state-building and the emergence of powerful bureaucracies (Huitema and Meijerink 2014). Different sectors requiring or impacting water were managed separately, leading to a 'silo effect' (Xie 2006).

Ideas around the need to protect, preserve and conserve water resources emerged during the nineteenth century with the beginning of the conservation movement in the United States (Agyenim 2011: 31). The persistence and aggravation of a range of water-related problems in the second half of the twentieth century (e.g. Wat&San, energy, food production, transboundary conflicts) increased awareness about the need for a new water resources management approach (Agyenim 2011: 32). Global climate change has brought further attention to this issue. Nevertheless, water crises are often portrayed as issues of physical scarcity rather than inadequate governance (Biswas and Tortajada 2010) or a deeply political challenge from global to local ('glocal') level (Gupta et al. 2013a). This position reinforces technocratic approaches based on standardized large-scale technological solutions at the expense of more sustainable technologies and practices (Farrelly and Brown 2011; GTT 2014). However, persistent and worsening water challenges reveal the limitations of engineering to solve these problems (Li et al. 2015). Consequently, there is growing consensus that the global water crisis is first and foremost a 'crisis of governance' (Castro 2007; Tortajada 2008; Biswas and Tortajada 2010; Vörösmarty et al. 2010; Gupta et al. 2013a).

The concept of IWRM gained global attention during the International Conference on Water and the Environment in Dublin in 1992, when participants defined the *Dublin Principles*, which call for integrated, participatory management (ICWE 1992). It was developed by practitioners from on-the-ground experience and promoted as early as at the first global water conference in Mar del Plata in 1977 (WWAP 2009). The concept has since been adopted by international agencies such as the World Bank (Abers 2007; Abers and Keck 2013), contributing to its influence among academic and professional circles worldwide. The Hague Forum, in 2000, pushed IWRM further by including a wide range of water management stakeholders from around the world, as opposed to the Mar del Plata and Dublin events, where participants exclusively consisted of intergovernmental participants and experts (Rahaman and Varis 2005). During this forum, participants suggested applying equity criteria in water management and called for collaboration and partnerships at all levels, meaningful participation and cooperation with international organizations and the UN (Rahaman and Varis 2005). In 2001, the International Conference on Freshwater in Bonn suggested IWRM as the best approach for meeting the water-related needs of the poor and for promoting decentralization and new partnerships. It also recommended prioritizing efforts towards better governance, mobilizing financial resources, building capacity and sharing knowledge (Rahaman and Varis 2005). The concept was also integrated into the European Union's Water Framework Directive of 2000 and subsequent programmes such as the EU Water Initiative (Wallington et al. 2010).

Although it is not clearly apparent in the description of the IWRM principles, privatization and public–private partnerships are often considered to go hand in hand with IWRM, which can be considered as part of the neo-liberal trend (GWP-TAC 2000; Rahaman and Varis 2005; Gupta and Pahl-Wostl 2013). In theory, privatization can bring many benefits to Wat&San services. It can increase the predictability of Wat&San investments and services, ensuring that they are properly maintained. The rationale is straightforward: Wat&San services are costly, long-term projects with few immediate political gains. By delegating services to private parties, governments can bring in capital they do not have, diminish costs by tapping into private efficiency and ensure that investments will be maintained despite potential political changes as companies are less subject to political influence (Wolff and Palaniappan 2004). Nonetheless, privatization may also lead to fragmentation, single-purpose planning and management and a lack of transparency, and make water unaffordable for the poor in developing regions through full cost-recovery (Rahaman and Varis 2005).

Integrated River Basin Management (IRBM) is a subset of IWRM focused on the river basin scale. IWRM is a broader concept with an administrative logic, where the national level also plays a crucial role. A literature review of IWRM papers focusing on the main trends within the period 2000–2011 showed that 44.4% of papers were dominated by a focus on the river basin management unit (Gallego-Ayala 2013). Within IRBM, the basin is the main arena for building relations of collaboration, cooperation and coordination at different levels and scales of organization (Burgos and Bocco 2014). The purpose is to match the system's natural sphere, the "hydrologically-defined basin", to the management scale, thereby allowing for comprehensive problem analysis and the internalization of otherwise externalized problems (i.e. downstream effects of upstream activities) (Pahl-Wostl et al. 2008; Lindborg et al. 2017). A number of governments and institutions (e.g. EU Water Framework Directive) have adopted this principle (Molle 2009a). In practice, water resources management at the basin level is carried out through River Basin Organizations (RBOs) often composed of representatives of a variety of interests and can go involve several states and nations. Participatory processes that include marginalized groups may promote greater equity in water resources management, including in access to water (Swallow et al. 2006).

4.2.2 Main Characteristics

The IWRM and IRBM literature sheds light on their goals, the spatial scale at which they are implemented, their approach to water resources, key actors and their mandates, water users, water's 'inputs' and 'outputs' (see Table 4.1).

Goals: As water resources management shifted towards IWRM/IRBM, its goals moved from a pure resource exploitation ethic to include social equity and environmental sustainability (Hooper 2005). IWRM and IRBM aim to integrate water uses for people (i.e. public/urban use), food production (i.e. irrigation), nature (i.e. maintaining ecosystems), industry and other uses (i.e. energy, commerce) ('GWP Integrated Urban Water Management', no date).

Approach: Conventional water resources management paradigm has a linear approach, focusing on supply augmentation rather than demand management, and addressing water resources concerns separately from land use, planning and environmental management (Xie 2006). IWRM/IRBM consider water systems as a whole, including surface (e.g. wetlands) and groundwater, upstream and downstream, as well as quantity and quality concerns through both supply and demand management (Chenoweth et al. 2001; Jønch-Clausen and Fugl 2001; Jones et al. 2006; Savenije and Van der Zaag 2008; Molle 2009a; Butterworth et al. 2010). IWRM/IRBM integrate

Table 4.1 Main characteristics of IWRM/IRBM

Main characteristics	IWRM/IRBM
Scale	Basin (for IRBM) and national (for IWRM)
Goals	Integrate water uses for people, food, nature, industry and others. Balance economic, environmental and equity goals
Approach	Integrate all water resources and related concerns within the basin. Integrate sectors and multiple interests
Actors	Basin organizations (for IRBM), national and state/provincial departments for (IWRM)
Mandates	Water allocation and bulk supply; wastewater/water quality; flood mitigation; participatory planning; monitoring/evaluation; conflict resolution; community development
Users	All users within a river basin. Agriculture is often the biggest user
Water 'inputs'	IWRM/IRBM aim to consider all water resources in the basin (including within urban areas), but focus is on blue water
Wastewater/stormwater 'outputs'	Concern for impacts downstream communities and ecosystems. Efforts to regulate and reduce pollution (industrial and urban, and diffuse pollution). Climate change adaptation increasingly considered (focus on droughts)

Source Author

both the natural and human systems through a holistic institutional approach (Jønch-Clausen and Fugl 2001; Abdullah and Christensen 2004; Molle 2009a), considering links between land and water resources, social and economic activities, and environmental management functions such as pollution control, development planning and biodiversity conservation (Abdullah and Christensen 2004; Medema et al. 2008; Savenije and Van der Zaag 2008; Molle 2009a; Molle and Mamanpoush 2012; Koç 2015). This involves integrating objectives, viewpoints and sectoral responses (Jønch-Clausen and Fugl 2001; Medema et al. 2008), and emphasizing stakeholder participation and cross-agency coordination for equitable allocation and ecosystem protection (Jønch-Clausen and Fugl 2001; Hooper 2005; Agyenim 2011; Gain et al. 2013). Furthermore, IWRM/IRBM treat water as an economic good, pricing it at levels that promote cost recovery and allocation to the most beneficial uses (Xie 2006). Responses involve both hard and soft measures that go beyond technical solutions (Chenoweth et al. 2001; Abdullah and Christensen 2004).

Spatial scales: IWRM/IRBM emphasize the river basin unit as the ideal management level,[2] although IWRM also has a spatial national focus through the integration of its principles in national-level legislation, policies and institutions (Abdullah and Christensen 2004; Watson 2004; Hooper 2005; Rahaman and Varis 2005; Jones et al. 2006; Butterworth et al. 2010). While the river basin level dominates, IWRM can be implemented at local, transboundary and global levels (Butterworth et al. 2010). Moreover, some catchments' large size and population mean that management at the scale of polders, wetlands or aquifers may be more logical (Butterworth et al. 2010). Transformations of the natural environment (e.g., irrigation schemes, inter-basin transfers) and different administrative boundaries further complicate basin management (Butterworth et al. 2010; Moss and Newig 2010).

Organization: Conventional WRM is hierarchical, centralized and top-down, with central governments often directing water supply development, service provision and water use regulation (Xie 2006). IWRM/IRBM embraces (a) decentralization, with preference for the (sub-)basin level (Xie 2006; Molle 2009a); (b) transparency and participatory decision-making (Rahaman and Varis 2005; Jones et al. 2006); and (c) implementation through special purpose organizations (e.g., RBOs) rather than general purpose government layers (Molle 2009a; Huitema and Meijerink 2014). Diverse actors can be involved, including user associations, environmental NGOs and affected communities.

Mandates: Conventional WRM's mandates are limited to river basin planning and policy "with no enforcement, regulatory or water-resource development functions, and no direct representation of local interests and stakeholders, other than through state agencies" (Molle 2009a: 490). IWRM/IRBM mandates concern all basin-related activities (e.g. basin-wide evaluation, planning, strategy implementation, monitoring,

[2] The boundaries of surface watersheds are generally delineated from topographic maps (Vaughan et al. 2016). The rain and snow that falls within this area is stored, filtered, seeped or drained into a common water body through a network of streams that drain the surface area and the groundwater that contribute to those streams.

water allocation mechanisms, water quality maintenance, basin guidelines, negotiation/ dispute resolution mechanisms, flood warning and mitigation, and community development) (Chenoweth et al. 2001). These mandates are shared between actors, including RBOs and other multi-stakeholder platforms, but are often blurred (Jønch-Clausen and Fugl 2001; Grigg 2008). IWRM/IRBM supporters argue that adequate implementation requires clearly delimited responsibilities that match to authority to capacities for action (Jønch-Clausen and Fugl 2001). In some cases, government bodies are responsible for regulating water uses but lack the administrative and political power needed to oppose traditional line agencies (Molle 2009a; Agyenim 2011). In other cases, quantitative functions (e.g. water supply, irrigation, flood mitigation) are managed separately from qualitative functions (e.g. water pollution, land use), while a single organization could facilitate a more integrated approach (Abdullah and Christensen 2004).

Users: IWRM/IRBM considers all users within a river basin, including domestic, agricultural, industrial and recreational users (Savenije and Van der Zaag 2008). Agriculture is generally the largest user of water resources, followed by the energy sector (i.e. hydropower), making the water-food-energy nexus crucial for integrated water management, particularly in the context of climate change (Biswas 2008; OECD 2016). Domestic and recreational users represent a relatively small portion of total water use. In river basins with low population densities, ecosystems such as grasslands or woodlands may be major water users (Harrington et al. 2009). However, in heavily urbanized basins, water resources' allocation can be a source of competition and conflict between urban and rural users (Butterworth et al. 2010; Molle and Mamanpoush 2012; OECD 2016). Downstream users are particularly vulnerable to variations in the water regime occurring upstream (Molle 2009b). Users also differ in their access to natural or financial resources and in their political power (Swyngedouw et al. 2002; Molle and Hoanh 2007; Molle 2009b). Effective and legitimate norms and instruments for water allocation are therefore crucial in heavily urbanized basins, but are often lacking, especially in cases with large informal economies (Watson 2004; Butterworth et al. 2010; OECD 2016).

Inputs and outputs: IWRM/IRBM consider the hydrological cycle as a unitary whole and are therefore concerned with all water resources within a river basin (including rainwater and snowmelt) (Savenije and Van der Zaag 2008; Harrington et al. 2009). They call for awareness on how different types of water (e.g., blue water, grey water) interact and how this interaction impacts equitable use and the distribution of costs and benefits (Savenije and Van der Zaag 2008). However, IWRM/IRBM focus on 'blue water' (i.e. surface and groundwater), ignoring 'grey water' (i.e. water after basic treatment) and 'green water' (i.e. water in the unsaturated zone of the soil and plants which has critical ecosystem services, and is sensitive to land degradation) (Jewitt 2006; Molle and Mamanpoush 2012; Hayat and Gupta 2016). 'Rainbow water' (i.e. atmospheric moisture) is not included in IWRM/IRBM, yet perturbations in moisture recycling, often linked to changes in land use, can affect precipitation in distant locations. Water availability in one river basin can therefore be modified due to activities (i.e. deforestation, irrigation) on the other side of the planet (Wang-Erlandsson et al. 2018). Furthermore, IWRM/IRBM do not address virtual water, i.e., the import of items, such as food, which reduces the need to use local water resources for their production (Dalin et al. 2017).

The linear approach of conventional WRM contributed to water contamination downstream. Especially following World War II, the focus was on economic development, and water quality was not a priority for the industrial and agricultural sectors (Xie 2006). IWRM and IRBM call for protecting water resources and ecosystem services through adequate wastewater management (i.e. collecting and treating wastewater before discharge, wastewater recycling, and waterway restoration) (Martinez-Santos et al. 2014). This may involve (de)centralized infrastructure, but also regulating and reducing diffuse pollution, most of which derives from agricultural activities (Xepapadeas 2011).

4.2.3 Drivers, Institutions and Instruments

The rise of IWRM and IRBM was in part spurred by drivers of water resources challenges such as increasing competition for water resources through population growth and economic development, climate change and variability, and land use changes due to human interventions or natural processes (Hooper 2005; Gallego-Ayala 2013; Wang-Erlandsson et al. 2018). Major shifts in societal behaviour further contributed to a shift in WRM (Hooper 2005).

The 'Dublin Principles' defined four IWRM principles that aim to guide concerted action from the global to the local level. The first defines fresh water as a "finite and vulnerable resource, essential to sustain life, development and the environment" (ICWE 1992: Guiding Principles). The second principle emphasizes that water governance should be participatory and that decisions should be taken at the lowest appropriate level with public consultation and involvement of users (Ibid). The third principle recognizes the central role that women play in water governance and aims to empower them so that policies can better answer their needs. The final principle defines water as an economic good and suggests that giving water an economic value will prevent wasteful and harmful uses. From these principles, IWRM/IRBM frameworks derive a variety of instruments (see Table 4.2). Key among these are water (re-)allocation policies. In particular, *who* has priority to use water resources in water scarce environments (in many cases, the priority is granted to domestic water use) (Butterworth et al. 2010; Agyenim 2011). Other influential instruments include regulatory and economic instruments to manage pollution (e.g. wastewater discharge fees) for Wat&San utilities, industries and other bulk water users), the identification and licensing of polluters (i.e. permits for wastewater discharge), effective monitoring systems, identification and responses to infringements of laws, regulations and permits, and regulations and incentives to preserve areas with important ecosystems (i.e. conservation areas, compensation to municipalities for conservation efforts, PES programmes). Other crucial instruments, especially in water scarce areas, are those that establish rules for water diversions from rivers, lakes and groundwater

Table 4.2 Instruments of IWRM/IRBM

Types of instruments	IWRM/IRBM
Economic	Bulk water use fees; PES programmes; pollution fees, and economic and financial tools for investment, cost recovery and behaviour change (e.g. subsidies for water efficient technology and infrastructure, royalties)
Regulatory	Bulk water use permits; Transfers of bulk water use permits; land use regulation within basin (restricting development); Licensing (of polluting activities); Conservation measures (e.g. easements and zoning)
Suasive	Anti-pollution campaigns; Public awareness and education programmes on environmental protection; Water resources assessment and information
Coordination	Pollution monitoring systems; Meteorological monitoring systems; Participatory spaces for negotiation and conflict-resolution on water resources management

Source Author

(Agyenim 2011). IWRM/IRBM instruments interfere at national and basin levels, and focus more on environmental and economic dimensions than social and relational goals.

Political, legal and institutional factors also shaped the shift towards IWRM/IRBM (Tortajada 2001; Grigg 2008; Agyenim 2011). Institutions rooted in centralized structures, sectoral fragmentation, weak local capacity, low political awareness or will and inadequate data and information management hinder the successful IWRM/IRBM implementation (Jønch-Clausen and Fugl 2001; Medema et al. 2008; Molle 2008). Weak implementation of IWRM/IRBM in the Global South may be explained by the fact that it requires a stable political and institutional environment (Agyenim 2011; Cameron and Katzschner 2017). Property rights regimes may be different from those in the Global North and stable financing and administrative capacity is more likely to be lacking, as is the ability to enforce laws and regulations (Agyenim 2011). On the other hand, in the Global South, IWRM can be compatible with local water rights that are participatory, self-regulatory, sensitive to the vulnerable and able to contain conflicts (Agyenim 2011; Agyenim and Gupta 2012). Latin American countries that adopted IWRM/IRBM principles shared strong support for decentralization policies within broader political reforms (Lee 2000; Tortajada 2001).

IWRM/IRBM institutions are often designed at basin level, under the premise that they "fit" their context and are therefore more effective (Young et al. 2008; Agyenim 2011). Nonetheless, IWRM/IRBM has developed at multiple levels. At global level, International Financial Institutions (IFIs) have financed large-scale water resources development projects in the Global South (Lebel 2005). The national level is crucial for IWRM through the design and implementation of laws, norms, regulations and rules (Gallego-Ayala 2013). National governments set standards for water quality and water allocation. Some nations also possess a national water regulatory agency (OECD 2011). Hydropower planning and management are often at national level as this enables better distribution and economies of scale. When water becomes an issue of national security (i.e. during drought), governments may take control over water resources (Gupta et al. 2013b). In federal countries, significantly more water policy responsibilities are devolved to subnational governments. States usually have sovereignty over water resources within their territory and establish the basic rules about how individuals and groups can access these (Gupta et al. 2013b). However, water policy roles and responsibilities across national and sub-national levels vary significantly, making it impossible to capture a "national model" with comprehensive institutional mapping (OECD 2011). Furthermore, multiple institutions are sometimes superimposed onto each other (e.g. communal irrigation institutions and modern water use allocation systems) (Lebel 2005). The multiple actors involved have differing objectives, responsibilities, and interests, which can lead to conflicts (Koç 2015). The institutional integration expected of IWRM may not be realistic, and fostering coordination, cooperation and collaboration may be more adequate (Biswas 2004; Koç 2015).

4.3 IWRM/IRBM for Sustainable and Inclusive Development

4.3.1 Steps Forward

The shift from conventional WRM approaches towards IWRM/IRBM creates conditions for more inclusiveness. By emphasizing the blending of different viewpoints, IWRM/IRBM enable a more holistic and inclusive approach to water-related issues (Grigg 1999, 2008; Medema et al. 2008). Decentralization and participation, by integrating actors with knowledge of the local environment and local users' needs, may enhance equity and efficiency in resource management (Andersson and Ostrom 2008). IWRM/IRBM strives for a fair allocation of water resources among upstream and downstream users, between current and future generations, and for both human and ecosystem needs (Harrington et al. 2009; Molle 2009b; Molle and Mamanpoush 2012). IWRM/IRBM can also challenge centralized expertise and decisions when water-related challenges require adaptive, polycentric (Sneddon et al. 2002; Molle 2009a) and contextual solutions (Jønch-Clausen and Fugl 2001; Xie 2006). IWRM/IRBM may enhance sustainability by recognizing the role of ecosystems in WRM (Molle 2009a) and promoting demand management (Grigg 2008).

4.3.2 Lingering Obstacles

Continued environmental degradation in areas where IWRM/IRBM has been implemented highlights the limitations of these principles for fostering ecological sustainability. Moreover, the focus of IWRM/IRBM has been on 'blue water', to the detriment of 'grey water', 'rainbow water', 'green water', and the 'embedded water' or ‘virtual water’ in imported food products and other goods (Molle and Mamanpoush 2012; Hayat and Gupta 2016; Dalin et al. 2017; Wang-Erlandsson et al. 2018; Brandeler et al. 2019). An understanding of ‘green water’ flows can lead to more efficient irrigation for instance. There is also no consensus on whether IWRM/IRBM enhances adaptability, which is problematic in a context of climate change (Medema et al. 2008; Foster and Ait-Kadi 2012; Gain et al. 2013).

In addition, the principle of water as an economic good is frequently criticized as it may foster the notion of water as a commodity, shifting perception away from water as a common good that brings shared duty and responsibility (Rahaman and Varis 2005). Pushing for cost recovery in areas with large pockets of poverty, such as nations in the Global South, can eliminate vital subsidies for basic water infrastructure and services (Rahaman and Varis 2005).

Furthermore, there is no consensus on how to implement IWRM/IRBM. While the idea of moving from a fragmented to an integrated and coordinated approach has become mainstream, its implementation is still lacking (Chenoweth et al. 2001; Medema et al. 2008; Jacobi et al. 2015). There are several reasons for this. The integration goals are ambitious and assume political will and the adequate availability and sharing of data and knowledge (Allan 2003; Molle 2008). They are challenged by uncertainty (Agyenim 2011), tensions between basin-wide approaches and more decentralized resource management[3] (Miller and Hirsch 2003), trade-offs (Efficiency, Equity and Environmental sustainability goals) (Molle 2008), and the jurisdictional complexity of river basins that cross national or even state borders (in the case of federal regimes) (Choudhury and Islam 2015).

Also, the Dublin statement on IWRM does not provide suggestions on how to implement effective participation mechanisms (Rahaman and Varis 2005: 16). Integrating different views requires significant resources and time-commitments (i.e. time spent in meetings, travel time and costs of attending meetings), and participants lacking expert knowledge often struggle to follow and intervene in discussions (Brandeler 2013). Decision-making spaces are often still dominated by state-affiliated experts and traditional power relations, and the broad goals of IWRM/IRBM can lead some groups to hijack these to legitimize their own agendas (Wester and Warner 2002; Molle 2008). For instance, urban elites and state bureaucracies may use these spaces to facilitate upstream interventions by downstream stakeholders (Molle 2009b). Marginalized voices and interests struggle to gain representation and negotiate on equal footing (Swyngedouw et al. 2002; Molle 2009b; Brandeler et al. 2014; Varis et al. 2014). This especially concerns indigenous and rural communities as well as the urban poor (Boelens 2009; Brandeler et al. 2014). Furthermore, RBOs often struggle with limited enforcement, regulatory or water-resource development functions (Molle 2009a). IWRM/IRBM cannot be established in contexts of war and conflict and require stable democratic processes and institutions (Hooper 2005), but beyond that it is not clear how to address unequal power relations (Jønch-Clausen

[3] Although the river basin management approach has often been framed as one of decentralization, it represents a scaling up from municipalities (Cohen and Davidson 2011).

and Fugl 2001; Brandeler et al. 2019). This may require broader political reforms beyond the scope of water management institutions.

4.4 Scalar Limitations of IWRM/IRBM

4.4.1 The "Naturalness" of the River Basin Scale as a Social Construction

The IRBM and, to a large extent, the IWRM literature, have framed the spatial scale of the river basin as the ideal unit for water resources planning and management (see Sect. 4.2.2). However, this simplifies hydrological dynamics. First of all, river basins are generally formed by multiple watersheds nested into each other, as each tributary forms a sub-division, so deciding where to draw the boundaries for a management area can be a challenge in itself (Cohen and Davidson 2011; Perreault 2014).[4] These boundaries are constantly shifting as our understanding of surface and groundwater flows and GIS technology improve (Cohen and Davidson 2011). Second, the hydrological cycle is not restricted to river basin boundaries. Land–atmosphere feedback, such as local and regional land use change and moisture recycling patterns, can have effects on remote precipitation and, thereby, on distant river flows (Wang-Erlandsson et al. 2018). Model simulations suggest that complete deforestation of Central Africa could decrease February precipitation by 35% in the Great Lakes region of the U.S. (Avissar and Werth 2005), while irrigation in India may support up to 40% of the precipitation in arid regions of eastern Africa (Vrese et al. 2016). Third, river basins usually share the same boundaries as granular, unconfined aquifers (i.e. just below the surface), but the recharge areas of karstic and confined aquifers mostly do not match those of river basins (Demiroğlu 2017). Aquifer boundaries may be a more appropriate spatial framework for groundwater management and protection in the latter case (deep aquifers in arid regions) (Garduno et al. 2006; Foster and Ait-Kadi 2012).

Moreover, the view of river basins as closed ecological systems characterized by stability and predictability has lost meaning in the context of human interventions (e.g. climate change, pollution, loss of wetlands) that lead to increased uncertainty, change, complexity and conflict (Holling 1978; Watson 2004). As Europe's river basins are now linked by canal networks, inland navigation no longer relies on natural waterways and is coordinated nationally and regionally rather than at basin scale (Muller 2018). River flows have also been altered to allow for water supply in expanding human settlements through inter-basin transfers, the drilling of deep wells and desalination (Cohen and Davidson 2011; Perreault 2014; Muller 2018). Unintended human interventions, such as the accidental introduction of invasive

[4] For instance, the US Geological Survey (USGS) divides the United States into 21 regions, 222 sub-regions, 370 basins, 2,270 sub-basins, approximately 20,000 watersheds and around 100,000 sub-watersheds (USGS 2013).

species into a river basin, can also have devastating biological and economic impacts (Cohen and Davidson 2011).[5] As these transformations, as well as the governance systems that respond to these, transcend across spatial scales, effective solutions are beyond the reach of any single actor (Watson 2004; Varis et al. 2014). An exclusively basin-scale approach may therefore limit flexibility and lead to transboundary problems.

Ultimately, the choice of scale for water resources management is socially produced and inherently political, emerging from social practices, perceptions and relationships over time (Molle 2009a; Perreault 2014). Treating the river basin as the 'natural' scale can lead to a technical approach that ignores political struggles and power relations (Perreault 2014). On the other hand, governance units based on conventional electoral boundaries (i.e. municipal, provincial, state, or national boundaries) rarely correspond to hydrological boundaries, leading to a mismatch between institutional and hydrological logics that blurs the allocation of responsibilities across multiple scales and complicates the relationships between elected representatives, local authorities, water agencies, resource managers and end users (Bahri 2012; Söderbaum and Granit, 2014; OECD 2015). Government participants in river basin initiatives are more likely to respond to the concerns of the jurisdictions where they were elected and to which they are accountable (Cohen and Davidson 2011). This creates obstacles for coordinating different actors, particularly for financing (Newig and Fritsch 2009) and land use management (Tucci 2007), and may contribute to management failures within water governance, such as a lack of cooperation, participation and transparency (OECD, 2015). Additionally, the socioeconomic drivers and power relations that shape WRM are not necessarily contained within the basin (Molle 2009a).

4.4.2 IWRM/IRBM Ignore the Urban

While basin-level management has become the dominant paradigm within the water governance literature, River Basin Organizations (RBOs) struggle in heavily urbanized environments, raising doubts about the adequacy of the basin as the spatial scale for addressing urban water challenges (Formiga Johnsson and Kemper 2005; Jacobi and Francalanza 2005; Rahaman and Varis 2005; Alvim 2006; Roggero and Fritsch 2010; Brandeler 2013; Freitas and Fracalanza 2013). The IWRM and IRBM literature do not clearly address how water management at basin scale can be reconciled with the urban scale. A search within ScienceDirect for articles on IWRM published between 1970 and 2015 revealed that only around 14% contained any of the terms 'urban', 'city', 'cities', 'megacity', 'megacities' or 'metropolitan' in their title, abstract or keywords (Brandeler et al. 2019). The same search within IRBM

[5] The introduction of zebra mussels into the Great Lakes, from ship ballast water, is estimated to have cost the North American economy over US$ 100 million and wreaked havoc on food webs and water supply systems (Cohen and Davidson 2011).

articles gave a result of 10%. Many of the IWRM/IRBM publications that do refer to the urban context have a narrow focus (e.g. urban demand management, pollution abatement tools, leakage control, wastewater recycling, public–private partnerships), and ignore IWRM's broader dimensions (Rees 2006; Bahri 2012). They also do not address the particular challenges for IWRM implementation in large metropolises (e.g. coordinating across urban jurisdictions that may have competitive or conflictive relations).

A straightforward explanation for this is that in most basins most of the water is used for agricultural (i.e. irrigation). WRM entities have often largely focused on regulating flows for irrigation (70% of the world's water withdrawals, even reaching 90% in countries of the Global South) and power generation, with little influence over water quality management (Lee 2000; Molle 2008). In addition, irrigation's heavy water use is often also inefficient, with over half of the water deliveries percolating into groundwater or returning to streams without watering crops, evaporating or otherwise lost (Molle 2008). However, when demand within a basin exceeds supply, surface and groundwater resources become over-exploited, which affects development prospects (Tortajada 2008; Barrios et al. 2009). As basins urbanize and industrialize, and rural and urban water users compete for increasingly limited water resources, appropriate supply and demand management and water resources allocation mechanisms are crucial (Butterworth et al. 2010; Molle and Mamanpoush 2012).

Furthermore, to implement IWRM/IRBM in heavily urbanized river basins, integrated decision-making processes must address trade-offs and externalities between rural and urban areas. One key trade-off concerns the allocation of scarce water resources between public/urban users, other sectors as well as the environment. This is especially pertinent in basins where all available water is committed, and new developments require reallocation (Rees 2006). As domestic water use typically has priority over other uses, this translates into reallocation from agricultural to urban uses (Rees 2006). Importing water from other basins allows actors and users within the recipient basin to avoid this contentious, and potentially conflictual decision. Although this may increase the economic efficiency of water use, such transfers also imply economic, social and political costs to donor basins (Rees 2006). Large cities around the world are increasingly relying on neighbouring basins.

The IWRM/IRBM literature promotes integration between water and land resources and management. However, land use changes in rural areas (e.g. deforestation) and diffuse pollution from agriculture are difficult to regulate and their impacts on cities cannot easily be traced to specific sources. The expansion of human settlements and intensification of their use of land and water resources in urban areas strengthens interlinkages between spatial planning and water management (Hartmann and Spit 2014). Nevertheless, there seems to be no comprehensive analytical framework for assessing the coordination of IWRM/IRBM with urban and territorial planning (Neto 2016). RBOs often have some (limited) influence over land and water uses upstream and across the basin, but little leverage over policy fields that directly impact water quantity, quality and risks, such as land use planning, agriculture, hydropower, stormwater management, conservation, economic development and more (Moss 2004: 85). The mismatch between hydrological and political-administrative borders may further lead to poor cooperation, particularly in basins with large metropolises and megacities, and many stakeholders and interests (Aguilar 2008; Feiock 2009; Sorensen 2011; Li et al. 2015). In addition, climate variability and change can also lead to droughts or floods, impacting both river basins and the cities within them. If IWRM/IRBM measures do not take steps to prepare for and adapt to these risks, cities may be left struggling to cope, or they may be overruled by urban water frameworks, as major cities gain influence in water management. Droughts affect water supplies, and in basins with large cities even short droughts can have significant impact.

4.5 Inferences

This chapter reviewed the water resources management literature to gain insights on how it understands and addresses metropolitan water challenges. It explored this in relation to two knowledge gaps identified in Sect. 1.2.3, namely (a) how the literature relates to the local context and integrates urban water concerns such as water services and water-related risks, (b) how the literature relates to the metropolitan context as a multi-jurisdictional and often heterogeneous urban environment. The IWRM/IRBM literature review identified Several factors that may facilitate or hinder sustainable and inclusive development.

Sustainable and inclusive development

First, the literature review highlights a shift from WRM based on modernist ideals towards IWRM, and its subset of IRBM, that embrace sustainable and inclusive development principles. IWRM and IRBM aim for decentralization to the river basin level and stakeholder participation, which can lead to the design of institutions and instruments that foster the inclusion of multiple viewpoints and interests, greater efficiency in resource management for multiple users, the protection of ecosystems and the promotion of water demand management. The review identified aspects of IWRM/IRBM that could hinder sustainable and inclusive development, such as limited considerations for types of water other than 'blue water' and for water as a hazard (water is defined as a resource). Participatory processes may struggle to include diverse, and particularly under-represented voices, as they require significant time commitments, finance (Anggraeni et al. 2019) and as expert knowledge dominates to the detriment of local, tacit knowledge. Elected representatives involved in participatory processes are also likely to defend their constituents' interests, even at the expense of the basin-wide interest. The approach to water as an economic good can also lead to the exclusion of the poor and marginalized. The chapters on São Paulo and Mexico City evaluate whether such measures were part of the responses to metropolitan water challenges, and if so, how effective they were.

The river basin scale may not always be ideal

One of IWRM/IRBM's core principles is the idea that the river basin is the ideal unit for water resources management, as this allows for overcoming the mismatch between hydrological and administrative scales. However, the literature review points out important limitations when parties take for granted the superiority of the river basin as a planning unit. This book recognizes that choosing the basin as a unit for water resources planning and management is a social construction and attempts to consider other relevant scales (for drivers, institutions and instruments) into a broader framework for understanding metropolitan water challenges. This includes scales related to elements of the hydrological cycle (e.g. groundwater dynamics, atmospheric water, green water) and scales that gain relevance due to human interventions on the 'natural' environment. For this purpose, a governance framework that addresses metropolitan water challenges must therefore go beyond the main characteristics of IWRM/IRBM identified in Sect. 4.2.2.

Influence of basin entities over specifically urban drivers and institutions

IWRM/IRBM have a strong focus on agriculture and rural areas, even though they promote balancing the needs and goals of all users. However, in heavily urbanized basins, a significant or even dominant proportion of water uses are for urban uses, and water-related challenges are different (e.g. water quality challenges, such as contamination from solid waste and domestic and industrial wastewater, reductions in aquifer recharge due to soil sealing, tensions between users). For IWRM/IRBM institutions and instruments to have an impact on these challenges, they must address the relevant drivers and actors. This is even more complex in metropolitan regions, due to the larger number of actors involved and their sometimes conflicting political and administrative interests.

References

Abdullah D, Christensen B (2004) Integrated river basin management. Bullet 22:21–23

Abers RN (2007) Organizing for governance: building collaboration in Brazilian river basins. World Dev 35(8):1450–1463. https://doi.org/10.1016/j.worlddev.2007.04.008

Abers R, Keck ME (2013) Practical authority: agency and institutional change in brazilian water politics—excerpts. In: American Political Science Association 2013 Annual Meeting. Available at: https://papers.ssrn.com/sol3/papers.cfm?abstract_id=2302112. Accessed: 14 Oct 2018

Aguilar AG (2008) Peri-urbanization, illegal settlements and environmental impact in Mexico City. Cities. https://doi.org/10.1016/j.cities.2008.02.003

Agyenim JB (2011) Investigating institutional arrangements for integrated water resource management in developing countries: the case of white Volta Basin. Vrije Universiteit Amsterdam, Ghana

Agyenim JB, Gupta J (2012) IWRM and developing countries: Implementation challenges in Ghana. Phys Chem Earth Parts a/b/c 47–48:46–57. https://doi.org/10.1016/J.PCE.2011.06.007

Allan T (2003) IWRM/IWRAM: a new sanctioned discourse? Available at: https://www.soas.ac.uk/water/publications/papers/file38393.pdf. Accessed: 9 Jan 2019

Alvim ATB (2006) A contribuição do Comitê do Alto Tietê à gestão da bacia metropolitanaentre 1994 e 2002. Pós Revista Do Programa De Pós-Graduação Em Arquitetura e Urbanismo Da FAUUSP 19:26–44. https://doi.org/10.11606/issn.2317-2762.v0i19p26-44
Andersson KP, Ostrom E (2008) Analyzing decentralized resource regimes from a polycentric perspective. Policy Sci 41:71–93. https://doi.org/10.1007/s11077-007-9055-6
Anggraeni M, Gupta J, Verrest HJLM (2019) Cost and value of stakeholders participation: a systematic literature review. Environ Sci Policy. https://doi.org/10.1016/J.ENVSCI.2019.07.012
Avissar R, Werth D (2005) Global hydroclimatological teleconnections resulting from tropical deforestation. J Hydrometeorol 6:134–145. Available at: https://doi.org/10.1175/JHM406.1. Accessed: 1 Dec 2018
Bahri A (2012) Integrated urban water management: briefing note. Technical Committee Background Paper 16. Stockholm
Barrios JE, Rodríguez-Pineda JA, de la Maza Benignos M (2009) Integrated river basin management in the Conchos River basin, Mexico: a case study of freshwater climate change adaptation. Climate Dev 1(3):249–260
Biswas AK (2004) Integrated water resources management: a reassessment. A water forum contribution. Water Int 29(2):248–256. https://doi.org/10.1080/02508060408691775
Biswas AK (2008) Integrated water resources management: is it working? Int J Water Resour Dev 24(1):5–22. https://doi.org/10.1080/07900620701871718
Biswas AK, Tortajada C (2010) Future water governance: problems and perspectives. Int J Water Resour Dev 26(2):129–139. https://doi.org/10.1080/07900627.2010.488853
Boelens R (2009) The politics of disciplining water rights. Dev Change 40(2):307–331. https://doi.org/10.1111/j.1467-7660.2009.01516.x
Burgos A, Bocco G (2014) La gestión del agua y el aporte de la geografía al enfoque de cuencas hidrográficas en México. In: Pérez Correa F (ed) Gestión pública y social del agua en México. Universidad Nacional Autónoma de México, México
Butterworth J et al (2010) Finding practical approaches to integrated water resources management. Water Alternat 3(31):68–81. Available at: www.water-alternatives.org
Cameron R, Katzschner T (2017) Every last drop—the role of spatial planning in enhancing integrated urban water management in the city of cape town. S Afr Geogr J 99(2):196–216. https://doi.org/10.1080/03736245.2016.1231622
Castro JE (2007) Water governance in the twentieth-first century. Ambiente Sociedade 10(2):97–118. https://doi.org/10.1590/S1414-753X2007000200007
Chenoweth JL, Malano HM, Bird JF (2001) Integrated river basin management in the multi-jurisdictional river basins: the case of the Mekong river basin. Int J Water Resour Dev 17(3):365–377. https://doi.org/10.1080/07900620120065147
Choudhury E, Islam S (2015) Nature of transboundary water conflicts: issues of complexity and the enabling conditions for negotiated cooperation. J Contempor Water Res Educ 155(1):43–52. https://doi.org/10.1111/j.1936-704X.2015.03194.x
Cohen A, Davidson S (2011) The watershed approach: challenges, antecedents, and the transition from technical tool to governance unit. Water Alternat 4(1), pp 1–14. Available at: www.water-alternatives.org. Accessed: 11 Jan 2019
Custódio V (2012) Escassez de Água e inundações na Região Metropolitana de São Paulo. Humanitas/Fapesp, São Paulo
Dalin C et al (2017) Groundwater depletion embedded in international food trade. Nature 543(7647):700–704. https://doi.org/10.1038/nature21403
Dellapenna JW, Gupta J (2009) The evolution of the law and politics of water. Springer, Dordrecht. https://doi.org/10.1007/978-1-4020-9867-3_1
Demiroğlu M (2017) Identifying the groundwater basin boundaries, using environmental isotopes: a case study. Appl Wat Sci 7(3):1161–1167. https://doi.org/10.1007/s13201-016-0516-y
Farrelly M, Brown RR (2011) Rethinking urban water management: Experimentation as a way forward? Glob Environ Chang. https://doi.org/10.1016/j.gloenvcha.2011.01.007

Feiock RC (2009) Metropolitan governance and institutional collective action. Urban Affairs Rev 44(3):356–377. https://doi.org/10.1177/1078087408324000

Formiga Johnsson RM, Kemper KE (2005) Institutional and policy analysis of river basin management: the Alto-Tietê River Basin, São Paulo, Brazil. 3650. Available at: http://econ.worldbank.org

Foster S, Ait-Kadi M (2012) Integrated water resources management (IWRM): how does groundwater fit in? Hydrogeol J 20:415–418. https://doi.org/10.1007/s10040-012-0831-9

Freitas S, Fracalanza AP (2013) GESTÃO DA ÁGUA E METRÓPOLE: DESAFIOS DA BACIA HIDROGRÁFICA DO ALTO TIETÊ'. In: 4° Encontro Internacional da Governança da Água. São Paulo, Brazil

Gain AK, Rouillard JJ, Benson D (2013) Can integrated water resources management increase adaptive capacity to climate change adaptation? A critical review. J Water Resour Protect 5:11–20. https://doi.org/10.4236/jwarp.2013.54A003

Gallego-Ayala J (2013) Trends in integrated water resources management research: a literature review. Water Policy 15(4):628–647. https://doi.org/10.2166/wp.2013.149

Garduno H et al (2006) Groundwater dimensions of national water resource and river basin planning. Washington, DC

Grigg NS (2008) Integrated water resources management: balancing views and improving practice. Water Int 33(3):1–20. https://doi.org/10.1080/02508060802272820

Grigg NS (1999) Integrated water resources management who should lead, who should pay? J Am Water Resour Assoc 35(3):527–534. https://doi.org/10.1111/j.1752-1688.1999.tb03609.x

GTT (2014) Integrated Urban Water Resources Management. In: Humanity, Gulbenkian think tank on water and the future of humanity- revisiting water security. Springer International Publishing, pp 109–132

Gupta J, Pahl-Wostl C, Zondervan R (2013a) Glocal water governance: a multi-level challenge in the anthropocene. Curr Opin Environ Sustain 5:573–580. https://doi.org/10.1016/j.cosust.2013.09.003

Gupta J, Pahl-Wostl C, Zondervan R (2013b) "Glocal" water governance: a multi-level challenge in the anthropocene. Curr Opin Environ Sustain 5(6):573–580. https://doi.org/10.1016/j.cosust.2013.09.003

Gupta J, Pahl-Wostl C (2013) Global water governance in the context of global and multilevel governance: its need, form, and challenges. Ecol Soc 18(4):art53. https://doi.org/10.5751/ES-05952-180453

GWP-TAC (2000) Integrated water resources management. Stockholm. Available at: www.gwpforum.org. Accessed: 30 May 2019

GWP integrated urban water management' (no date)

Harrington L et al (2009) Cross-basin comparisons of water use, water scarcity and their impact on livelihoods: present and future. Water Int 34(1):144–154. https://doi.org/10.1080/02508060802661584

Hartmann T, Spit T (2014) Frontiers of land and water governance in urban regions. Water Int 39(6):791–797. https://doi.org/10.1080/02508060.2014.962993

Hayat S, Gupta J (2016) Kinds of freshwater and their relation to ecosystem services and human well-being. Water Policy 18(5):1229–1246. https://doi.org/10.2166/wp.2016.182

Holling CS (ed) (1978) Adaptive environmental assessment and management. Wiley, Chichester

Hooper B (2005) Integrated river basin governance: learning from international experience. International Water Association Publishing, London

Huitema D, Meijerink SV (2014) The politics of river basin organisations : institutional design choices, coalitions and consequences. Edward Elgar Publishing Limited, Cheltenham. Available at: https://www.e-elgar.com/shop/eep/preview/book/isbn/9781782549222/. Accessed 11 June 2018

ICWE (1992) The Dublin statement on water and sustainable development'. International Conference on Water and the Environment, Dublin. Available at: http://www.wmo.int/pages/prog/hwrp/documents/english/icwedece.html. Accessed: 23 June 2018

Jacobi P, Fracalanza AP, Silva-Sánchez S (2015) Governança da água e inovação na política de recuperação de recursos hídricos na cidade de São Paulo. Cadernos Metropole 17(33):61–81. https://doi.org/10.1590/2236-9996.2015-3303

Jacobi PR, Francalanza AP (2005) Comitês de bacias hidrográficas no Brasil: desafios de fortalecimento da gestão compartilhada e participativa. Desenvolvimento e Meio Ambiente 11(0). https://doi.org/10.5380/dma.v11i0.7816

Jewitt G (2006) 'Integrating blue and green water flows for water resources management and planning. Phys Chem Earth Parts a/b/c 31(15–16):753–762. https://doi.org/10.1016/J.PCE.2006.08.033

Jønch-Clausen T, Fugl J (2001) Firming up the conceptual basis of integrated water resources management. Int J Water Resour Dev 174:501–510. https://doi.org/10.1080/07900620120094055

Jones T, Newborne P, Phillips B (2006) Applying the principles of integrated water resource and river basin management—an introduction

Koç C (2015) A study on the role and importance of irrigation management in integrated river basin management. Environ Monitor Assess 187(8). https://doi.org/10.1007/s10661-015-4647-7

Lebel L (2005) The politics of scale, position, and place in the governance of water resources in the Mekong Region. Ecol Soc 10(2). https://doi.org/10.5751/ES-01543-100218

Lee TR (2000) Urban water management for better urban life in Latin America. Urban Water 2(1):71–78. Available at: www.elsevier.com/locate/urbwat. Accessed: 13 Dec 2018

Li E, Endter-Wada J, Li S (2015) Characterizing and contextualizing the water challenges of megacities. J Am Water Resour Assoc 51(3):589–613. https://doi.org/10.1111/1752-1688.12310

Lindborg R et al (2017) How spatial scale shapes the generation and management of multiple ecosystem services. Ecosphere 8(4):1–15

Lopez-Gunn E (2009) Agua Para Todos: A New regionalist hydraulic paradigm in Spain. Water Alternat 2(3):370–394. Available at: www.water-alternatives.org. Accessed: 30 May 2019

Martinez-Santos P, Aldaya MM, Llamas MR (2014) Integrated water resources management in the 21st century: revisiting the paradigm. CRC Press/Balkema, Leiden

Medema W, Mcintosh BS, Jeffrey PJ (2008) From premise to practice: a critical assessment of integrated water resources management and adaptive management approaches in the water sector. Ecol Soc 13(2):29. Available at: http://www.ecologyandsociety.org/vol13/iss2/art29/

Miller F, Hirsch P (2003) Civil society and internationalized river basin management, AMRC Working Paper Series. Sydney

Mithen S (2010) The domestication of water: water management in the ancient world and its prehistoric origins in the Jordan Valley. Philos Trans R Soc a: Math Phys Eng Sci 368(1931):5249–5274. https://doi.org/10.1098/rsta.2010.0191

Molle F (2008) Nirvana concepts, narratives and policy models: Insights from the water sector. Water Alternat 1(1):131–156

Molle F (2009a) River-basin planning and management: The social life of a concept. Geoforum 40:484–494. https://doi.org/10.1016/j.geoforum.2009.03.004

Molle F (2009b) Water, politics and river basin governance: repoliticizing approaches to river basin management. Water Int 34(1):62–70. https://doi.org/10.1080/02508060802677846

Molle F, Mamanpoush A (2012) Scale, governance and the management of river basins: a case study from Central Iran. Geoforum 43(2):285–294. https://doi.org/10.1016/j.geoforum.2011.08.004

Molle F, Hoanh CT (2007) Implementing integrated river basin management: lessons from the Red River Basin, Vietnam. Colombo, Sri Lanka. Available at: www.iwmi.org. Accessed: 30 May 2019

Moss T (2004) The governance of land use in river basins: prospects for overcoming problems of institutional interplay with the EU water framework directive. Land Use Policy 21(1):85–94. https://doi.org/10.1016/J.LANDUSEPOL.2003.10.001

Moss T, Newig J (2010) Multilevel water governance and problems of scale: setting the stage for a broader debate. Environ Manage 46(1):1–6. https://doi.org/10.1007/s00267-010-9531-1

Muller M (2018) Scale and consequences – the limits of the river basin as a management unit. Wat Sci Technol Water Supply 1–10

Neto S (2016) Water governance in an urban age. Utilities Policy 43. https://doi.org/10.1016/j.jup.2016.05.004

Newig J, Fritsch O (2009) Environmental governance: participatory, multi-level—and effective? Environ Policy Govern 19(3):197–214. https://doi.org/10.1002/eet.509

OECD (2015b) Water and cities: ensuring sustainable futures. OECD Publishing, Paris. https://doi.org/10.1787/9789264230149-en

OECD (2011) Water governance in OECD countries: a multi-level approach. OECD studies on Water, OECD Publishing

OECD (2016) Water governance in Cities. Paris

Pahl-Wostl C, Gupta J, Petry D (2008) Governance and the global water system: a theoretical exploration. Global Govern 14(4):419–435. Available at: https://www.jstor.org/stable/pdf/27800722.pdf?refreqid=excelsior%3A810836c39ba3eb599a2f922ea3bc60c7. Accessed: 14 Oct 2018

Pahl-Wostl C et al (2007) Managing change toward adaptive water management through social learning. Ecol Soc 12(2):18. Available at: https://www.jstor.org/stable/26267877. Accessed: 4 Dec 2018

Perreault T (2014) What kind of governance for what kind of equity? Towards a theorization of justice in water governance. Water Int 39(2):233–245. https://doi.org/10.1080/02508060.2014.886843

Rahaman MM, Varis O (2005) Integrated water resources management: evolution, prospects and future challenges. Sustain Sci Pract Policy 1(1). Available at: http://ejournal.nbii.org

Rees, J. A. (2006) *Urban Water and Sanitation Services; An IWRM Approach.*

Roggero M, Fritsch O (2010) Mind the costs: rescaling and multi-level environmental governance in Venice Lagoon. Environ Manag 46(1):17–28. https://doi.org/10.1007/s00267-010-9449-7

Savenije HHG, Van der Zaag P (2008) Integrated water resources management: concepts and issues. Phys Chem Earth 33:290–297. https://doi.org/10.1016/j.pce.2008.02.003

Sneddon C et al (2002) Contested waters: conflict, scale, and sustainability in aquatic socioecological systems. Soc Nat Resour 15(8):663–675. https://doi.org/10.1080/08941920290069272

Söderbaum F, Granit J (2014) The political economy of regionalism: the relevance for international waters and the global environment facility. Washington, DC. Available at: https://www.thegef.org/sites/default/files/council-meeting-documents/GEF.STAP_.C.46.Inf_.02_The_Political_Economy_of_Regionalism_-_The_Relevance_for_International_Waters_and_the_Global_Environment_Facility_April_25_2014_rev_4.pdf. Accessed: 8 Oct 2018

Sorensen A (2011) Megacity sustainability: urban form, development, and governance. In: Sorensen A, Okata J (eds) Megacities: urban form, governance, and sustainability. Springer, Tokyo, pp 397–418. https://doi.org/10.1007/978-4-431-99267-7_17

Swallow B et al (2006) The challenges of inclusive cross-scale collective action in watersheds. Water Int 31(3):361–375. https://doi.org/10.1080/02508060608691938

Swyngedouw E (1999) Modernity and hibridity: nature, regeneracionismo, and the production of the Spanish waterscape, 1890–1930. Ann Assoc Am Geogr 89:443–465

Swyngedouw E, Kaïka M, Castro E (2002) Urban water: a political-ecology perspective. Built Environ 28(2):124–137. Available at: https://www.jstor.org/stable/pdf/23288796.pdf?refreqid=excelsior%3A33d4175e87e690aec72bb0ffb516a88b. Accessed: 10 Jan 2019

Tortajada C (2001) Institutions for integrated river basin management in Latin America. Int J Water Resour Dev 17(3):289–301. https://doi.org/10.1080/07900620120065084

Tortajada C (2008) Challenges and realities of water management of megacities: the case of Mexico City metropolitan area. J Int Aff 61(2):147–166

Tucci CEM (2007) Urban flood management. Porto Alegre, Brazil

USGS (2013) Federal Standards and Procedures for the National Watershed Boundary Dataset (WBD) (4th ed.): U.S. Geological Survey Techniques and Methods 11–A3. Available at: https://pubs.usgs.gov/tm/11/a3/pdf/tm11-a3_4ed.pdf. Accessed: 11 Jan 2019

van den Brandeler F (2013) Towards inclusive and resilient water governance? A case study on the implications of deliberation and participation in Guarulhos. University of Amsterdam, Brazil van den Brandeler F et al (2014) Decentralization, participation and deliberation in water governance: a case study of the implications for Guarulhos, Brazil. Environ Urban 26(2):489–504. https://doi.org/10.1177/0956247814544423

van den Brandeler F, Gupta J, Hordijk M (2019) Megacities and rivers: scalar mismatches between urban water management and river basin management. J Hydrol 573:1067–1074. https://doi.org/10.1016/j.jhydrol.2018.01.001

Varis O, Enckell K, Keskinen M (2014) Integrated water resources management: horizontal and vertical explorations and the "water in all policies" approach Integrated water resources management: horizontal and vertical explorations and the "water in all policies" approach. Int J Water Resour Dev 30(3):433–444. https://doi.org/10.1080/07900627.2014.912130

Vaughan S et al (2016) Groundwater report: lake Superior North Watershed. Saint Paul, MN. Available at: www.pca.state.mn.us. Accessed: 11 Jan 2019

Vörösmarty CJ et al (2010) Global threats to human water security and river biodiversity. Nature 467(7315):555–561. https://doi.org/10.1038/nature09440

Vrese PD, Hagemann S, Claussen M (2016) Asian irrigation, African rain: remote impacts of irrigation. Geophys Res Lett 43:3737–3745. https://doi.org/10.1002/2016GL068146.1

Wallington TJ et al (2010) Knowledge-action systems for integrated water management: national and international experiences, and implications for South East Queensland. Available at: http://www.griffith.edu.au/. Accessed: 4 Jan 2019

Wang-Erlandsson L et al (2018) Remote land use impacts on river flows through atmospheric teleconnections. Hydrol Earth Syst Sci 22(8):4311–4328. https://doi.org/10.5194/hess-22-4311-2018

Watson N (2004) Integrated river basin management: a case for collaboration. Int J River Basin Manag 2(4):243–257. https://doi.org/10.1080/15715124.2004.9635235

Wester P, Warner J (2002) River basin management reconsidered. In: Turton A, Henwood R (eds) Hydropolitics in the developing world: a Southern African perspective. African Water Issues Research Unit, Centre for International Political Studies, University of Pretoria, pp. 61–71. Available at: https://library.wur.nl/WebQuery/wurpubs/123299. Accessed: 9 Jan 2019

Wittfogel KA (1955) Developmental aspects of hydraulic states. In: Irrigation civilizations: a comparative study. Pan-American Union, Social Science Monographs, Washington, DC

Wolff GH, Palaniappan M (2004) Public or private water management? Cutting the Gordian Knot. J Water Resour Plan Manag 130(1):1–3. https://doi.org/10.1061/(ASCE)0733-9496(2004)130:1(1)

WWAP (2009) Third world water development report. Preface xii

Xepapadeas A (2011) The economics of non-point-source pollution. Ann Rev Resour Econ 3(1):355–373. https://doi.org/10.1146/annurev-resource-083110-115945

Xie M (2006) Integrated water resources management (IWRM)-introduction to principles and practices 1. Available at: http://pacificwater.org/userfiles/file/IWRM/Toolboxes/introductiontoiwrm/IWRMIntroduction.pdf. Accessed: 5 Jan 2019

Young OR, Schroeder H, King LA (2008) Institutions and environmental change. The MIT Press

Chapter 5
The Implementation of IWRM/IRBM in São Paulo

5.1 Introduction

This chapter examines how different drivers and institutions at multiple levels of the river basin governance regime shape water-related challenges experienced in the Metropolitan Region of São Paulo (MRSP). It uncovers the causal links behind water challenges and the effectiveness of existing policy instruments. It reviews the relevant historical and geographical context of Brazilian river basin governance and the driving forces on the river basin from the local to the global level (see Sect. 5.2); explores how IWRM actors and institutions at multiple levels address water challenges at the basin scale (see Sect. 5.3); analyses the related instruments according to their stated mandates, their effect on actors' behaviour and their impacts on inclusive and sustainable water governance (see Sect. 5.4) and draws conclusions on how more appropriate instruments could be (re)designed for the São Paulo (see Sect. 5.5).

5.2 Context and Drivers of São Paulo's River Basin Challenges

5.2.1 Context in Relation to the River Basin

Although Brazil is water abundant, with about 12% of the world's surface water resources,[1] water is unevenly distributed (Formiga Johnsson and Kemper 2005).

[1] Per capita water availability reaches 40,000 m^3/inhab./year.

F. van den Brandeler, *Scalar Mismatches in Metropolitan Water Governance*, Water Governance - Concepts, Methods, and Practice,
https://doi.org/10.1007/978-3-031-08061-6_5

Water resources are concentrated in the humid and sparsely populated Amazon rainforest, while most Brazilians live within 300 km of the coast. Brazil also has extremely dry regions such as the Northeast.

São Paulo is located near the Tietê River's springs, in the Alto-Tietê Basin (ATB),[2] a sub-basin of the Tietê River. Despite being only 22 km from the Atlantic Ocean, the coastal escarpment forces the river to flow inwards (see Map 2.1). The river crosses the MRSP and flows 1136 km until it joins the Paraná River.[3] The ATB is a heavily urbanized sub-basin that virtually overlaps with the MRSP. Due to the MRSP's population size, water is relatively scarce (Interviews-B4/B5/B28). Water availability per capita here is just above 130 m^3 per year[4] (SABESP 2017) and is aggravated by severe contamination.

5.2.2 *Main Drivers of São Paulo's River Basin Challenges*

Many driving forces at multiple levels directly or indirectly shape water-related challenges in the MRSP and its river basin (see Table 5.1 for overview).

Climate

São Paulo has a humid sub-tropical climate. Heavy summer precipitation causes frequent floods. However, winters are dry, and the city regularly experiences longer dry spells. Between 2013 and 2015, Brazil's Southeast, including São Paulo, experienced a record-breaking drought. Unusually high temperatures and low precipitation levels contributed to an almost complete exhaustion of the metropolis' water reservoirs (Nobre et al. 2016). In 2020–2021, Brazil faced another, more widespread, drought—the worst in almost a century—with parts of the State of São Paulo under "extreme" and "exceptional" drought conditions (ANA 2021). MapBiomas, a collaborative Brazilian land use and land cover mapping initiative, found that Brazil has lost nearly 16% of its surface water over the past three decades (MapBiomas 2021). Scientists expect these events to become more frequent as a result of global anthropogenic climate change. Since the early twentieth century, the country has warmed on average (Nobre 2001; Marengo et al. 2007). Several climate change scenarios indicate warming trends up to 4–6 °C in parts of the country, particularly in the Amazon region, by the end of the century (Nobre 2001; Gosling et al. 2011). Precipitation in the Southeast (where São Paulo is located) is projected to decrease overall (up to 5%) but with more intense and irregular rains (São Paulo Legislative Assembly 2009; Gosling et al. 2011; Angelo and Feitosa 2015). This could aggravate water scarcity as well as floods and mudslides.

[2] Alto-Tietê means Upper-Tietê.

[3] The waters of the Tietê River discharge into the Atlantic Ocean in the Paraná Delta north of Buenos Aires.

[4] Regions face absolute water scarcity if renewable water resources are below 500 m^3/inhab./year (FAO 2012).

Table 5.1 Multi-level drivers of water-related challenges on the river basin

	Direct		Indirect	
	Local	Regional/global	Local	Regional/global
Land use change	Urbanization (especially informal)	Urbanization		
		Agriculture		
Demographic	Population growth in urban periphery	Growing water demand for public supply	Rapid population growth during twentieth century	Rural/urban migration
	Growing water demand for public supply			Population growth in nearby regions
Economic development	Industrial (water use and contamination)			
	Mining (contamination)			
	Water demand for hydropower			
Climate			Heat island effect	Climate variability and change
			2013–2015 drought	

Source Author

Although climate change was acknowledged as a risk multiplier, public managers remained sceptical (Interviews-B5/B8/B24/B33). Scientific models estimate that deforestation in the Amazon rainforest contributes to higher temperatures and decreases in precipitation (São Paulo Legislative Assembly 2009; Nobre 2014). Nonetheless, São Paulo's State Secretary for Sanitation and Water Resources dismissed this link citing the heavy rains and floods of 2016 and blaming natural variability (Watts 2017).

Demographics

The Brazilian population increased exponentially from 51.9 million inhabitants in 1950 to 190.8 million in 2010 (IBGE 2010). National population growth has slowed down to 0.8% a year, and projections indicate that this trend will continue in the coming decades (IBGE 2018). Parallel to this, Brazil's urban population rose from 31.24% in 1940 to 84.36% in 2010, pushed by rural to urban migration (IBGE 2010). By 2005, 42.8% of Brazilians lived in metropolitan regions (Braule Pinto 2007). São Paulo also experienced explosive population growth during the twentieth century (see Sect. 2.2.2). The Municipality of São Paulo currently grows 0.55% a year, and the metropolis as a whole, 0.73% (Seade 2019). Moreover, the population has also grown beyond the MRSP, leading to the definition of the Macro-metropolitan region of São Paulo.

Land Use Changes

Before the region was transformed by Portuguese settlers, it was an ecotone—a transition area between several biomes—composed of Atlantic rainforest, grasslands, wetlands, and vegetation typical of the South, the coastal area and the Central West region of Brazil (Cardim 2018). This led to a rich variety of native flora and fauna. It is estimated that 80 km of wetlands extended along the rivers that crossed today's MRSP (Del-Rio et al. 2015). Most of this original landscape disappeared and, within the small pockets of vegetation that are left, 90% of the plants are non-native (Cardim 2018). São Paulo urbanized rapidly and without planning,[5] leading to increased inequalities, including in the distribution of risks derived from precarious construction (Jacobi and Sulaiman 2017). Expansion at the margins of the metropolis and irregular occupation of the green belt and areas of springs led to water contamination (Interviews-B4/B5/B14/B19/B33). While rapid, unplanned urban growth significantly deteriorated the river basin's water quality and decreased water availability for public supply, water demand grew exponentially (Jacobi et al. 2015). The relative scarcity of water further drove water use conflicts between public/urban use, industrial use, agriculture and energy production within the MRSP and with neighbouring regions.

In addition, regional land use changes such as deforestation, agricultural expansion or intensification and unplanned urbanization (in particular inadequate drainage and sewage discharge[6]) caused or contributed to erosion, which deteriorates soils and leads to siltation, ultimately aggravating risks of mudslides and floods (Jacobi et al. 2015; FABHAT 2016) (Interviews-B17/B19). The lack of adequate soil conservation in rural areas further aggravates these risks (Interview-B17).

Economic Development

In colonial times, the Tietê River was strategic for exploration and served as an integrating element within the state (Paganini 2008). The need for water resources for people and industry, as well as the need for space for urban expansion, led to a transformation of the regional hydrology, as rivers were channelled, rectified, diverted and/or buried. Many of the Tietê's tributaries were buried (e.g., Tamanduateí and Anhangabaú Rivers), becoming invisible and forgotten (Gouveia and Moroz-Caccia 2016).

Although the public-urban and industrial sectors dominate the ATB's water use, irrigation accounts for 65% of water use at state level (DAEE 2009). However, WRM in the region first developed for energy generation. The Tietê River's hydropower potential was harnessed in the late nineteenth century through dam construction (Paganini 2008). The Serra do Mar project that began in 1927, reversed the natural flow of the Pinheiros River, a tributary of the Tietê River and the second most important river in São Paulo (see Fig. 5.1). This transformed the river into a canal diverting

---

[5] São Paulo's first master plan dates from 1972, when the MRSP had over 8 million inhabitants (São Paulo City, no date; Saconi and Entini 2013).

[6] Erosion can be caused by large volumes of effluents that are concentrated and discharged in one location, in a stream that is not channelled (Interview-B17).

Fig. 5.1 Pinheiros River in São Paulo and the Traição pumping station in central São Paulo. *Source* Author

water to the newly built Billings dam. From there, water flows towards the coastal cliffs, generating power through a 750 m drop (Braga et al. 2006). This increased the region's energy capacity from 16 to 480 MW in the early 1950s, supporting the MRSP's rapid urbanization and industrialization (Braga et al. 2006; Paganini 2008).

5.3 The Institutional Framework for IWRM/IRBM in São Paulo

This section examines which formal institutions and organizations at multiple levels shape river basin governance in the MRSP.

5.3.1 Global Level

International institutions finance mega-projects. The World Bank (partially) funded the 'Programa Mananciais' (Headwaters Program), which aims to preserve water sources and surroundings ecosystems.[7] The Inter-American Development Bank

[7] Although this programme was on hold for lack of funding, it received USD $10 million in 2016 and 5 million in 2017 (SABESP 2016, 2017).

(IDB) funded the 'Tietê project', together with SABESP, which aims to clean the Tietê River by expanding sewage treatment services in the MRSP (see Box 6.1).

5.3.2 Transboundary Level

The MRSP is not located within a transboundary river basin or aquifer. However, the Guaraní aquifer, measuring over one million km^2, reaches approximately 200 km from the city of São Paulo (DAEE, no date). Municipalities in the western part of São Paulo State heavily rely on the aquifer. It is the world's largest and is shared by Brazil, Argentina, Paraguay and Uruguay. While an integrated management agreement was signed in 2010 by the four nations, it did not include institutional arrangements for its implementation (Silva and Barbosa Pereira 2018).

5.3.3 National Level

During the military dictatorship that started in 1964, Brazil experienced a process of centralization that eliminated elections at state and municipal levels, and reduced their fiscal autonomy (Abers and Keck 2004). Water management was fragmented between national line ministries and between government departments at state level, while also centralized, technocratic and top-down (Abers and Jorge 2005; Kellas 2010; OECD 2015).

The democratic shift of the 1980s promoted political and financial decentralization and civil society participation, which became key characteristics of new Brazilian federalism (Eghrari 2012). The 1988 Constitution engendered widespread reforms across sectors and substantial revenues were transferred from federal to state and local governments (Abers and Keck 2004). It also states that all waters within the national territory are public, belonging to the nation if they cross state borders, and to a specific state if they are contained within its borders (Arts. 20 and 26, Federal Constitution). The Federal government is responsible for instituting a national WRM system and defining water allocation criteria (Art. 21). This only concerns surface water resources as groundwater is under state domain (Art. 26).

Enacted in 1997, the National Water Law (NWL) (Federal Law nº 9.433) instituted the National Water Resources Management System (SINGRH), based on São Paulo State's 1991 Water Law (see Sect. 5.3.4). It considers the multiple uses of water and suggests a multilevel governance system, including integration, decentralization to the basin level and stakeholder participation (see Table 5.2). River basin committees are responsible for managing basins through inclusive arrangements for stakeholder participation (equal representation of state government, municipalities and organized civil society) (Brazil 1997, Art. 1) (see Sect. 5.3.5).[8] The Law recognizes water as a

[8] Committees are created for both river basins of federal and state basin domain.

Table 5.2 Main aspects of the National Water Law of 1997

Approach to water	Public good Limited natural resource with economic value WRM provides for the multiple uses of water Water for human and animal consumption have priority in times of scarcity River basin as the territorial unit for policy implementation Decentralization of WRM and participation of the Public Authority, users and communities
Goals	Ensure current and future generations the necessary water availability in quality standards appropriate to their uses Rational and integrated use of water resources, including waterway transport, aiming for sustainable development The prevention and defence against critical hydrological events of natural origin or arising from inappropriate uses of natural resources
Key actors	The National Council of Water Resources (CNRH) The National Water Agency (ANA) The River Basin Committees for national rivers The State and Federal District Councils on Water Resources The federal government, state, municipal and federal district bodies whose responsibilities relate to WRM

Source Federal Law n° 9.433, 1997

public good with economic value and a limited natural resource—contrary to Brazil's traditional vision of water's inexhaustibility (Benjamin et al. 2005). This framework thus promoted a shift towards IWRM and introduced new instruments (see Sect. 5.4).

The SINGRH sets national water management objectives, which it pursues through several key entities (see Annex E: Main Actors in São Paulo's Metropolitan Water Governance). The National Council on Water Resources (CNRH) is a multi-stakeholder platform for river basins of Federal domain. The National Water Agency (ANA) regulates bulk water use and water use permits for water bodies under federal jurisdiction. Although this model of WRM is over two decades old, power asymmetries remain and maintain a centralized power structure at the state level (Porto and Porto 2017). Environmental politics have suffered from the deterioration of the relationship between the State and civil society organizations (CSOs) and a focus on mega-infrastructure projects and expanding natural resources extraction (de Castro and Motta 2015).

5.3.4 State Level

In 1991, São Paulo State adopted IWRM through State Law 7.633 (São Paulo State Water Resources Policy), inspiring the 1997 NWL. The policy includes principles of decentralization, participation and integration. Integration concerns surface and groundwater, quantity and quality, and stakeholders (users, the Public Authority and Civil Society). The State Water Law created the State Council on Water Resources, which defines the water resources policy for water bodies under State domain and is responsible for its supervision and regulation. Other key state entities are the DAEE, SABESP, CETESB, EMAE and the Public Prosecutor's office.

The DAEE (Department of Water and Hydropower), a parastatal agency of the State Department of Sanitation and Water Resources (SSRH), implements the State's WRM policy. Created in 1951, it is an autonomous agency that oversees quantitative water management aspects (i.e. water supply dams, water use permits, macro-drainage and flood control) (DAEE 2018). It allocates water by defining criteria for water use, issuing water use permits and regulating water uses (see Sect. 5.4.2). It is seen as technocratic, focusing on large engineering works to control water, dominated by aging engineers and lacking in innovation (Interview-B6).

SABESP is the state Wat&San (Water and Sanitation) company, providing services to most MRSP municipalities and around half of those in São Paulo State (see Sect. 6.3.3). Due to its size and dominance within the MRSP, it also plays a key role in WRM. It has built multiple reservoirs that constitute the 'Integrated Metropolitan System' (see Sect. 5.4.1).

The Environment Department (SIMA) is the state agency responsible for environmental policy, including water quality. CETESB (São Paulo State Environment Agency), an autonomous branch of SIMA, monitors and licenses activities that could cause pollution. Its activities include the 'green agenda' (vegetation clearing), 'grey agenda' (pollution sources, sewage and water treatment plants, landfills) and 'blue agenda' (areas of springs). CETESB works with municipalities to facilitate licensing and pollution control (CETESB, no date). It focuses on contamination with little attention to issues such as climate change (Interview-B8), although SIMA must present a new State climate adaptation plan by late 2022.

The State office of the Public Prosecutor aims to defend the rights of citizens and the public's interest (MPSP, no date). It acts through special groups on issues such as environmental crime and irregular land parcelling. It has been particular active on the issue of untreated wastewater discharge (Interviews-B22/B23).

Finally, the EMAE (Metropolitan Company for Water and Hydropower) is a state company created in 1998 that operates several hydropower plants (EMAE, no date).[9] It collaborates closely with the DAEE and SABESP (Interview-B19).

5.3.5 *River Basin Level*

The 1991 State Water Resources Policy promoted the creation of basin committees across São Paulo State. The ATB committee was established in 1994 (SIGRH, no

[9] The EMAE originated in 1899 in the creation of the São Paulo Railway, Light and Power Company, a Canadian company that provided public transportation services and lighting São Paulo City (Custódio 2012).

date) (see Annex E: Main Actors in São Paulo's Metropolitan Water Governance).[10] This basin overlaps almost entirely with the MRSP's boundaries, leading many to refer to it as the 'metropolitan committee' (van den Brandeler 2013). The area's complexity justified the formation of six sub-committees to manage the ATB's sub-basins (Campos 2009).

Box 5.1 Alto-Tietê Basin Plan

The basin plan must have a 12-year planning timeframe, with short, medium and long-term goals (Deliberation 146, CBH-AT, 2012). The plan in place at the time of fieldwork (2017–2018) was for 2009–2012, and was disproportionally focused on flood risks and piece-meal environmental education programmes (Interviews-B6/B7/B17/B24). Water quality was not granted significant attention (Interview-B7). The basin plan was under revision at the time of fieldwork through a more inclusive and multi-disciplinary process. This included developing the plan 'in-house' so that committee members would be familiar with it and 'own' it (Interview-B7). It was discussed with a wider audience, including through public hearings (Interviews-B6/B7/B20).

Finally approved in 2018, the new plan includes a wider range of concerns, including climate change, water use reduction, water losses, wastewater recycling, (more emphasis on) water quality, land use and development, and lessons from the water crisis (although no contingency plan) (FABHAT 2016). It considers Wat&San services and supports municipalities in developing these (Interviews-B11/B12). The plan aims to go beyond diagnostics and provide concrete planning and actions, including for the protection of areas of springs (see Sect. 5.4.4) and for flood risk management (see Sect. 6.4.2) (Interviews-B6/B7). The new plan identifies the interconnection of water supply systems within and beyond the basin as a potential, short-term, solution to water shortages, and it supports further research on the potentiality of aquifers (FABHAT 2016). Sub-basin committees develop plans separately, and these are not always aligned with the main basin plan.

The basin committee has no direct influence on water allocation. However, it sets the fees for bulk water use and can therefore affect water demand (see Sect. 5.4.3). It also has limited say on the planning and construction of large-scale infrastructure for WRM. The committee emphasizes the role of municipalities in containing urban expansion in areas that are still undeveloped, particular those that contain springs (FABHAT 2016). For this purpose, it supports local actors in appropriating knowledge, information and actions developed at regional level (Interview-B6). It has limited powers, as its decisions are not binding.

[10] The first was the Piracicaba-Capivari-Jundiaí committee, a region with a history of strong WRM participation.

The basin agency (FABHAT) leads the elaboration of the basin plan, which establishes priorities for actions and projects (see Box 5.1), the water resources situation report, which monitors water quantity and quality indicators in the basin, and provides grants according to the criteria established by the committee (SIGRH, no date). These grants are available through the FEHIDRO fund (State Fund for Water Resources) (see Sect. 5.4.3).

5.4 Instrument Analysis

5.4.1 Inter-basin Water Transfers

Design

The MRSP mostly uses surface water resources. As water availability within the ATB became insufficient (or too contaminated), SABESP's unofficial strategy for 'water security' shifted to importing water. Construction began in 1966 of an inter-basin water scheme (the Cantareira System) from the neighbouring PCJ (Piracicaba-Capivari-Jundiaí) river basin (Braga et al. 2006). Of the 81 m^3/s of water used in the MRSP, 33 m^3/s were imported from the Cantareira System (see Table 5.3) (FABHAT 2016; SABESP 2019).

Table 5.3 SABESP water supply systems within and outside the Alto-Tietê Basin

System	Approximate production (m^3/s)	Percentage of total production (%)	Population supplied (in millions)	River basin
Cantareira System	33.0[a]	40.62	7.5	Piracicaba-Capivari-Jundiaí
Alto-Tietê	15.0	18.46	4.20	Alto-Tietê
Guarapiranga	15.0	18.46	3.70	Alto-Tietê
Rio Grande	5.5	6.77	1.20	Alto-Tietê
Rio Claro	4.0	4.92	2.06	Alto-Tietê
Alto Cotia	1.2	1.48	0.36	Alto-Tietê
Baixo Cotia	1.05	1.29	0.42	Alto-Tietê
Ribeirão da Estiva	0.1	0.12	0.04	Alto-Tietê
São Lourenço	6.4	7.88	1.4	Ribeira de Iguape e Litoral Sul
Total	81.25	100.00	20.9	

Source Adapted from SABESP (2019)

[a]Approximately 8 m^3/s are used within the PCJ basin

The São Lourenço system, inaugurated in 2018, has contributed another 6.4 m^3/s to the MRSP from a river basin southwest of the ATB. In 2018, SABESP also interconnected the Cantareira system and the Jaguarí dam in the Paraíba do Sul river basin, located Northeast of the MRSP (SABESP 2017). This has allowed for diverting up to 8.5 m^3/s to the Cantareira system or 12.2 m^3/s to the Paraíba do Sul system, depending on which is in greater need.

Inter-basin transfers require water use permits (see Sect. 5.4.2). The water transfers from the PCJ and the Paraíba do Sul basin spread across multiple states,[11] and their permits involve agreements at federal level (SABESP 2015) (Interviews-B2/B4). Water imports depend on sufficient water availability within reservoirs and are paid through water use fees (see Sect. 5.4.3).

Effectiveness on Actors in Terms of Mandated Goals

The MRSP water supply systems interlink to form SABESP's Integrated Metropolitan System. Different parts of the MRSP can be supplied by two or more systems, depending on the reservoirs' water availability (Interview-B4). Reliance on inter-basin transfers and the interconnection of water systems accelerated after the 2013–2015 water crisis, and according to SABESP representatives, this increased water security through increased reservoir capacity, but also by adding redundancy and flexibility to the system (Interviews-B4/B5/B33).

However, many non-state actors have argued that these large-scale works are a temporary fix and that insufficient emphasis has been given to reducing water use, contamination and losses, or prioritizing certain uses (São Paulo (Estado) 2017: 178) (Interviews-B12/B14/B30/B31/B32/B33/B34). The ATB plan was not coordinated or aligned with plans from donor basins, limiting knowledge sharing and potential synergies (Interview-B20). SABESP's strategy also ignores climate change and population growth. As the MRSP expands, so will water demand and the need to search for more distant water sources at increasing costs. Climate variability and change, meanwhile, could lead to increased water demand due to rising temperatures, and decreased availability with higher evaporation in dams and more frequent droughts. In the long-term, exclusive reliance on this strategy is unsustainable.

Impact on Inclusiveness and Sustainability

Inter-basin transfers increased water supply in the short-term, but also caused environmental degradation while affecting aquatic species from changes in water flows. In July 2018, after beginning operations the Jaguarí dam dropped by 4 m in 20 days (Lira 2018). Land use changes around the dams, such as deforestation, affect the springs that recharge them. The construction and maintenance of basin transfers was also expensive, but no cost–benefit analysis about this approach versus alternatives was publicly available. Moreover, since the Cantareira System was built, water demand for public/urban use and irrigation within the PCJ basin increased significantly, raising tensions over water allocation between the two basins (Braga et al.

[11] The Paraíba do Sul basin spreads across the states of São Paulo, Rio de Janeiro and Minas Gerais.

2006; Anazawa 2018). It also represented a challenge for maintaining minimum environmental flows in drier periods.

The MRSP's economic power and its dependence on water imports led the DAEE and the ANA to prioritize it in inter-basin transfer negotiations to guarantee continued supply, often at the expense of other urban areas (Interviews-B4/B34). Such interference may also be inevitable due to the two basin's contrasting interests, which make decentralized management of such transfers unrealistic (Interviews-B2/B4/B5/B11/B12/B22/B23). Such interference may also centralize power and sideline the basin committee and local governments (Interviews-B22/B23/B32). An example of how the imbalance in power relations affects water allocation is that the Cantareira water use permit—which guarantees a water transfer of 30 m^3/s to the ATB—was renewed in the middle of the 2013–2015 water crisis in a top-down manner with little transparency or consideration for environmental impacts (Interviews-B32/B33). The water use fees (see Sect. 5.4.3) that the PCJ basin received in exchange for the transfer did not compensate for the lost economic opportunities and environmental costs (Demajorovic et al., no date).

SABESP's scale and resources have enabled its plans to effectively become the State's Wat&San policy and to solidify its conceptualization of water security as a supply-side problem (Interviews-B6/B11/B14/B22/B23/B31/B32/B33/B34). The basin committee considers SABESP's plans for its basin plan, but not vice versa, and no other WRM plan at any level is used by the committee, municipalities, utilities or state actors (Interview-B16).

5.4.2 *Water Use Permits*

Description

The 1997 NWL aimed to ensure quantitative and qualitative control over water resources by introducing water use permits and wastewater discharge permits (for the latter, see Annex G: Additional Instruments). Water use permits concern derivations or abstractions from surface or groundwater for various uses. Permits are granted if the volume in question can be sustainably abstracted from a certain water body.[12] This process requires data on water availability in terms of quantity and quality, as well as on users upstream and downstream of the abstraction point (Braga et al. 2008). Permits are granted by the DAEE or ANA, depending on whether the water resources are of state or federal domain (Brazil 1997, Art. 14). They are registered to a specific holder, abstraction point and type of use, and are subjected to inspections. Permits can be transferred to other holders for an administrative fee (DAEE 2017). They are valid for up to 35 years but can be renewed, and they are subject to the water use priorities' ranking established in the state water resources plan. Each state defines its own criteria for issuing water use permits. São Paulo State applies the

[12] Sustainable abstraction depends on the effect on the water stock or flow.

'Q7,10' method to define the minimum environmental flow: the drought flow over a period of seven consecutive days that occurs approximately once every 10 years (São Paulo State Legislative Assembly 1994). Holding a permit is a prerequisite for obtaining an environmental license from CETESB for activities or enterprises that affect waterways (i.e., through abstraction or discharge). The environmental licensing process involves an environmental impact assessment (EIA). The basin committee and agency have no role in granting water use permits (Interview-B20).

Effectiveness on Actors in Terms of Mandated Goals

Water permits have allowed the State to control water use, in contrast to the private appropriation of water resources with no accountability (Menezes Da Costa and Tybusch 2015). Permits must be compatible with the state's water resources plan, which has ensured consistency with water use priorities. Moreover, as permits are a prerequisite for obtaining an environmental license, WRM could be coordinated with environmental management.

Nevertheless, despite relative water scarcity, there were no quantitative restrictions to obtain a permit beyond the 'Q7,10' method in the ATB.[13] São Paulo State had areas with permit restrictions where groundwater extraction is more widespread. Nonetheless, there were no restrictions in the ATB (Interviews-B16/B20). Agricultural users do not have water metres, encouraging wasteful irrigation practices, and the DAEE lacks capacity to adequately monitor users and ensure compliance with allocated volumes (Interviews-B5/B6/B8/B16). There were also no substantial qualitative or environmental criteria for granting water use permits (Interviews-B5/B6/B29).[14] Although the new basin plan also aimed to reduce water demand, this topic was still marginal within the basin committee's discussions (Interview-B14). In certain circumstances, such as environmental degradation, a permit could be suspended (Brazil 1997, Art. 15). However, the DAEE eased regulation by only requiring users to have the necessary paperwork required for a permit and not necessarily verifying it (Interview-B16).

Impact on Inclusiveness and Sustainability

Although the water permit system gave the DAEE greater control, it did not reduce overall water allocation. In the MRSP and surroundings, the majority of water permits were for surface water, and groundwater use was often dismissed as irrelevant by respondents (Interview-B19).[15] However, it was estimated that around 11 m^3/s of groundwater were extracted across the MRSP, and clandestine groundwater use was estimated to be significantly above registered use (FABHAT 2016; OECD 2017)

[13] However, the issuing of new permits was temporarily suspended during the 2013–2015 water crisis.

[14] A SABESP official specified that users will not receive permits for water of low quality, but criteria around this are not formally defined (Interview-B5). Further, there were no criteria that restrict groundwater use near potential pollutant sources such as gas stations.

[15] The Guaraní Aquifer is too far from the MRSP to be used (with current infrastructure and technologies), while the aquifer below the MRSP had limited capacity and is susceptible to contamination.

(Interviews-B4/B5/B20/B28).[16] Registered and unregistered wells increased significantly during the water crisis, particularly by residential and commercial buildings in the (wealthier) centre of the city, with often no quality regulation, exposing users to potential public health risks (Interviews-B6/B7/B19).

Municipal utilities could obtain permits for water resources beyond their borders, but this required agreements with the state government and the municipalities where water would be abstracted, as well as large infrastructure investments to transport this water (Interviews-B22/B23). As a result, if they needed to import water, they bought it from SABESP. According to some respondents, this created a dependency on the state and pushed municipalities to delegate water services to SABESP (Interview-B36).

5.4.3 Water Use and Wastewater Discharge Fees

Description

River basin committees can charge bulk water use fees to users holding water use permits within the basin (i.e. user-payer policy), except those of the energy sector (CEDE 2015).[17] Water use fees aim to: (a) recognize water as an economic good and give users an indication of its value; (b) incentivize rational water use; and (c) obtain financial resources to fund programmes and interventions defined in the basin plan (Brazil 1997, Art. 19). The fees should not significantly increase costs but be sufficient to promote behavioural changes inducing rational water use and constitute a financial reserve for actions within the basin (CEDE 2015).

After years of debate, this system was implemented in the ATB in 2013/2014.[18] The fees include an abstraction, consumption and discharge component[19] and reflect the relationship between availability and demand (MMA 2010). Their value is set within the basin committee and the collected fees go to the State Fund for Water Resources (FEHIDRO). The basin committee members decide how the funds are used, based on the goals outlined by the basin plan, reviews by the technical boards and voting in plenary sessions (Interview-B20). All members can propose projects, which are then reviewed by the committee's technical boards.

The PCJ Committee charges an additional fee (0.02 Real/m^3) for water not returned to the basin—as is the case with inter-basin transfers. However, SABESP negotiated a 50% discount on this fee for imports from the Cantareira System to the

[16] A SABESP official claimed there were around 5000 registered groundwater use permits for an estimated 12,000 wells (Interview-B5). Unregulated wells could be at risk of contamination (e.g. by gas stations) (Interview-B10).

[17] The energy sector pays a "financial compensation for water use for electricity generation", which aims to indemnify states, the Federal District and municipalities for liabilities caused by this activity.

[18] The PCJ basin was the first to implement it, and many basins are yet to follow suit (Interviews-B5/B29).

[19] The abstraction fee is 0.01 Real/m^3, the consumption fee is 0.02 Real/m^3 and the discharge fee is 0.10 Real/kg of Organic Water Pollutant (Interview-B20).

MRSP in a 2006 agreement (when the PCJ implemented water use fees) (Consórcio PCJ 2018). SABESP's aim was to limit the impact of the transfer fee on water tariffs in the MRSP.

Effectiveness on Actors in Terms of Mandated Goals

The implementation of water use fees has slightly increased awareness of water's scarcity but limited increase in rational water use. However, collecting the fees is difficult. When first implemented, it was estimated that the total collected fees would sum up to USD 10 million per year, but in 2015 the collected total amounted to just over half of that (FABHAT 2016). Users must send proof of their water meter reading and, if they fail to comply, they are charged the full amount they are authorized to use according to their permit (FABHAT 2017). They therefore lack incentives to declare uses above their permitted use. Water utilities and industries pay the same fees. If utilities have a "Water Losses Master Plan", and industries a rational water use programme, they pay 20% less (São Paulo State 2010). However, agricultural users do not pay any fees (Interview-B6). Agricultural use is relatively low in the ATB, as most of the land is urbanized or protected, so urban and industrial users were the main contributors (Interviews-B1/B6/B10).[20] Due to high levels of urban and industrial use, the ATB collected higher fees than most basins, and therefore had a more substantial budget (Interview-B25).

However, the value charged was too low to lead to substantial reductions in water use (e.g. for large industrial users it was often not even 1% of their advertising costs) (Interviews-B7/B11/B25/B29/B32). Estimates indicate that, in 2017, SABESP paid around USD 19 million in bulk water fees across the state out of its USD 3.6 billion revenue (Interviews-B11/B29). Moreover, the rate per cubic metre decreases as the volume used increases, which incentivizes wasteful practices (Interview-B29). In recent years, the PCJ Committee has tried to renegotiate the fee for inter-basin transfers and strengthen the shared management of the Cantareira System (Consórcio PCJ 2018). These funds would support the economic sustainability of municipalities upstream of the Cantareira system, and the protection of springs.

Another challenge concerns uncertainty around the cost of treating wastewater to the quality standards required. With better knowledge of these costs, a more adequate system of fees can be designed that could decrease water use or increase wastewater treatment (Da Silva and Rios Ribeiro 2006). Some industries were reported to have reduced their permit's allocated volume of water, reduced their water use and adopted measures such as wastewater reuse, although these cases still seemed marginal (Interviews-B7/B20/B34).

Impact on Inclusiveness and Sustainability

The implementation of these fees has increased the committee's budget tenfold and revived participants' interest in the basin committee (Interview-B16). While the fees charged were not large for SABESP, they can strengthen the committee and allow for

[20] A member of the committee and civil society representative estimated that SABESP and large industrial users contributed approximately 60% to the budget (Interview-B35).

capacity-building and investments in protecting water resources (Interviews-B4/B6). However, the FEHIDRO fund cannot be used to address land use and housing-related issues in areas of protected springs that significantly contribute to water-related challenges. Such issues are outside the basin committee's mandates (Interview-B15). Moreover, at the time of fieldwork, the basin plan, which determines the priorities for actions financed by the FEHIDRO fund, was widely criticized for being inadequate and outdated (Interview-B29).[21] Potential recipients (e.g. municipal departments) often lacked the technical capacity to elaborate projects that complied with the criteria (CEDE 2015). There was also disagreement within the basin committee on whether to spend FEHIDRO funds on investments in sanitation, which is the mandate of Wat&San utilities. Furthermore, tensions increased between the ATB and PCJ committees, as the latter claimed greater compensation for the use and transfer of its water resources. The importance of the MRSP at national level has contributed to top-down intervention by SABESP, the DAEE and the ANA to ensure its access to subsidized bulk water.

5.4.4 Areas of Protection and Rehabilitation

Description

In 1975–76, São Paulo State adopted legislation for the protection of areas surrounding springs, streams, reservoirs and other water bodies of interest in the MRSP, with the aim to protect water production services by regulating land use and land occupation, Wat&San services,[22] and natural resources management (São Paulo State 1997) (Interview-B10). However, as the city grew, real estate and political interests in developing these areas were high, and these laws were not adequately enforced, resulting in informal occupations (FABHAT 2016) (Interviews-B5/B7/B14/B17/B19/B33/B36). There was no public policy to address regional inequalities and unequal urbanization (Interview-B33), nor metropolitan-level planning for these areas.

Hence, in 1997, a new State Law (nº 9.866) was adopted setting guidelines and norms for the protection and restoration of areas of springs of regional interest in São Paulo State (1997). This led to the creation of four Areas of Protection and Recovery of Springs (APRMs) within the ATB, corresponding to 55% of its territory (FABHAT 2016) (see Map 5.1). APRMs aim to (1) preserve and rehabilitate springs of regional interest for public supply; (2) harmonize these actions and environmental protection more broadly with land use and socio-economic development; (3) promote participatory management; (4) decentralize planning and management;

[21] A large proportion of funds went to small, scattered environmental education projects or other local projects, with insufficient transparency on project selection, spending and impact assessment (Interviews-B5/B17/B29).

[22] This concerns water treatment, stormwater drainage, flood control, solid waste management, sewage management and the transmission and distribution of electric energy.

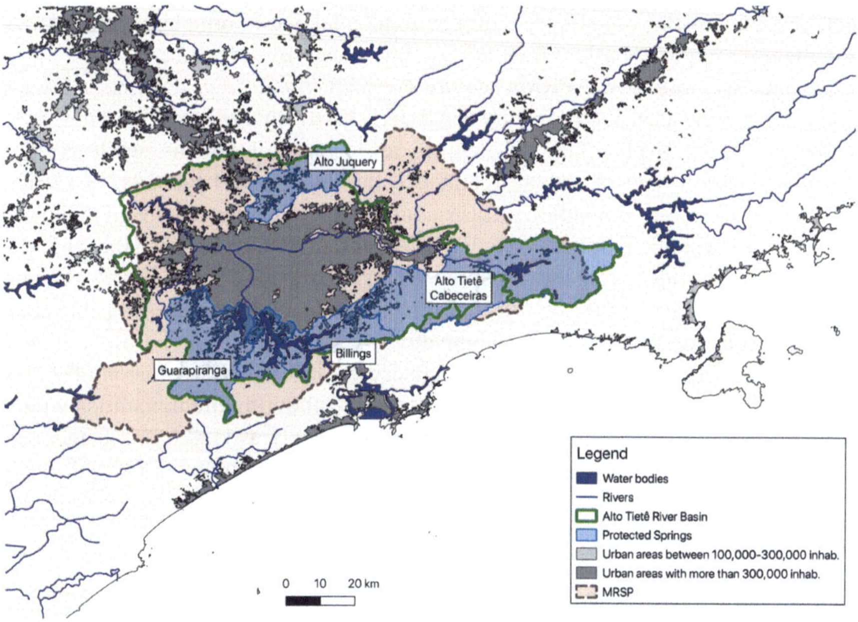

Map 5.1 Expansion of urbanization towards areas of protected springs in the MRSP. *Source* Author

and (5) integrate housing policy and environmental protection (São Paulo State 1997). Sub-basin committees are responsible for developing specific plans for each APRM, aligned with the basin plan (Interview-B6). These plans establish policy guidelines on housing, transportation, environmental management and infrastructure that interferes with water quality.

Within the APRMs, there are three categories of 'Intervention Areas': 'Areas of Restriction to Occupation', with the highest level of protection; 'Areas of Directed Occupation', which allow for urban or rural uses if these comply with environmental preservation criteria, and; 'Areas of Environmental Rehabilitation', which compromise the areas of springs and require corrective intervention (São Paulo State 1997). Each of these areas have guidelines and norms.

Effectiveness on Actors in Terms of Mandated Goals

The focus on areas of springs is a more pragmatic approach that allows for regularization and slum upgrading and for prioritizing areas that are more important for protecting springs and reservoirs. Installing sewage infrastructure is possible even in the most restrictive 'Intervention Areas', but must be approved by CETESB, and the discharged effluents must be compatible with the classification of the receiving water body (São Paulo State 1997). The CETESB addresses environmental issues rather than land occupation, which limits the integration between the two policy areas (Interview-B15). The lack of integration between different mandates and goals perpetuates the status quo, with environmental actors reluctant to regularize settlements or approve basic infrastructure (Interviews-B5/B15/B20).

Despite legal reforms, APRMs require constant monitoring and interventions, and municipalities lack such capacity (FABHAT 2016) (Interview-B14). They were blamed for a lack of planning and control over local developments, while utilities were blamed for the lack of sewage treatment (Interviews-B5/B10/B22/B23). Even if they can install sewage collection infrastructure, connecting households to treatment plants remains a challenge (Interview-B10). Municipal master plans and the APRM's specific laws were also often incompatible. Comparing zoning maps of APRMs with municipal zoning maps reveals different delimitations and land uses for the same areas (Nascimento 2019). Many municipalities did not update and harmonize their master plans with the APRM legislation (Interview-B30). This confusion obscures and facilitates inadequate land use. In addition, neighbouring municipalities within APRMs often did not coordinate preservation adequately as land use management is a municipal mandate (Interviews-B20/B30). This made common policies difficult to agree on.

Impact on Inclusiveness and Sustainability

Some municipalities of the MRSP have large portions of their territory within APRMs,[23] which restricts their ability to obtain environmental licenses from CETESB for development, condemning them to poverty and informality (Interview-B7). The new legislation is more permissive and has allowed for regularization in some areas and thereby the installation of basic urban infrastructure and sanitation services. Between 2008 and 2015, over 1000 km of sewage pipes were installed in the APRMs (FABHAT 2018). Despite this, the effects on slum upgrading and the reservoirs' water quality have been subpar (e.g. phosphorus levels in effluents remain significantly above targets). In 2013/2014 studies reported cases of water-borne diseases within the Guarapiranga and Billings APRMs and showed that contamination levels remained high above the target levels (FABHAT 2016). In 2015, CETESB registered high mortality rates in fish and other aquatic fauna (FABHAT 2016). Moreover, urban growth is higher within the APRMs than in the rest of the basin (FABHAT 2016), indicating that not enough is done to prevent new informal settlements in these areas.

5.5 Instrument Assessment and Redesign

The instruments employed in river basin governance in the MRSP show mixed results, both in terms of their ability to change actors' behaviour and to foster more sustainable and inclusive water governance (see Table 5.4).

[23] 17 municipalities of the MRSP have at least 50% of their territory inserted in an APRM, and 6 are entirely contained within one (FIA, no date).

Table 5.4 Assessment of IWRM/IRBM policy instruments in the MRSP

	Instrument design	Effect on actors in relation to mandate	Impact on sustainability and inclusiveness
Inter-basin transfer	[+] Aims to increase water supply and redundancy through integration Donor basin is compensated	[+] Increases supply and integrates supply systems Disincentivizes water demand measures Not a long-term solution considering drivers of CC and urban growth	Ecol: [−−] Externalities are transferred to donor basin Soc: [+] Access to drinking water is high Econ: [−] Positive for MRSP but negative for donor basin, and costs are transferred to future generations Rel: [−−] Donor basin has little voice in the process and its own development is affected
Water permits	[+] Dependent on water availability, considering minimum environmental flow and priorities ranking	[−] As a prerequisite, environmental licenses encouraged users to obtain water permits However, criteria to obtain one are too lax. Enforcement is weak, especially for agricultural users	Ecol: [−] Greater quantitative control of water resources, but no reduction of overall water allocation. Irregular (groundwater) use is high Soc: [−] Lack of qualitative control, with health risks Eco: [−] Irregular uses have short-term costs (i.e. loss of fees) and long-term costs (i.e. depletion) Rel: [−] The MRSP exports its water scarcity

(continued)

Table 5.4 (continued)

	Instrument design	Effect on actors in relation to mandate	Impact on sustainability and inclusiveness
Water use and wastewater discharge fees	[++] Reflect availability, promote rational use and are redistributed within basin	[+] Effects were limited by the fees' low value (stronger effect for industrial users) The general trend gives ground for optimism, as fees will be increased progressively	Ecol: [+] Limited impact, but as the fees' value rises it is expected that users will reduce use (assuming improved enforcement) Soc: [++] Utilities are charged lower fees than industries. The funds from the fee can also be used to support poorer municipalities Econ: [++] The basin committee's funds have increased, allowing it to invest more Rel: [++] Interest in the basin committee rose with funds, giving it more regional influence
APRMs	[+] Pragmatic approach to preservation and rehabilitation and more inclusive of marginalized residents	[0] Has enabled basic services provision to certain areas, although differences in plans and mandates between sectors and levels of government limit progress	Ecol: [+] Legalizing areas, allowing for sewage infrastructure, may improve water quality more than full protection (with no compliance) Soc: [+] If implemented adequately, residents in these areas will have greater access to services Econ: [+] Legalisation adds paying water consumers. If pollution is reduced, local water bodies become usable, reducing need for imports Rel: [+] It has potential for better integrating marginalized areas within the MRSP

Relative assessment scores ++ Very positive; + Positive; 0 Neutral; – Negative; –– Very negative (see Sect. 2.4)

Redesign

Based on the instrument evaluation, the following redesign options are recommended.

No additional inter-basin transfers: To be sustainable, the conceptualization of water security must integrate demand-side measures and the preservation of ecosystems. As SABESP depends on water sales for its revenue, it lacks incentives to reduce these. If bulk water in other basins becomes more expensive, through higher bulk water use and inter-basin transfer fees, users may have an incentive to reduce imports and improve the quality of nearer water sources. Cancelling SABESP's discounted inter-basin fee—over the course of several years to allow it to adjust—could incentivize water conservation and investments in alternative strategies (e.g. wastewater reuse, rehabilitation of the ATB's water bodies by expanding wastewater collection and treatment), while increasing revenue for the PCJ basin committee preserving the Cantareira System and sustainable economic activities around it. The PCJ Committee and other stakeholders from donor basins require a greater voice in decision-making processes that involve water transfers, to negotiate outcomes that benefit both donor and receiving basins.

Inter-basin transfers lock water users in for a long time. To achieve sustainable and inclusive development, alternative measures to additional inter-basin transfers could be explored. Water demand is expected to increase with urban growth and climate change, this allows for a slow transition while other measures such as wastewater treatment and reuse and water saving equipment are progressively scaled up.

More scrutiny on the issuing of water permits: Most water in the ATB is allocated to public supply. Still, the allocation of permits to intensive industries—a lower priority use than public supply—could undergo more rigorous scrutiny of costs and benefits. Incentives could be put in place to nudge industries towards water reuse. For instance, flexible permits could be introduced so that users could still receive part of their water through the public supply, subsidies could be granted for investments in water saving technology, and reuse could be incorporated into certifications or labelling schemes recognizing greener industries. Industries could also establish partnerships with sewage treatment plants, if located nearby (e.g. receiving treated wastewater at a discounted price, or for free, in exchange for their water allocation). These measures require adequate risk assessments.

In addition, unregistered water abstractions must be addressed. A starting point could be irregular use by industries and commercial or large residential buildings, as they are more likely to use groundwater. More reliable and easily accessible quantitative and qualitative data is also necessary to prevent over-abstraction through more effective monitoring. Key actors virtually ignore groundwater use despite estimates of significant irregular use, and this could receive greater attention if groundwater management was effectively included within water resources management. Furthermore, despite reliance on inter-basin transfers, water use permits continue to be granted to new users. The DAEE could introduce a restriction zone within the ATB, increasing requirements for obtaining a permit (e.g. through environmental and social impact assessments, compensation mechanisms), and potentially creating a moratorium on new permits, where users can still transfer existing permits.

Expand water use fees (progressively): Returning water use fees to the basin committee, to fund projects within the basin, according to priorities that were set in a collaborative manner, has strengthened participatory basin management. The fees are still too low to have a significant effect and nudge users towards conservation or reuse. A sudden increase in rates would create difficulties for smaller users. However, a progressive, annual increase (as was done in other cases, such as in French basin agencies) would allow users time to adjust, for instance by adopting water saving technologies or switching to water reuse. To prioritize water for public supply, an increase in water use fees can be part of a differentiation between types of users. Large, lucrative industrial and commercial water users could be charged a higher fee. Agricultural users could be charged a fee, albeit significantly lower, to discourage wastewater irrigation practices. Efforts could also be made to increase the proportion of water users that pay these fees, which is linked with enforcing water permits.

Integrating APRMs in regional, inter-sectoral planning: The APRMs represent a more pragmatic approach to preserving BESS within and around the mega-city. Although concrete results are still limited, with proper implementation this instrument has potential to keep uncontrolled urban growth in check. One obstacle is the incoherence of plans and mandates at different levels of governance and across sectors. For example, municipal actors must reconcile state-level environmental policies with their constituents' demands for housing and basic infrastructure and services. The new basin plan aims to support municipalities to make their municipal Wat&San plans compatible with the basin plan, and thereby with the APRMs. As an incentive, municipalities that do so could receive points for applying for FEHIDRO funds, which would increase their qualification for receiving funds. If municipal Wat&San plans are coordinated with land use management and urban planning, they could identify parts of informal settlements to legalize, upgrade with certain services or relocate (e.g. when in areas at risk or where no form of sanitation can be installed).

Moreover, regional planning could identify areas to direct urban growth and densification, with shared efforts towards the protection of environmentally valuable areas so that it does not fall disproportionally on the shoulders of low-income peripheral municipalities. Such regional spatial planning has greater potential when linked to the granting of environmental licenses, and thereby to water use and discharge permits. The Integrated Urban Development Plan for the Macro-metropolis (see Sect. 6.3.4), under development at the time of research, aims for regional planning on land use, Wat&San, environmental matters, water resources, housing and more. It remains to be seen whether it will develop instruments promoting a shared regional vision and a fairer distribution of the costs of preserving crucial ecosystem services. As of May 2021, reports indicated that this plan was yet to be approved and converted into law (CNM 2021).

Missing Instruments

In addition, the suggestions below could be incorporated into the instrument mix.

The research found that no significant suasive instruments were in place. The basin committee and other organizations had supported environmental education initiatives and awareness campaigns, but these tended to be small-scale and piecemeal, and they had not been assessed to examine results. On the other hand, during the water crisis, media reports, documentaries and other sources were effective in spreading awareness about the need to reduce water use. In Spain, campaigns to promote rational water use and the use of water-saving technologies have had major impact on reducing water consumption (Tortajada et al. 2019). The challenge in São Paulo is the state government's own reluctance to discuss water scarcity. The basin committee could collaborate with multiple NGOs (and even the private sector) to develop a more comprehensive, large-scale awareness campaign.

PES programmes have been implemented in some parts of Brazil, but not in the ATB. Their effective implementation presented several challenges, such as continuity if they depended on donations. However, with stable funding, they could help preserve areas of springs. They can link users to the ecosystems that they rely on and help make this relationship more visible and valued. If FEHIDRO funds increase, through the collection of water use fees, a portion could be allocated in this way and be linked to the APRMs.

The NWL originally included financial compensation to municipalities (for land use and development restrictions due to environmental and zoning regulations) as one of its instruments, but this was ultimately vetoed. Several respondents mentioned that a financial compensation instrument would be fair, considering the disproportionate negative externalities put on peripheral, poorer municipalities for benefits in the urban core. This could be linked to wastewater discharge fees or to taxes on real estate developments and industries, and it could also make polluting more costly and incentivize polluters to invest in sewage treatment, reuse and other measures.

References

Abers R, Jorge KD (2005) Descentralização da gestão da água: por que os comitês de bacia estão sendo criados? Ambiente Soc 8(2):99–124. https://doi.org/10.1590/S1414-753X2005000200006

Abers RN, Keck ME (2004) Muddy waters: decentralization, coordination and power struggle in the Brazilian water management reform. In: Conference of the Latin American studies association, Las Vegas, Nevada

ANA (2021) Monitor de Secas. Available at: Monitor de Secas do Brasil (ana.gov.br)

Anazawa TM (2018) A escassez hídrica na Região Metropolitana de Campinas entre 2013–2015: a perspectiva de um desastre socialmente construído. Cad Metrop São Paulo 20(42):347–369

Angelo C, Feitosa C (2015) País poderá viver drama climático em 2040, indicam estudos da Presidência—Observatório do Clima. Observatório do Clima. Available at: http://www.observatoriodoclima.eco.br/pais-podera-viver-drama-climatico-em-2040/. Accessed 31 Jan 2019

Benjamin AH, Marques CL, Tinker C (2005) The water giant awakes: an overview of water law in Brazil. Tex Law Rev 83(7):2185–2244. Available at: https://papers.ssrn.com/sol3/papers.cfm?abstract_id=874819. Accessed 2 June 2019

Braga BPF, Porto MFA, Silva RT (2006) Water management in metropolitan São Paulo. Int J Water Resour Dev 22(2):337–352. https://doi.org/10.1080/07900620600649850

Braga BPF et al (2008) Pacto federativo e gestão das águas. In: Revista Estudos Avançados: Dossiê Água. Instituto de Estudos Avançados da USP 33(63):17–42. Available at: http://www.scielo.br/scielo.php?script=sci_issuetoc&pid=0103-401420080002&lng=pt&nrm=iso

Braule Pinto SG (2007) Regiões metropolitanas: por que não cooperam? Federal University of Rio de Janeiro. Available at: http://scholar.google.com/scholar?hl=en&btnG=Search&q=intitle:REGIõES+METROPOLITANAS:+porque+não+cooperam?#0%5Cnhttp://scholar.google.com/scholar?hl=en&btnG=Search&q=intitle:REGIõES+METROPOLITANAS:+por+que+não+cooperam?%230

Brazil (1997) National water law. Available at: http://www.planalto.gov.br/ccivil_03/leis/l9433.htm. Accessed 19 Apr 2018

Campos VNDO (2009) Estruturação e implantação da gestão compartilhada das águas: o comitê de bacia hidrográfica do Alto Tietê. In: Jacobi PR (ed) Atores e Processos na Governança de Água no estado de São Paulo. Annablume, São Paulo, Brazil, pp 13–34

Cardim R (2018) As grandes árvores remanescentes da Mata Atlântica. Olhares, São Paulo. Available at: http://www.esquina.net.br/2018/11/23/as-grandes-arvores-remanescentes-da-mata-atlantica/. Accessed 1 Feb 2019

CEDE (2015) Instrumentos de gestão das águas. Brasília

CETESB (no date) CETESB Apresentação. Available at: https://cetesb.sp.gov.br

CNM (2021) CNM alerta: seis anos do Estatuto da Metrópole e maioria das regiões metropolitanas não possui planos de desenvolvimento urbano. In: Confederação Nacional de Municipios, 26 May. Available at: https://www.cnm.org.br/comunicacao/noticias/cnm-alerta-seis-anos-do-estatuto-da-metropole-e-maioria-das-regioes-metropolitanas-nao-possui-planos-de-desenvolvimento-urbano

Consórcio PCJ (2018) Municípios de cabeceiras das Bacias PCJ querem mais recursos da cobrança pelo uso da água para investimentos em conservação. Tratamento de Água, 11 Jan 2018. Available at: https://www.tratamentodeagua.com.br/pcj-cobranca-uso-da-agua/. Accessed 10 Aug 2019

Custódio V (2012) Escassez de Água e inundações na Região Metropolitana de São Paulo. Humanitas/Fapesp, São Paulo, Brazil

Da Silva SC, Rios Ribeiro MM (2006) Enquadramento e cobrança da água no rio Pirapama—PE. Eng Sanit Ambient 11(4):371–379. Available at: http://www.scielo.br/pdf/esa/v11n4/a10v11n4.pdf. Accessed 27 Apr 2019

DAEE (2009) A Irrigação e o uso racional da água, a trajetória de um profissional, por Dirceu D´Alkmin Telles, Memória do DAEE. Available at: https://memoriadodaee.wordpress.com/2009/12/15/a-irrigacao-e-o-uso-racional-da-agua-a-trajetoria-de-um-profissional-por-dirceu-d´alkmin-telles/. Accessed 4 June 2019

DAEE (2017) PORTARIA DAEE n° 1.630, de 30 de maio de 2017. Available at: www.daee.sp.gov.br. Accessed 2 June 2019

DAEE (2018) DAEE Institucional. Departamento de Águas e Energia Elétrica. Available at: http://www.daee.sp.gov.br/

DAEE (no date) O Aquífero Guarani. Available at: http://www.daee.sp.gov.br/index.php?option=com_content&view=article&id=960%3Ao-aquifero-guarani&catid=51%3Aconsultas&Itemid=55. Accessed 9 Oct 2018

de Castro F, Motta R (2015) Environmental politics under dilma: changing relations between the civil society and the state. LASA Forum 46(3):25–27. Available at: https://www.researchgate.net/profile/Renata_Motta/publication/281101310_Environmental_Politics_under_Dilma_Changing_Relations_between_the_Civil_Society_and_the_State/links/55d4e1ea08ae43dd17de4c44.pdf. Accessed 30 Sept 2019

Del-Rio G, Rêgo MA, Silveira LF (2015) A multiscale approach indicates a severe reduction in Atlantic Forest wetlands and highlights that São Paulo Marsh Antwren is on the brink of extinction. In: Moreira F (ed). PLoS ONE 10(3):e0121315. https://doi.org/10.1371/journal.pone.0121315

Demajorovic J, Caruso C, Jacobi PR (no date) Cobrança do uso da água e comportamento dos usuários industriais na bacia hidrográfica do Piracicaba, Capivari e Jundiaí. Rev Adm Públ-Rio de Janeiro 49(5):1193–1214. https://doi.org/10.1590/0034-7612137792

Eghrari S (2012) Metropolitan governance in Brazil: application and practices in public consortia and participatory budgeting. In: Governing the metropolis: powers and territories. New directions for research, Paris, 28–30 Nov 2012. Available at: https://www.researchgate.net/publication/245024772. Accessed 16 Jan 2019

EMAE (no date) Missão, Visão e Valores. Empresa Metropolitana de Águas e Energia S.A. Available at: http://www.emae.com.br/conteudo.asp?id=Missao. Accessed 2 June 2019

FABHAT (2016) Relatório—I Plano de Bacia Hidrográfica do Alto Tietê. São Paulo, Brazil

FABHAT (2017) COMUNICADO AOS USUÁRIOS DE RECURSOS HÍDRICOS, n. 213. Available at: http://fabhat.org.br/site/index.php?option=com_content&task=view&id=127&Itemid=74. Accessed 29 June 2019

FABHAT (2018) Plano de Bacia Hidrográfica do Alto Tietê: Oficina Técnica no 1—Socioeconomia & Uso e Ocupação do Solo. São Paulo

FAO (2012) Coping with water scarcity: an action framework for agriculture and food security. Rome. Available at: http://www.fao.org. Accessed 11 Dec 2018

FIA (no date) Áreas de Proteção aos Mananciais na RMSP. Fundação FIA. Available at: http://www.fundacaofia.com.br/gdusm/apm.htm. Accessed 5 June 2019

Formiga Johnsson RM, Kemper KE (2005) Institutional and policy analysis of river basin management: the Alto-Tietê River Basin, São Paulo, Brazil, 3650. Available at: http://econ.worldbank.org

Gosling SN et al (2011) Climate: observations, projections and impacts: Brazil. Devon, UK. Available at: http://eprints.nottingham.ac.uk/2040/7/Brazil.pdf. Accessed 23 Jan 2019

Gouveia ICM-C, Moroz-Caccia IC (2016) A cidade de São Paulo e seus rios: uma história repleta de paradoxos. Confins. Théry, Hervé 27. https://doi.org/10.4000/confins.10884

IBGE (2010) Taxa de Urbanização. Censo demográfico 1940–2010

IBGE (2018) Projeção da população do Brasil. Available at: https://www.ibge.gov.br/apps/populacao/projecao/. Accessed 1 June 2019

Jacobi PR, Sulaiman SM (2017) Governança Ambiental Urbana face às Mudanças Climáticas. In: Jacobi P, Fracalanza AP, Empinotti V (eds) Governança da água no contexto da escassez hídrica, 1st edn. IEE-USP, UFABC and GovAmb, São Paulo, Brazil

Jacobi P, Fracalanza AP, Silva-Sánchez S (2015) Governança da água e inovação na política de recuperação de recursos hídricos na cidade de São Paulo. Cadernos Metropole 17(33):61–81. https://doi.org/10.1590/2236-9996.2015-3303

Kellas H (ed) (2010) Inclusion, collaboration and urban governance: Brazilian and Canadian experiences. The University of British Columbia, Vancouver. Available at: www.pucminas.br/editora. Accessed 30 Jan 2019

Lira N (2018) Após interligação de represas, queda brusca no nível do reservatório do Rio Jaguari preocupa Prefeitura e pescadores de Santa Isabel | Mogi das Cruzes e Suzano, Globo. Available at: https://g1.globo.com/sp/mogi-das-cruzes-suzano/noticia/2018/07/25/apos-interligacao-de-represas-queda-brusca-no-nivel-do-reservatorio-do-rio-jaguari-preocupa-prefeitura-e-pescadores-de-santa-isabel.ghtml. Accessed 23 Jan 2019

MapBiomas (2021) Superfície de água no Brasil reduz 15% desde o início dos anos 90. Available at: https://mapbiomas.org/superficie-de-agua-no-brasil-reduz-15-desde-o-inicio-dos-anos-90

Marengo JA et al (2007) Eventos extremos em cenários regionalizados de clima no Brasil e América do Sul para o Século XXI: Projeções de clima futuro usando três modelos regionais. São Paulo, Brazil. Available at: http://www.grec.iag.usp.br/link_grec_old/outros/ambrizzi/relatorio5.pdf. Accessed 1 June 2019

Menezes Da Costa JV, Tybusch JS (2015) Uma Abordagem Crítica Sobre a Outorga dos Direitos de Uso dos Recursos Hídricos no Brasil. An Sem Acad FADISMA ENTREMENTES 12. Available at: http://sites.fadisma.com.br/entrementes/anais/wp-content/uploads/2015/08/uma-abordagem-critica-sobre-a-outorga-dos-direitos-de-uso-dos-recursos-hidricos-no-brasil.pdf. Accessed 26 Apr 2019

MMA (2010) Relatório técnico parcial 7. Brasília

MPSP (no date) MPSP Institucional. Ministério Público do Estado de São Paulo. Available at: http://www.mpsp.mp.br/portal/page/portal/home/home_interna. Accessed 2 June 2019

Nascimento M (2019) Metodologia para Compatibilização entre os Planos Diretores Municipais e as Leis Específicas das Áreas de Proteção e Recuperação aos Mananciais. São Paulo. Available at: http://www.sigrh.sp.gov.br/public/uploads/documents/CBH-AT/14091/res-142_2018_metodologia-de-compatibilizacao_eg_is_ea.pdf. Accessed 28 Apr 2019

Nobre CA (2001) Mudanças climáticas globais: possíveis impactos nos ecossistemas do país. Parcerias Estratégicas 6(12):239–258. Available at: http://seer.cgee.org.br/index.php/parcerias_estrategicas/article/view/186. Accessed 31 Jan 2019

Nobre AD (2014) The future climate of Amazonia: scientific assessment report. São José dos Campos, Brazil. Available at: http://d2ouvy59p0dg6k.cloudfront.net/downloads/the_future_climate_of_amazonia_report.pdf. Accessed 1 June 2019

Nobre CA et al (2016) Some characteristics and impacts of the drought and water crisis in southeastern Brazil during 2014 and 2015. J Water Resour Prot 8(2):252–262. https://doi.org/10.4236/jwarp.2016.82022

OECD (2015) Water resources governance in Brazil. OECD Studi. OECD Publishing, Paris. https://doi.org/10.1787/9789264238121-en

OECD (2017) Cobrança pelo Uso de Recursos Hídricos no Brasil: Caminhos a seguir. Éditions OCDE, Paris. Available at: http://arquivos.ana.gov.br/institucional/sag/CobrancaUso/Estudos/CobrancaPeloUsoRecursosHidricos_01082014.pdf

Paganini WDS (2008) A identidade de um rio de contrastes o tietê e seus múltiplos usos. ABES. Available at: https://books.google.nl/books/about/A_identidade_de_um_rio_de_contrastes.html?id=TF3JQwAACAAJ&redir_esc=y. Accessed 31 Jan 2019

Porto MF, Porto PSDS (2017) Ecologia política da água: conflitos ambientais no Brasil e defesa dos comuns. In: Jacobi P, Empinotti V, Fracalanza AP (eds) Governança da água no contexto da escassez hídrica. IEE-USP, UFABC e GovAmb, São Paulo

SABESP (2015) Crise Hídrica, Estratégia e Soluções da SABESP para a Região Metropolitana. São Paulo. Available at: http://site.sabesp.com.br/site/uploads/file/crisehidrica/chess_crise_hidrica.pdf. Accessed 17 Feb 2019

SABESP (2016) Relatório de Sustentabilidade. São Paulo. Available at: http://site.sabesp.com.br/site/uploads/file/sociedade_meioamb/relatorio_sustentabilidade_2016.pdf. Accessed 4 June 2019

SABESP (2017) Relatório de Sustentabilidade. São Paulo. Available at: http://site.sabesp.com.br/site/uploads/file/asabesp_doctos/relatorio_sustentabilidade_2017.pdf. Accessed 23 Jan 2019

SABESP (2019) Mananciais—Região Metropolitana. Available at: http://site.sabesp.com.br/site/interna/Default.aspx?secaoId=31. Accessed 23 Jan 2019

Saconi R, Entini CE (2013) Como era São Paulo sem Plano Diretor. O Estado de São Paulo, 20 Sept 2013

São Paulo (Estado) (2017) Plano Estadual de Recursos Hídricos (PERH): 2016/2019. CRH/CORHI, São Paulo. https://doi.org/10.1902/jop.2010.104002

São Paulo City (no date) População Recenseada: Região Metropolitana de São Paulo e Municípios (1950–2010), Prefeitura de São Paulo—Urbanismo e Licenciamento. Available at: https://www.prefeitura.sp.gov.br/cidade/secretarias/upload/urbanismo/infocidade/htmls/7_populacao_recenseada_1950_10552.html. Accessed 31 Jan 2019

São Paulo Legislative Assembly (2009) Política Estadual de Mudanças Climáticas. Available at: https://www.al.sp.gov.br/repositorio/legislacao/lei/2009/lei-13798-09.11.2009.html. Accessed 12 May 2018

São Paulo State (1997) LEI ESTADUAL Nº 9.866, DE 28 DE NOVEMBRO DE 1997. Available at: http://licenciamento.cetesb.sp.gov.br/legislacao/estadual/leis/1997_Lei_Est_9866.pdf. Accessed 12 May 2018

São Paulo State (2010) DECRETO Nº 56.503, DE 9 DE DEZEMBRO DE 2010. Available at: http://fabhat.org.br/site/images/docs/decreto56.503.pdf. Accessed 30 June 2019

São Paulo State Legislative Assembly (1994) Lei nº 9.034, de 27 de dezembro de 1994. Available at: https://www.al.sp.gov.br/repositorio/legislacao/lei/1994/lei-9034-27.12.1994.html. Accessed 3 June 2019

Seade (2019) Perfil dos Municípios Paulistas, Fundação SEADE. Available at: http://www.perfil.seade.gov.br/. Accessed 1 June 2019

SIGRH (no date) SigRH Apresentação. Available at: http://www.sigrh.sp.gov.br/fabhat/apresentacao. Accessed 12 May 2018

Silva DM, Barbosa Pereira MC (2018) O AQUÍFERO GUARANI E A BUSCA PELA GESTÃO INTEGRADA DO SISTEMA DE ÁGUAS TRANSFRONTEIRIÇAS: PROPOSTA DE IMPLEMENTAÇÃO DO ACORDO DO AQUÍFERO. Observatório das Águas. Available at: http://www.observatoriodasaguas.org/publicacoes/id-725026/o_aquifero_guarani_e_a_busca_pela_gestao_integrada_do_sistema_de_aguas_transfronteiricas__proposta_de_implementacao_do_acordo_do_aquifero. Accessed 9 Oct 2018

Tortajada C et al (2019) Water demand management strategies for water-scarce cities: the case of Spain. Sustain Cities Soc 45:649–656. https://doi.org/10.1016/j.scs.2018.11.044

van den Brandeler F (2013) Towards inclusive and resilient water governance? A case study on the implications of deliberation and participation in Guarulhos, Brazil. University of Amsterdam

Watts J (2017) The Amazon effect: how deforestation is starving São Paulo of water. The Guardian, 28 Nov 2017

Chapter 6
The Implementation of UWM in São Paulo

6.1 Introduction

This chapter examines how different drivers and institutions at multiple levels of the urban water governance regime shape water-related challenges in the Metropolitan Region of São Paulo (MRSP). It uncovers the causal links behind these water challenges and the effectiveness of existing policy instruments. It reviews the relevant context of Brazilian urban water governance and its main drivers (see Sect. 6.2); analyses the driving forces according to their scalar level; explores which formal actors and institutions shape Urban Water Management (UWM) (see Sect. 6.3); analyses the instruments of UWM according to their stated mandates, their effect on actors' behaviour and their impacts on inclusive and sustainable water governance (see Sect. 6.4). Finally, the chapter summarizes the main empirical findings and considers how more appropriate instruments could be (re)designed for São Paulo in relation to UWM (see Sect. 6.5).

6.2 Context and Drivers of São Paulo's Urban Water Challenges

6.2.1 Context in Relation to the Metropolitan Region

São Paulo was founded in 1554, but remained a small and relatively unimportant settlement until the nineteenth century, when the country attained independence and the Southeast developed the coffee industry (Bógus and Véras 2000).[1] In the 1930s, São Paulo transitioned into a new period of industrialization and the city

[1] São Paulo's population was 30,000 in 1872 when the first census was conducted (São Paulo, no date).

F. van den Brandeler, *Scalar Mismatches in Metropolitan Water Governance*,
Water Governance - Concepts, Methods, and Practice,
https://doi.org/10.1007/978-3-031-08061-6_6

experienced exponential urban growth (see Sect. 2.2.2). With few natural barriers, the city expanded horizontally, absorbing surrounding towns and transforming into one of the world's largest metropolises. The MRSP is composed of 39 municipalities, occupying an area of 8050 km^2 (Kellas 2010).

6.2.2 Main Drivers of São Paulo's Urban Water Challenges

Table 6.1 provides an overview of the main direct and indirect drivers at multiple levels of urban water challenges in the MRSP.

Urbanization

Rapid urban growth, mainly through rural to urban migration due to the modernization of agriculture and the expansion of industrialization, meant that cities could not meet the demand for housing. In addition, cities lacked spatial planning and affordable housing policies. These factors, combined with real estate speculation and the inadequate enforcement of land use regulations, led to the occupation of vacant plots and abandoned buildings in inner cities and the expansion of informal settlements

Table 6.1 Multi-level drivers of water-related challenges in the city

	Direct		Indirect	
	Local	Regional/global	Local	Regional/global
Land use change	Urbanization (especially unplanned, informal)	Urbanization		
Demographic	Population growth in urban periphery	Growing water demand for public supply	Rapid population growth during twentieth century	Rural/urban migration
	Growing water demand for public supply			Population growth in nearby regions
Economic development	Insufficient investments in sanitation and solid waste management		Economic centre of Brazil	Economic crisis in Brazil (2012–ongoing)
Environmental awareness			Lack of environmental awareness	
Climate			Heat island effect	Climate variability and change
			Drought of 2013–2015	

Source Author

within and around cities (Jacobi 2004; Monteiro et al. 2017) (Interviews-B15/B19). Unplanned urbanization in the MRSP has contributed to soil-sealing[2] and the occupation of hillsides, aggravating the risk of floods and mudslides (Jacobi et al. 2015; FABHAT 2016; Bis 2017) (Interviews-B14/B17/B19). These risks are compounded by inadequate sanitation and solid waste management, and poor citizen awareness, causing the accumulation of waste in streams and rivers that consequently clog drains and stormwater channels (Jacobi et al. 2015; Bis 2017) (Interviews-B16/B17/B19). Informal settlements spread to the margins of the MRSP, and especially to the East and South, around springs and water supply reservoirs, as these areas were vacant due to their protected status (see Sect. 5.4.4) (Alvim and Kato 2011; Rolnik et al. 2015). Low-income residents were pushed away from the urban core, in precarious housing, and without access to land tenure nor adequate infrastructure and services (Denizo 2009; Monteiro et al. 2017).

From the 1970s, housing programmes were designed by all levels of government to address the growing low-income housing crisis.[3] These mainly aimed to fulfil the need for shelter, with less consideration for urban planning and basic infrastructure and services (e.g. drainage, sanitation, lighting, and other public services) or environmental impacts (Interview-B30). Housing projects were mainly developed in the MRSP's periphery (Rolnik et al. 2015), disregarding proximity to jobs or the loss of social fabric that can come from uprooting communities (Denizo 2009). Housing programmes were also not coordinated with other social policies, such as education and jobs, nor did they consider the target population's limited ability to pay taxes and maintenance and utility costs. Eventually, these programmes stopped targeting the poorest, due to the high default rate on subsidized loans offered for the acquisition of a home (Monteiro et al. 2017). At federal level, the focus remained on promoting housing programmes rather than developing housing policies that include slum upgrading and integration with land use management and urban planning (Marguti 2018) (Interview-B30).

Economic Development

Municipalities lack the financial and human capacity to contribute adequately to investments in Wat&San services. This motivates many to delegate service provision to the state water company, SABESP or to a private company. Municipalities, especially small ones, often rely heavily on financial support from the federal government. Economic crises over the last decade further reduced budgets at all levels of government for sanitation and risk prevention.

[2] It is estimated that 37% of the land within the ATB is impermeable (Jacobi et al. 2015).

[3] The State is responsible for the housing needs of those that cannot access the formal housing market (Denizo 2007). It operates through the State Housing Department and the Housing and Urban Development Company of São Paulo State (CDHU). The CDHU attends families that earn from one to 10 minimum wages (CDHU, no date).

Heat Island Effect

Furthermore, large-scale urbanization has triggered urban heat islands. Combined with water shortages, these are believed to have caused a severe dengue epidemic during the 2013–2015 water crisis (Clorosur 2015) (Interview-B33). The heat island effect also leads to a decrease in atmospheric pressure causing heavier precipitation (Goldenstein 2017).

6.3 Institutional Framework

Actors and institutions of UWM at multiple levels shape metropolitan water-related challenges and responses to these in the MRSP (see Fig. 6.1).

6.3.1 Global Level

Global level actors have played a crucial role in UWM through the financing of Wat&San infrastructure in the MRSP. The World Bank has granted loans to SABESP, for programmes aimed at reducing leaks and connecting areas using irregular connections to the public network (SABESP 2018).

6.3.2 National Level

During the 1970s, Brazil went through a period of major investments in Wat&San infrastructure, with the implementation of the PLANASA in 1971 (National Sanitation Plan). This centralized, top-down process, led by the military regime, focused on expanding water supply infrastructure through the creation of powerful state water companies (Saiani and Toneto Júnior 2010; Jacobi et al. 2015). The Federal government established policy guidelines for state companies, and municipalities were given a merely passive role in the sector (Sousa and Costa 2016). Investments dropped sharply in the late 1970s, and the plan was abandoned in 1992, after the democratic transition. As the centralized state remains associated with the military dictatorship to this day, there is broad support for decentralization as a mechanism for empowering disadvantaged groups (Abers and Keck 2004). Although the Constitution gave municipalities the responsibility for Wat&San provision, a policy vacuum remained at national level until 2007 (Saiani and Toneto Júnior 2010).

In 2007, the Federal Law No. 11.445 for basic sanitation established a legal framework for Wat&San services (Brazil 2007). These services include drinking water for public supply, the collection, treatment and discharge of wastewater, urban drainage,

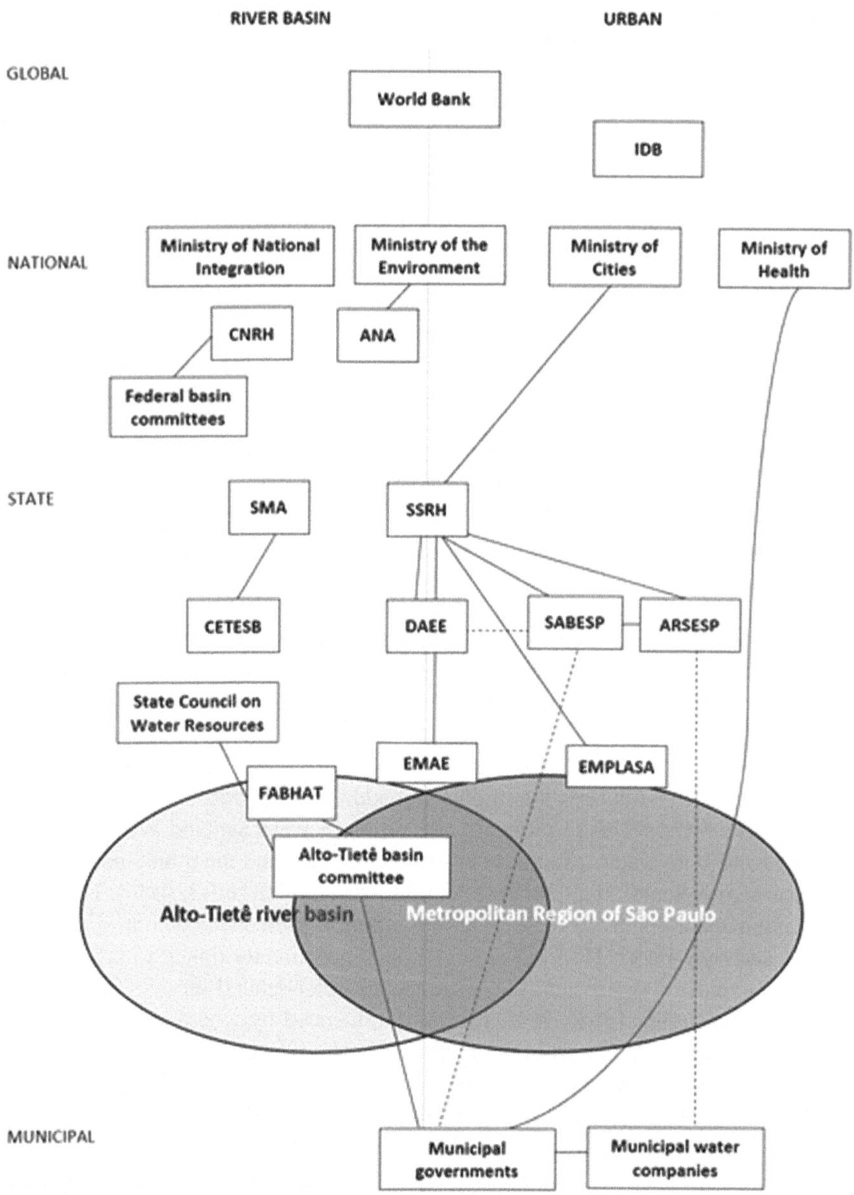

Fig. 6.1 Basin and urban water stakeholders in the MRSP. *Source* Author

Table 6.2 Main aspects of the PLANSAB

PLANSAB	
Approach to Wat&San	A citizen's right, fundamental to the improvement of their quality of life
Goals	Universalization, combined with fair prices and tariffs for: – Access to drinking water for 99% of the population by 2033 – Wastewater management (collection, treatment and disposal) for 90% of the population by 2033 – Urban solid waste management (collection, treatment and disposal) – Adequate urban stormwater management (and thereby flood control)
Key actors	Ministry of the Cities, through the National Secretariat for Environmental Sanitation (SNSA)

Source Adapted from Federal Law nº 11.445, 2007

stormwater management and solid waste management.[4] This legal framework led to the National Basic Sanitation Plan (PLANSAB) (see Table 6.2), based on the principles of universality, equity, integration (i.e. Wat&San services should be provided together), sectoral integration, economic efficiency and sustainability, and alignment with public health and environmental concerns. The National Secretariat for Sanitation is responsible for implementing the PLANSAB in municipalities of more than 50,000 inhabitants (SNS 2019). Although the framework is a distinctive shift away from the sectoral approach to Wat&San services, investments may be too low to reach the goal of universalization by 2033 (Almeida 2017). In 2020, Brazil invested around USD $3.3 billion in Wat&San, just a fraction of the USD $93 billion estimated to be needed to reach the PLANSAB's goals (KPMG 2020; ITB 2022). The Law also explicitly stipulates that it does not address water resources management (Brazil 2007, Art. 4). This reinforces a separation in Wat&San and WRM policies, as federal and state water resources laws focus on users and the municipality is not given a role (dos Santos et al. 2019) (Interviews-B12/B15/B16/B32/B36). The Environmental Ministry is involved in urban water policy, such as flood control, areas of springs, and river parks. Housing policy, though not directly linked to urban water governance, plays a significant role, as access to water-related services and protection from water-related risks is significantly influenced by where and how people live. The 1988 Federal Constitution represented a turning point for housing policy, as it defined access to housing as a right and devolved greater responsibilities for social housing to states and municipalities (Souza et al. 2009; Santos and Duarte 2010). However, this was not combined with adequate financial mechanisms, which especially hampered municipalities' capacity to adequately respond to the housing demand (Santos and Duarte 2010).

In 2015, the 'Metropolitan Statute' established an institutional framework for metropolitan governance and created norms and guidelines for "public functions of common interest" (Casa Civil 2015). This Federal Law derived from a 2013 Supreme Court decision that stated that services of 'common interest' in metropolitan

[4] The Law defines national parameters such as minimum standards for drinking water and the promotion of incentives for conscientious water consumption (Paganine 2015).

regions should be managed jointly by the state and local governments (Costa and Góes 2013). Wat&San services fall under this category, which has important implications for UWM in the MRSP. In fact, ambiguity surrounding the mandates of Wat&San services in metropolitan regions prompted the debate behind this decision (see Sect. 6.3.4). The Metropolitan Statute requires all metropolitan regions in Brazil to develop Integrated Urban Development Plans (IUDPs) that harmonize municipal master plans.

6.3.3 State Level

State actors follow the PLANSAB's guidelines as there is no state-level legal framework for Wat&San. Municipal governments are responsible for Wat&San planning and must guide the service provider's actions (see Sect. 6.3.5) (Brazil 2007). Before the PLANSAB's adoption, state water companies often developed Wat&San plans, de facto shaping the sector's policies (e.g. defining priorities, tariffs) (Interviews-B4/B29/B30/B32/B34/B36). Key actors at state level are SABESP, ARSESP, the DAEE, EMAE and the Public Prosecutor's office.

SABESP, the São Paulo State's Wat&San Company, is the main actor for drinking water and sewerage services and operates in around half of the state's municipalities. The largest in South America, it serves approximately 28.6 million consumers, including in informal settlements (SABESP, n.d.).[5] It is a Government-sponsored, publicly-traded company (51% owned by the State and 49% owned by stockholders) and is listed on the stock exchanges of São Paulo and New York (van den Brandeler 2013). SABESP is autonomous but linked to the SSRH (State Department of Sanitation and Water Resources) (see Sect. 5.3.4), and was founded during the military dictatorship, when the federal government pushed for the establishment of powerful state companies. The company has been consistently profitable and 60% of its net revenue is concentrated in the MRSP (SABESP 2020) (Interview-B36). At the time of research, SABESP operated in 34 of 39 municipalities of the MRSP, and contracts were since then signed for its takeover of the operations in some remaining municipalities.

ARSESP (São Paulo State Regulatory Agency of Sanitation and Energy) is an independent regulatory agency, bound to the SSRH and created by the 2007 Federal Law on Wat&San. ARSESP regulates stated-owned sanitation services (i.e. SABESP's services), and some municipal and private companies' services. It promotes the expansion of these services at an affordable price, while meeting

[5] Initiatives such as the 'Legal Water' programme, initiated in 2016, aim to bring drinking water services to informal settlements (SABESP 2018). This contributes to universalizing water supply services and reduces water losses from irregular connections. It represents a shift in urban policy towards proactively addressing the challenges of informal urbanization by coordinating utilities and local governments (Pasternark and D'Ottaviano 2018). However, without land tenure, these interventions remain in a legal grey zone. Moreover, the physical layout can make the installation of sanitation infrastructure practically impossible.

commercial (e.g. billing, tariffs) and operational (e.g. water quality) targets. It also monitors whether utilities are meeting the targets established in municipal Wat&San plans, such as reducing leakages. It can punish the company or establish a Conduct Adjustment Commitment through which it converts a fine into an investment in the service. As the regulating agency was created long after SABESP and many private providers, it has had to negotiate its position and there ambiguities remain regarding the limits of its role (Interviews-B5/B11). Overall, its powers are dwarfed by those of SABESP.

The DAEE is involved in UWM as it is responsible for macro-drainage and flood control around major rivers (i.e. those that cross municipal boundaries) (see Sect. 6.4.2). This means it must prevent flooding of the Tietê and Pinheiros Rivers within the MRSP. The EMAE also has mandates related to urban flood management within the MRSP, as it is responsible for controlling the volume of water in the canalized Pinheiros River to mitigate flood risks after heavy rains (EMAE, no date).[6] Finally, the State office of the Public Prosecutor monitors, pressures and prosecutes polluting activities, and sets targets for municipalities towards the universalization of sewage collection and treatment services. This has incentivized municipalities to regularize informal settlements to allow for sewerage infrastructure to be installed, as removing populations is often practically impossible (Interviews-B22/B23). The Public Prosecutor's office was criticized for lacking technical knowledge and putting disproportionate blame on Wat&San utilities and local governments (Interviews-B5/B22/B23).

6.3.4 Metropolitan Level

The MRSP was created in 1973 by Federal law (Brasil 1973), as part of the centralized planning apparatus of the military regime. The 1988 Federal Constitution (Art. 25) delegated power to the states to institute metropolitan regions, urban agglomerations and micro-regions to bring municipalities together in the planning and implementation of public functions of common interest, such as Wat&San (Casa Civil 2015). The MRSP almost entirely overlaps with the ATB (Alto-Tietê Basin)—a significant opportunity for collective action and harmonized policies.[7] Nevertheless, metropolitan and basin institutions largely act separately (Interviews-B4/B15).

The state government is responsible for regional and metropolitan planning, including the development of the Integrated Urban Development Plans (IUDPs) required by the 2015 Metropolitan Statute (EMPLASA, no date). It formulates policies at macro-metropolitan level on land occupation issues and compatibility with the

[6] The EMAE and DAEE can use the Pinheiros River as a reservoir when too much water accumulates in the Tietê River. A movable dam was built in the 1990s between the two rivers for this purpose (Interview-B19).

[7] Only five municipalities of the MRSP are not part of the basin.

region's sustainable development.[8] Inter-municipal consortia may also play a role in water governance at metropolitan level. They are legal entities with an autonomous governance structure and their own budget. They unite different municipalities in joint actions that, if produced individually by these municipalities, would not reach the same results or would require more resources (Vaz 1997). Possible joint actions include public services (e.g. basic sanitation provision) and environmental protection. Some municipalities have formed consortia within the MRSP, with their success significantly dependent on coordination between local politicians and actors, as well as on available funding (Interviews-B8/B28/B22/B34/B36).

6.3.5 Local Level

The Brazilian Constitution requires that services of 'local interest' be managed by municipal governments (Constitution of Brazil 1988). This decentralized power and enhanced local governments' roles in various policy decisions, although limited resources often challenged the design of effective policies. Wat&San services (including drainage and waste management) were generally understood as services of local interest. Some municipal attributions relevant to water governance also include land use management, urban planning, basic health care centres, and local environmental issues (Formiga Johnsson and Kemper 2005: 24). They are also responsible for the areas of springs in their territory. Municipalities can thus significantly influence local and regional water resources in terms of quantity and quality.

Municipalities are also in charge of stormwater and flood control, although the State is involved in macro-level aspects. Municipal civil defence authorities must cope with the immediate consequences of flood risks, which especially affect informal settlements in floodplains and unstable hillsides (Interviews-B8/B17). Municipalities must develop master plans that combine these various functions. However, these often overlook the basin plan (Interview-B16). A lack of funding to facilitate coordination and of political will by local officials has hindered integration between the basin and city plans (Interview-B36).

Municipalities choose whether to provide Wat&San services through a municipal company, to contract a private company or to delegate the responsibility to a state company (i.e. to SABESP).[9] Most municipalities within the MRSP have delegated these services to SABESP, but those that have not still bought at least part (and usually most) of their water from SABESP, de facto connecting them to the Integrated Metropolitan System for bulk water management (Interview-B4). As groundwater was generally not considered a reliable or sufficient source, and many surface water bodies in the MRSP were contaminated, the reliance on water resources

[8] In 2010, the São Paulo macro-metropolis had over 30 million inhabitants (EMPLASA, no date).

[9] Solid waste management is typically partially or entirely outsourced, whereas drainage services are mostly carried out by municipal authorities as these have not found a way to monetize these services (Tucci 2009).

beyond municipal borders reduced local utilities' autonomy. Their sewage was sometimes also treated in a regional rather than municipal treatment plant (Interview-B4). Regardless of the service provider, the municipality remains responsible for planning and ensuring a minimum volume of water per person per day (Interview-B32). Although municipal Wat&San plans are mandatory, lack of human and financial capacity often led to plans of low quality or copied from those of neighbouring municipalities (Interviews-B4/B7/B12/B30/B32/B35).[10] These should be updated every four years, which is rarely done,[11] and often do not match their context's reality (i.e. population growth, priority projects and investments), leading water companies to carry out their own planning in practice (Interviews-B11/B22).

Figure 6.1 shows that the mesh of actors responsible for managing water in the MRSP.

After the approval of the 2007 Federal Wat&San policy, debate arose on the ambiguity surrounding *who* had the mandate for operating Wat&San services. The Law delegates the services to 'titleholders', but does not clarify who they are (Brazil 2007). The issue of the mandate is controversial, as the military regime had pressured municipalities to delegate these services to state companies that were not regulated and received federal funds (De Sousa and Costa 2016). In 2013, the Supreme Court ruled that municipal governments were indeed responsible for Wat&San services but that, within metropolitan regions, Wat&San services must be jointly managed by municipalities and the State (Costa and Góes 2013). This decision was based on the principle that such services were of common, rather than local, interest within metropolitan regions (Costa and Góes 2013). This ruling led to uncertainty and significant debates on how to implement it in practice. The New Legal Framework for Sanitation approved in 2020 defines guidelines for regional planning and service provision, including for the creation of regional basic sanitation units with federal financial support (IAS 2020).

6.4 Instrument Analysis

The performance of policy instruments is examined in relation to water quantity, water quality, and adaptation to unpredictable and extreme water-related weather events, and their consistency across multiple levels of governance.

[10] An official from a municipality of the MRSP (Interview-B21) was unable to find the municipality's plan after significant searching during an interview. An ARSESP respondent claimed that mayors are sometimes convinced that their municipality does not have a Wat&San plan (Interview-B11).

[11] Even the municipality of São Paulo has not renewed its plan since 2009 (Interview-B12).

6.4.1 Water and Sanitation Tariffs

Design

The 2007 Federal Law for basic sanitation stipulates that utilities should ensure the financial sustainability of Wat&San services provision while striving for universal services and reasonable rates for consumers (Brazil 2007). Contracts for Wat&San services between a utility and a municipality must also include targets for rational water use (Art. 11 (§2)(II)). The regulatory agency is responsible for defining economic criteria for tariffs and potential subsidies (Ibid.) (Interview-28). The basin committee is not involved in decisions. SABESP applies the same tariffs in all municipalities it operates in by using cross-subsidies, through which smaller, poorer municipalities are subsidized by more profitable regions such as the MRSP (Interviews-B5/B11/B28). In municipalities with other water utilities, tariff rates vary.

SABESP uses 'increasing block tariffs'. For water consumption between 0 and 10 m^3, consumers pay a fixed rate (ARSESP, no date). The rate increases for each additional cubic metre and rises sharply above 20 m^3 to incentivize rational water consumption (Interview-B4). Most utilities charge one tariff for water supply and another for sewage collection and treatment, regardless of whether a household's wastewater was treated (Interview-B28). As collecting and treating sewage is costlier than providing drinking water, there are cross-subsidies between these (Interview-B29). A higher tariff is applied to large consumers, such as industrial and commercial consumers (ARSESP, no date). However, their rates per cubic metre decrease as consumption increases above the volume negotiated through a contract agreement, and they are fined if consuming less (IDS and Aliança pela Água 2017) (Interviews-B28/B29).[12]

SABESP separates domestic consumers into three tariff categories: 'regular', 'social' and 'favela' (the latter is only used in the MRSP) (Interview-B4). In 2019, in the MRSP, the social tariff was USD 2.40 and the favela tariff was USD 1.82 for water consumption up to 10 m^3, compared to USD 7.04 for regular residential consumers (see Annex H: Water Tariffs). The social or favela tariffs are only valid for the first 10 m^3 of water (Interviews-B4/B11/B28). Consumers are eligible for the social tariff if they meet at least one of the following criteria: A combined household income lower than three times the minimum wage, energy consumption below 170 kWh/month, a dwelling with up to 60 m^2 of surface area, or employment status (unemployed consumers have priority) (ARSESP 2009) (Interviews-B28/B29). Consumers paying social tariffs must reapply each year to prove their eligibility (ARSESP 2009). Some municipal utilities also apply such a tariff (van den Brandeler 2013) (Interview-B22).

[12] Between 500 and 1000 m^3, large consumers pay USD 3.8 per cubic metre. Above 40,000 m^3 they pay USD 2.5 (IDS and Aliança pela Água 2017).

Effectiveness on Actors in Terms of Mandated Goals

Drinking water provision is almost universalized within the MRSP (FABHAT 2016). Around 99% of the population in the municipality of São Paulo has access to drinking water through the piped network and this reaches 100% in some wealthy metropolitan municipalities. However, access falls heavily in smaller, poorer municipalities of the MRSP (e.g. only 63% in Salesópolis, where the Tietê River has its springs) (SNIS 2016). Residents in informal settlements often access drinking water through clandestine means (by connecting their home to the official network themselves, thus not paying for their consumption), and this population is included within data on access to drinking water, thereby masking inequalities in terms of quantity and quality.[13] Nonetheless, special programmes to install water supply infrastructure that require municipal authorizations have been implemented in recent years.[14] Sewage collection rates are above 90% in only seven municipalities of the MRSP (including São Paulo), whereas ten have rates below 50% (FABHAT 2016). Sewage treatment coverage varies from 0 to 100% across the MRSP. Consumers nevertheless pay for these services if they have access to drinking water (Interview-B19). While lack of land tenure in informal settlements is part of the explanation for the backlog in sewage collection and treatment, many formal neighbourhoods are lagging behind as well.

Designing a water tariff system that is affordable to all consumers is challenging in a context of severe inequalities, where consumers have vastly different abilities to pay. The inclusion of 'social' and 'favela' tariff rates helps, although fewer households receive this tariff than those who qualify for it (Interview-B29). The fixed tariff rate for consumption between 0 and 10 m^3, with rates increasing exponentially above that, encourages rational water use, but unfairly penalizes larger households for higher consumption even if their consumption per capita is reasonable (Interview-B11). The fixed bill for water consumption below 10 m^3 also reduces incentives to further conserve water (dos Santos et al. 2019) (Interview-B29).

Some local water utilities charged higher tariffs than SABESP (e.g. Guarulhos, East of São Paulo), even though many residents received water only every other day (van den Brandeler 2013) (Interview-B22). Higher tariffs could be due to reliance on SABESP for bulk water as they had insufficient water resources within the municipality's borders (Interviews-B12/B23). Some municipalities applied lower tariffs than SABESP, or did not even charge tariffs, due to widespread clientelism (Interviews-B4/B5/B28/B30). This was criticized as 'tariff populism' by a SABESP representative, as local utilities—often closely connected to the municipal government—could strengthen local support by highlighting that their tariffs were lower than those of SABESP (Interview-B4). These local utilities largely rely on SABESP for bulk water, but some do not pay for this service, which means that SABESP's own

[13] In Guarulhos, a municipality of approximately 1.3 million inhabitants neighbouring São Paulo, many residents have access to water every other day or less frequently (van den Brandeler 2013) (Interview-B23).

[14] Through its Legal Water Programme, SABESP aimed to install official water connections in 160,000 buildings, for 600,000 residents, in 2018. This corresponds to around 2.7% of the MRSP's population (SABESP 2017).

consumers ultimately subsidize water consumption in these municipalities (many of which are in relatively wealthier areas of the MRSP) (Interview-B4).

Regarding financial sustainability, SABESP was constantly profitable, but some local utilities were not. In 2018, SABESP's net profits were USD 760 million. Between 2003 and 2014, the company redistributed between 26 and 60% of its net profits to shareholders (Schapiro et al. 2018). Economies of scale allowed for state-wide cross-subsidies, helping expand services in rural and lower-income areas (Interviews-B4/B5). SABESP's tariffs were relatively low compared to rates across Brazil, and SABESP executives argued that higher rates on regular consumers would accelerate investments and increase access to services (Interviews-B4/B28/B29/B33). However, SABESP's profits could also indicate room for greater investments (e.g. to reduce leaks, increase sanitation services) or for expanding social tariffs, rather than paying dividends to shareholders (Interviews-B4/B16/B13/B22/B29/B32/B36).

Impact on Inclusiveness and Sustainability

The tariff structure does not incentivize rational water use and investments in wastewater management (Interviews-B4/B6/B13/B22/B29/B32/B36). For industrial and commercial consumers, the rate per cubic metre not only decreases as consumption increases, but this category of consumers must consume a minimum volume of water or pay a fine, which commodifies water and encourages wasteful practices (IDS and Aliança pela Água 2017) (Interview-B29). This perpetuates the reliance on inter-basin transfers to meet demand, instead of investments in alternative water sources (e.g. rainwater harvesting, wastewater recycling and reuse[15]) or in reducing water use (Interviews-B6/B29). Around 35% of treated water was lost (in leaks and clandestine connections) (Interviews-B20/B28/B33). ARSESP did not regulate the stability of water resources or the quality of wastewater treatment, further reinforcing disconnects between tariff structure and sustainable WRM (Interviews-B4/B29).

Another challenge was that the tariff-setting process was complex and lacked transparency. Tariffs were mainly set by SABESP, and ARSESP was sometimes described as a rubber-stamper (Interviews-B12/B29/B36). Although there are public hearings, these are difficult to follow for non-experts due to the topic's complexity (Interview-B36). In addition, although tariff revenues are the main source for investments in Wat&San, it is not clear who decides how revenue from these tariffs is reinvested (Interview-B29).

The lack of rational water use came to light during the 2013–2015 water crisis, partly due to SABESP's and the state government's slow reaction (Interview-B28). ARSESP does not have mandates over WRM (Interviews-B11/B29/B36). Nevertheless, during the water crisis, SABESP, with the approval of ARSESP, eventually implemented a system of bonuses and fines to incentivize consumers to reduce their

[15] Aquapolo, a large industrial wastewater reuse plant near the MRSP, produces up to 1000 L/s of reusable water for a petrochemical complex, a volume equivalent to the water supply of a city of 500,000 inhabitants (Aquapolo, no date). Despite significant potential demand for greywater, laws and norms have not been adjusted, and these practices needed to gain public acceptance (Interviews-B4/B34).

consumption. Fines were applied to monthly water bills that were more than 20% above of the consumer's average water bill of the past year, and bonuses for bills 20% below. Water consumption in the MRSP decreased from 169 l/day per person to 120 l in 2015 (SABESP 2017). By late 2017, consumption had rebounded to 130 l/day per person (SABESP 2017), but still remained below pre-crisis levels, indicating a potential last change in habits (Interviews-B4/B6/B7). In 2021, water consumption in the metropolis was still more than 10% below pre-crisis levels (Ribeiro 2021).

During the water crisis, the MRSP's periphery was significantly more affected by water shortages, especially neighbourhoods uphill and far from water supply dams (Leite 2014; Schmidt and Dezem 2014; Lobel et al. 2015) (Interview-B38). The State government claimed that there was no water rationing and that shortages were due to reductions in the system's pressure (i.e. to reduce leaks, which lowered water flow to the extremities of the system) (Schmidt and Dezem 2014; Martin 2015). Protests were organized by social movements that termed this situation 'selective water rationing' (Martin 2015).

While ARSESP focused on affordable tariffs and financial sustainability, it did not establish clear social or quality standards for services (e.g. water quality) (Interviews-B4/B29). Many households in informal settlements still obtained water through clandestine means (Interviews-B19/B32/B33). Although special tariffs helped some residents in low-income households or informal settlements, land tenure and slum upgrading were not addressed.

The fact that some local utilities did not pay for bulk water could have contributed to higher tariffs, as SABESP had to compensate its financial losses (Interviews-B4/B5/B28). Moreover, some municipal utilities receiving bulk water from SABESP were reported to sell it to industrial users (Interviews-B5/B30). The local utilities claim that SABESP's bulk water fees are excessive. The situation has led to conflicts and lawsuits, creating additional pressure for these municipalities to transfer the services to SABESP (Interviews-B5/B11/B20/B36).

6.4.2 Macro-Drainage

Design

Macro-drainage is defined as stormwater and flood control measures for inter-municipal rivers and is the responsibility of the State government. The Macro-drainage Master Plan of the Alto-Tietê Basin (MMP-ATB) is led by the DAEE, in collaboration with the State, municipalities and the basin committee. The MMP-ATB plan's first version, elaborated in 1998, aimed to restrict flows and store water (Abril 2016). The latest version aims for integrated stormwater and flood management by coordinating actors and policies through both structural (i.e. reservoirs) and non-structural measures (i.e. flood insurance, risk mapping, zoning, monitoring and warning systems, emergency plans) (DAEE 2012; FABHAT 2016). Structural measures include mainly grey but also green infrastructure, such as the development of linear parks along floodplains and the expansion of permeable areas (DAEE 2012). The plan is based on a 20-year planning timeframe and the basin as the planning unit (DAEE 2012). It strives for economically viable engineering solutions with an emphasis on environmental rehabilitation.

The plan points to urbanization and soil sealing as drivers of flood risks and promotes coordination with municipal stormwater plans. The MMP-ATB was incorporated into the ATB basin plan to promote an integrated regional approach to water-related risks implemented in 12 Drainage Districts, based on sub-basins divisions (FABHAT 2016) (Interview-B6). Local governments must adjust their stormwater plans to be coherent with the MMP, while state entities use it to guide regional investments. It is mainly funded through municipal and state budgets (DAEE 2012). The ATB committee also funds certain projects through the FEHIDRO (FABHAT 2016). State and local governments can also obtain funding from international (e.g. World Bank, IDB), federal (e.g. the National Bank for Economic and Social Development) and state sources (e.g. FEHIDRO, public–private partnerships) (DAEE 2012). The MMP-ATB also fosters partnerships between the DAEE and municipalities for infrastructure works, and with the EMAE for flood control and information-sharing (Interviews-B4/B19). By integrating both urban and basin considerations, the MMP-ATB thereby links IRBM and UWM.

Effectiveness on Actors in Terms of Mandated Goals

In practice, the DAEE takes on a dominating role. There seems to be more coordination between the DAEE and individual municipalities than between neighbouring municipalities (Interviews-B8/B17). For instance, the DAEE sets flow quotas for municipalities, which are generally respected (Interviews-B24/B30). The lack of horizontal coordination is a challenge because municipalities are still largely responsible for mitigating water-related risks, yet they act in isolated and reactive ways. Many municipalities lack local stormwater plans (FABHAT 2016). They often have limited financial resources and are reluctant to spend these outside their own borders for regional measures, despite sharing rivers and streams with neighbouring municipalities (Interview-B17). Between municipalities of different political parties, cooperation is even more limited, and tensions can be high (Interviews-B17/B23/B24).[16]

The Macro-drainage plan is not integrated with Wat&San planning, even though large volumes of wastewater are diverted into stormwater drains and streams (Travassos and Momm-Schult 2013) (Interviews-B4/B5/B22/B24/B30/B32). Solid waste also aggravates flood risks, by blocking drains and channels, but this is not addressed in the Macro-drainage plan (DAEE 2012). Informal urbanization further presents a major obstacle to regional stormwater management, but the MMP-ATB

[16] The municipality of São Paulo developed dykes, drainage systems and green areas in an area along the Tietê River. The municipality of Guarulhos, on the other side of the riverbank, was not informed or involved in these works, leaving it more vulnerable as river flows could only spill to one side (Interview-B24).

cannot directly act on land use management and urban planning, as these are the responsibility of local governments. The latter have been unable to prevent the informal occupation of areas at risk, leading to increasing support for regularizing and re-urbanizing informal settlements (Interviews-B4/B17). Local-level and basin planning identify vulnerable areas and potential responses, but ultimately have limited powers to guide municipal urbanization and development (Travassos and Momm-Schult 2013).

While the third version of the MMP-ATB aims to contain deforestation, preserve riverbeds and retain waters upstream, most actions of the plan's first and second versions focused on storing and diverting excess stormwater, dredging and river channelling (FABHAT 2016) (Interview-B8). This reflects the legacy of the plan's conventional, linear approach to UWM. One of the main infrastructural measures was the installation of concrete reservoirs known as 'large pools' (*piscinões*), which receive and hold excess stormwater runoff during heavy rains (FABHAT 2016). The *piscinões* tended to accumulate sediments and require extensive maintenance, and this approach ignored the roots of the problem (e.g. erosion from land use changes upstream,[17] soil-sealing) and prevented ecosystem functions such as aquifer recharge and the transportation of sediments (Travassos and Momm-Schult 2013) (Interviews-B17/B30). In addition, while *piscinões* may address small and medium-sized flood events, they may not be sufficient for larger events (Travassos and Momm-Schult 2013).

Nature-based solutions for stormwater infiltration and water treatment measures are only sporadically implemented by municipal departments, without a basin logic, inter-sectoral coordination and maintenance (Machado 2017; Cavalcanti 2018). São Paulo City has expanded green infrastructure measures through its 2010 Municipal Master Plan for Stormwater Management, which envisioned "another relationship between the city and its rivers", combining the basin, land use and green infrastructure (Bis 2017) (Interview-B24). However, most municipalities of the MRSP lack local-level plans. A basin-wide project of the Macro-drainage plan is the Tietê linear park, which would extend along 75 km, from São Paulo to the springs of the Tietê River (DAEE 2012). A decade after its launch, the project has failed to materialize, likely due to the challenge of removing informal settlements from the riverbank.

Impact on Inclusiveness and Sustainability

Almost two decades after the implementation of the first Macro-drainage plan, flooding remains a recurrent challenge in the MRSP. According to the new basin plan, 50% of the state's municipalities and most of the ATB's municipalities were affected by floods (FABHAT 2016) (Interviews-B5/B17). Flooding events in the municipality of São Paulo increased from 736 to 1191 between 2007 and 2016 (Bis 2017).

Residents in informal settlements located in floodplains or steep hillslopes are the most exposed to water-related risks (Interviews-B8/B24). These areas often lack

[17] As erosion and siltation are cumulative problems that are not immediately apparent, they do not receive adequate attention (Interview-B17).

adequate infrastructure for sanitation, drainage and solid waste collection, which increase the risks, through contamination and clogged drains (Hordijk et al. 2016) (Interview-B24).

Many flood risk measures, such as dredging, are costly and address only the symptoms of the problems (Interviews-B17/B19). The total dredging of the Pinheiros River would require removing around two million cubic metres of accumulated sediment and other materials (Interview-B19). Due to contamination, the transportation and disposal of these sediments is an added challenge (Interview-B17). As opposed to services such as drinking water provision and sewage collection and treatment, stormwater management does not generate any revenue and often relies on federal and state funds (Interviews-B18/B24). In low-income and informal settlements, residents are often left to their own devices to prevent flooding, and improvise walls and other measures to cope (van den Brandeler 2013; Hordijk et al. 2016).

6.4.3 Integrated Sewage System

Design

As with water supply, SABESP developed an integrated system for sewage collection and treatment at the metropolitan scale as part of the Tietê Project (see Box 6.1). This was based on the premise that integrating sewage mains across the metropolis and building fewer but larger sewage treatment plants would enhance efficiency through economies of scale (Interviews-B5/B11). SABESP divides the basin into sewage discharge basins that follow hydrological boundaries and guide the spatial planning of sewage collection and treatment (FABHAT 2016). Six large treatment plants spread across the MRSP's core and three times as many smaller plants formed 'isolated systems' in the periphery, where location, low population density and other factors made this more cost-effective (FABHAT 2016). The number and capacity of treatment plants was expanding, but some peripheral municipalities still discharged their sewage *in natura*. Some municipalities with local utilities, particularly within the MRSP's core, transported sewage to SABESP's treatment plants (FABHAT 2016).

Box 6.1 The Tietê Project

In the early 1990s, after growing public outrage with the Tietê's contamination, SABESP launched the Tietê Project (SOS Atlantic Rainforest), and the added media attention from the Rio 1992 Conference (Interviews-B3/B31). The project aimed to expand sewage collection and treatment across the MRSP to prevent effluent discharge into the Tietê River or its tributaries. Critics argued that progress had been slow despite around USD 2.7 billion invested over 25 years (Mori 2017). The 2013–2015 water crisis was a further setback as investments were divested towards water supply, and the pollution of the river expanded from 65 to 130 km (Interview-B31). Renewed investments in basic sanitation, especially in informal settlements, reduced the Tietê's pollution spread back to 85 km by late 2021 (SABESP 2021).

One of Latin America's largest wastewater treatment plants is located in Barueri, a municipality of the MRSP downstream of São Paulo and treats a large proportion of the MRSP's sewage. The municipality of São Paulo is involved in investments and decision-making regarding the plant (Interviews-B11/B12).

Effectiveness on Actors in Terms of Mandated Goals

In 2010, 87% of households in the MRSP were connected to the sewage network, up from 81% in 2000 (FABHAT 2016). Peri-urban municipalities had lower sewage collection levels, despite many being located near important springs (FABHAT 2016). Not all collected sewage is treated, with much of it discharged in waterways (FABHAT 2016; Goldenstein 2017) (Interviews-B4/B7). Local utilities with high rates of collection and treatment are mainly located in relatively wealthy municipalities. The municipality of Guarulhos, with around 1.3 million inhabitants, had a municipal utility that collected around 89% of its sewage but only treated around 5–8% of it (Interviews-B22/B28). The rest flowed to stormwater drains and into streams, even though SABESP had a sewage treatment plant nearby (Interviews-B5/B22). This may be due to inter-municipal rivalries and Guarulhos avoiding the loss of future revenue opportunities (SOS Mata Atlântica 2017). Similarly, the Barueri treatment plant is surrounded by municipalities with low rates of sewage collection and treatment (FABHAT 2016). In part, this is due to the extensive network of pipes and sewage mains required. However, in many cases, sewage interceptors and treatment plants in central areas of the MRSP are in place, but only a low flow of wastewater reach treatment plants due to the difficulty of installing sewer mains in informal settlements (FABHAT 2016). This also concerns SABESP's isolated systems and those operated by local utilities in metropolitan municipalities (FABHAT 2016).

Part of the population in areas not served by the public network have self-built and unregulated sceptic tanks (D'agostino 2013) (Interview-B21). Adequately built and regulated sceptic tanks, or small-scale decentralized treatment plants, could enhance sewage collection and treatment in marginalized areas, but such options are not mainstreamed. Resistance from governments at different levels and SABESP, even for informal settlements, are likely due to the need for a significant paradigm shift among water sector professionals, politicians fond of large and visible public works (believed to bring more votes) and the practical challenges of maintaining and regulating multiple, dispersed plants (Interviews-B30/B32).

As mentioned before (see Sect. 6.3.5), municipalities with lower financial and human capacity, had non-existent, outdated or inadequate Wat&San plans. Municipalities require such plans to qualify for federal funding for sanitation. For ARSESP, inadequate plans were also problematic as the agency relies on them to evaluate whether utilities are fulfilling their responsibilities (Interview-B11). Despite interconnections through large-scale infrastructure and water flows, municipalities' Wat&San plans were not coordinated with those of neighbouring municipalities (Interviews-B12/B36). They were also not aligned with other sectoral plans at municipal level, leading to contradictions between sectoral goals (Interviews-B6/B23).

The lack of local level planning lead SABESP to develop its own, informal plans (see Sect. 6.3.3). Although cross-subsidies between municipalities allowed SABESP to expand services while charging the same tariff, poorer and smaller municipalities tended to be left behind. Mairiporã, a municipality of the MRSP with one of the lowest rates of sewage collection and treatment, had not seen investments in sanitation since the 1970s (Correio Juquery 2017). There is little transparency around SABESP's investment decisions in different municipalities.

Furthermore, unplanned urbanization hindered utilities' ability to install sewage infrastructure (Interviews-B22/B32). Municipalities with informal settlements in APRM's especially struggle to expand these services due to legal restrictions on development in protected areas (see Sect. 5.4.4). However, even some wealthy areas of the MRSP are not connected to sewage treatment plants (Interview-B31). The goals and mandates of Wat&San utilities, the environmental sector and municipal governments were set at different levels: the state level (through CETESB) regulated Protected Areas and water contamination, municipalities were responsible for providing Wat&San and managing land use, but infrastructure that prevents sewage contamination was mainly managed at state level (by SABESP) (Interviews-B4/B5/B15/B22/B32). The urban and environmental agendas were at odds, with some actors pushing for better protection of areas of springs and others supporting land tenure and upgrading of informal settlements in these same areas (Interviews-B4/B5/B25). This led to a deadlock, where informal settlements were neither relocated nor regularized or upgraded, and sewage continued to contaminate water bodies (Interviews-B4/B5/B6/B7/B15/B19/B32/B33). This reflected the lack of a shared vision of the metropolitan water cycle (Interviews-B4/B25).

Impact on Inclusiveness and Sustainability

The lack of sewage collection and treatment across the MRSP caused the contamination of waterways far downstream, decreasing water availability (Interviews-B7/B16/B17/B20/B36). The cost of the contamination of the Tietê River is estimated at more than USD 19 million per year, based on the volume of Organic Water Pollutant in the yearly discharge of sewage within the basin and the fees charged for wastewater discharge (SOS Mata Atlântica 2017). Sewage contamination causes visual pollution and discomfort from its smell, threatens public health (e.g. dengue, leptospirosis, and diarrhoea) and water systems, substantially reduces potential water uses and leads to the loss of commercial value of riverside areas (van den Brandeler 2013; Goldenstein 2017) apart from the damage to the ecosystems that are affected (see Fig. 6.2).

Industrial contamination had decreased with stricter regulations and the move of industries to neighbouring basins, but it remained a challenge within the MRSP (SOS Mata Atlântica 2017) (Interviews-B7/B31). A 2014 study estimated that 28% of industrial wastewater in the MRSP was discharged untreated into rivers and

Fig. 6.2 View of the Billings Dam where the Pinheiros River flows in. *Source* Author

streams (Oliveira et al. 2014). Monitoring illegal discharges remains a challenge (Interview-B31). Inadequate solid waste management, particularly in informal settlements, and insufficient environmental awareness and education also affected water quality (Interviews-B19/B20/B36). The EMAE removed around 10,000 tons of solid waste (and a few corpses) from the Pinheiros River each year (Interview-B19).

Water contamination and increased water demand have led to conflicts between users. Moreover, the extreme contamination of the Pinheiros River, and thereby of the Billings Dam, has threatened the reservoir's water supply potential. As a result, since 1992, water from the Pinheiros River is only released into the dam after heavy rains, as a flood control measure, and the dam's hydropower potential has been reduced (Interview-B19).

6.5 Instrument Assessment and Redesign

Although Wat&San, stormwater and flood risks are mainly local responsibilities, state-level actors played important roles in the MRSP's UWM. Informal urbanization and inadequate urban planning are major obstacles for all three of the analysed instruments (see Table 6.3).

Table 6.3 Assessment of UWM policy instruments in the MRSP

	Instrument design	Effect on actors	Impact
Water tariffs	[+] Differences between utilities, but most of the MRSP has tariffs designed to be affordable and financially sustainable	[++] Drinking water access increased, but millions have no access or highly precarious access. Sanitation is lacking despite a sewage fee Tariffs are mostly affordable and include low rates for low-income households, but not all qualifying households receive these tariffs, and large households are penalized by the fixed rate for consumption between 0 and 10 m^3 Cross-subsidies facilitate access in smaller, poorer towns Clandestine connections lead to losses but operating in informal areas is a challenge. SABESP is financially sustainable	Ecol: [−−] Tariff structure does not incentivize rational water use, nor link to water availability. Bonuses and fines during the water crisis helped reduce consumption but were discontinued Soc: [++] Overall affordable, but many households that qualify for social/favela tariffs do not receive these, even though SABESP has significant profits Econ: [−] SABESP's recurrent profits suggests to some that it could invest more heavily in sanitation and reducing losses. Ignoring this will lead to higher costs in the future Rel: [−] Lack of transparency in tariff calculation. The regulator is much weaker than SABESP
Macro-drainage plan	[++] Multi-stakeholder planning with structural and non-structural measures, managed at sub-basin level	[0/+] Effective coordination between DAEE and municipali-ties, but less between the latter Emphasis on structural, reactive measures over non-structural and preventive ones It does not address the main driver of floods: informal urbanization. Urban policy does not adequately consider water-related risks Not integrated with Wat&San and solid waste management	Ecol: [−] Hard engineering focus that ignores ecological functions. Upstream erosion and solid waste are ignored Soc: [−] Floods and mudslides have a heavy toll, disproportionately affecting poorer, marginalized inhabitants Econ: [−] Flood costs are high, enhanced by the focus on measures that address the symptoms rather than the causes Rel: [0] Mandates mostly at local level, so poorer municipalities are less prepared. However, basin approach provides them more support

(continued)

Table 6.3 (continued)

	Instrument design	Effect on actors	Impact
Integrated sewage system	[+] Planning within sewage discharge basins. Focus on centralized infrastructure. Small systems in peri-urban and rural areas	[0/+] Combining a central, integrated system, with surrounding isolated systems is an efficient approach. Remote and informal areas are under-serviced and local alternatives might be more suitable The lack of Wat&San planning hampers efforts	Ecol: [−] Environmental degradation Soc: [−] Waterborne diseases, smell and visual pollution Econ: [−−] Contamination was high despite large investments in sanitation. Water has to be imported Rel: [−] Marginalized residents are more exposed to polluted water. SABESP neglects peri-urban municipalities

Relative assessment scores ++ Very positive; + Positive; 0 Neutral; − Negative; −− Very negative (see Sect. 2.4)

Redesign

Based on the instrument evaluation below, the following redesign options are recommended.

Water tariffs: Current incentives push utilities to sell more water, leading to more water imports. Rational water use measures can be better integrated into the tariff system by eliminating the decreasing rates for water consumption by industrial and commercial consumers (i.e. currently, the more they consume, the lower the cost of an individual cubic metre), which could reduce their consumption. Water saving measures, such as fines for excessive consumption (based on average consumption) implemented during the water crisis, could immediately go into effect during water shortages, as is done in the energy sector. Incentives for installing/retrofitting water saving equipment and appliances (e.g. subsidies) or faster water use permit approvals, could further promote behavioural changes.

A significant obstacle is the disconnect between the tariff structure and water availability in the basin. SABESP's tariff benefits from a discounted fee for water imported from the Cantareira System (see Sect. 5.4.3), and some municipal utilities receive bulk water from SABESP without paying for this service. These local utilities are then more likely to disregard water availability when setting their tariffs. By including water availability considerations in tariff regulations, ARSESP would hold SABESP and other utilities in the MRSP accountable to higher standards (i.e. increasing rational use, reducing water losses), and achieve greater environmental sustainability. Furthermore, utilities could increase transparency and accountability around how sewage tariff revenues are invested in sanitation, as consumers currently pay this tariff even if they do not receive sewage collection and treatment services. In addition, SABESP could use a greater share of its (significant) profits to fund tariff subsidies for households that qualify but currently do not benefit from these, as well as programmes for expanding water supply services in informal settlements. This would also benefit shareholders by increasing the number of paying consumers. If sanitation services are expanded to informal settlements, the reduced contamination in water bodies would reduce costs of treating and using this water.

Macro-drainage plan: To be effective, the Macro-drainage plan must be coordinated with municipal stormwater plans and other sectoral plans at local and state levels (Interview-B36). With support from the basin committee and FEHIDRO funds, local governments could harmonize their stormwater, Wat&San and solid waste management plans with the Macro-drainage plan to ensure synergies and coherence. Although municipal governments generally coordinated stormwater policies and actions with the DAEE, they did not always coordinate with each other. Aligning local stormwater plans with macro-drainage can prevent negative effects from local decisions on neighbours. In addition, expanding green infrastructure (e.g. floodplain rehabilitation, raingardens) could provide relatively low-cost alternatives to stormwater reservoirs, dredging works and other standard measures, while providing various ecosystem services. Peri-urban areas may be more suited for such measures than the urban core due to limited available space. Existing green infrastructure projects indicate the need for regional and inter-sectoral coordination. Furthermore, climate change considerations are not incorporated into the Macro-drainage plan. This may be due to a lack of studies on expected local impacts. Such studies are crucial to develop resilient infrastructure and identify areas that could become more exposed to risks from extreme weather events.

Integrated sewage system: In comparison to water supply, where local authorities depend on water resources beyond their borders, requiring centralized management, wastewater can often be more effectively managed at smaller or intermediary scales (Interview-B30). Large sewage plants can foster economies of scale, but the MRSP's largest treatment plant operates under capacity as many of the surrounding areas consist of informal settlements with no sewage connections. Other areas also fail to connect to this plant due to the lack of large sewage mains. Therefore, small-scale, local sewage treatment has greater potential in informal areas, including those in the APRMs, and where building or biophysical characteristics make it almost impossible to connect households to the sewage network. In such cases, local governments and utilities can collaborate on alternative solutions. Local governments can indicate in their Wat&San plans where such alternatives would be viable. However, they also need greater support in enforcing land use restrictions and in upgrading informal settlements. Within the APRMs, this could be through technical and financial support (i.e. FEHIDRO funds) from the basin committee. Access to these funds should also be conditional on the updating and harmonizing of Wat&San plans with the APRM's plans.

Although there are smaller wastewater treatment plants across the basin, SABESP has a mega-plant downstream of the MRSP, along the Tietê River. The downstream location prevents storing and reusing treated effluents within the basin (for human uses or for ecosystems). Decentralized wastewater management at sub-basin level would lead to higher treatment capacity and the repurposing of treated effluents within the basin.

There have been suggestions for compensating the basin downstream of the ATB for the contamination of the Tietê River that it receives by redirecting (part of) the revenue from wastewater discharge fees to the neighbouring basin committee. This would be fair, but as the ATB committee currently receives these fees it is not clear whether it would incentivize greater investments in sanitation by SABESP and other utilities in the MRSP. However, it would reduce the ATB committee's funding, including for sanitation projects.

Missing Instruments

Regional policies: In the absence of an official state policy for Wat&San and the weakness of municipal Wat&San planning, SABESP has taken a dominant role in shaping the sector (Interviews-B4/B29/B30/B32/B34/B36). This has led to a regionally centralized and infrastructure-oriented approach to Wat&San for the MRSP, with the development of Integrated Metropolitan Systems. Local governments, the basin committee and other state actors have little influence. The SSRH or DAEE could be the appropriate actor to develop a state-level Wat&San policy that oversees regional sanitation planning and harmonizes utilities' plans. Water services could then also be better coordinated with the state water resources policy, as the DAEE has significant mandates in both areas (Interview-B36).

The development of an Integrated Urban Development Plan (IUDP) at state level (see Sect. 6.3.4), could potentially strengthen integrated planning, including for Wat&San, at the metropolitan and macro-metropolitan level. The advantage is that the macro-metropolitan scale allows for considering interlinkages between basins for bulk water supply. ARSESP could oversee the implementation of the IUDP's Wat&San plans, as this complements its evaluation of municipal plans. Knowledge and data-sharing between municipalities and compensation mechanisms (e.g. for downstream municipalities that receive contaminated waters) within the IUDP's regional planning could further enable coherent regional planning. Furthermore, coordinating a regional Wat&San plan with regional plans for environmental preservation, land use, urban planning and housing could lead to a more sustainable and inclusive strategy for informal urbanization (i.e. which informal settlements to regularize, to relocate, and to upgrade)—the greatest challenge for expanding Wat&San and addressing water-related risks.

Climate change approach: Climate adaptation was not high on the agenda for most municipal or state actors, despite frequent floods and recent record-breaking droughts, with more focus on the heat island effect (Interviews-B7/B8/B24/B32). A State Policy for Climate Change was adopted in 2009, establishing the State's commitment towards climate change mitigation and adaptation, emphasising sustainable development, the polluter-payer principle, civil society participation and multi-level cooperation, among its main principles. It consolidates existing policies relevant for climate change mitigation and adaptation across departments (São Paulo Legislative Assembly 2009). These focus mostly on limited mitigation measures (e.g. emissions inventories and information campaigns) (Interviews-B7/B8/B9).

Efforts could be made to mainstream climate change adaptation into sectoral plans and the river basin plan. Although this has costs, current strategies focused on increasing water supply through inter-basin transfers and mitigating floods through stormwater reservoirs will become increasingly expensive and prone to fail, as climate change is expected to lead to less but more intense precipitation. Water demand management, floodplain rehabilitation and environmental preservation would build resilience to these changes, thereby reducing future costs. However, for regional adaptation strategies to consider interlinkages between areas (e.g. in terms of ecosystem services, large-scale infrastructure, upstream/downstream effects), a key obstacle is coordination between actors and political will. This could be addressed by making climate adaptation a cross-cutting objective of the various sectoral plans of the IUDP. In addition, task forces could be set up between agencies at state and municipal levels to build on common challenges and goals and identify synergies. For instance, developing a network of linear parks along waterways and green areas in flood-prone areas brings together agencies for stormwater management, risk management, sanitation, urban planning, and housing, where their diverse interests converge around one project.

Integrating urban and water policies: Addressing the MRSP's water-related challenges requires greater coordination between water and urban policies to jointly discuss how to address the challenge of informal settlements. They could identify areas to legalize, facilitating the provision of basic infrastructure and services and thereby reducing wastewater contamination in nearby water bodies. This concerns long-established settlements where eviction and relocation are not socially acceptable or financially viable options, but it cannot include occupations within floodplains or areas at risk (Interviews-B4/B6/B22). The latter must be relocated to social housing projects, ideally nearby to be less disruptive.[18] A regional strategy must identify areas available for these housing projects across the MRSP, including by revitalizing inner-cities. Municipal master plans must be harmonized with Wat&San plans and with the IUDP (Interviews-B4/B5). A regional strategy requires developing policies and plans that address regional inequalities across the MRSP and its rural hinterlands, as these put disproportionate pressure on peripheral municipalities with little capacity and that must preserve APRMs (Interviews-B4/B14/B15/B33). A regional strategy should develop financial instruments to support access to social housing for those who need to be relocated, such as cross-subsidies between high-end developments and low-income/social housing, and compensation mechanisms for municipalities that face restrictions on their development in order to preserve areas with crucial ecosystems.[19]

References

Abers RN, Keck ME (2004) Muddy waters: decentralization, coordination and power struggle in the Brazilian water management reform. In: Conference of the Latin American studies association, Las Vegas, Nevada

Abril MJ (2016) 54 PISCINÕES NAS VÁRZEAS DO ALTO TIETÊ O projeto de retenção de água pluvial na Bacia do Rio Aricanduva. In: Seminario Internacional de Investigación en Urbanismo. VIII Seminario Internacional de Investigación en Urbanismo, Barcelona. Available at: https://upcommons.upc.edu/bitstream/handle/2117/101507/49CAM_JulianaMarta.pdf. Accessed 27 Apr 2018

Almeida H (2017) Saneamento: Atraso do Plansab justifica revisão e aperto regulatório. Química, 27 May

Alvim ATB, Kato V (2011) Integração Entre Políticas Urbanas e Ambientais na Região Metropolitana. In: XIV Encontro Nacional da ANPUR, Rio de Janeiro, 23–27 May 2011

Aquapolo (no date) Sobre o Aquapolo. Available at: http://www.aquapolo.com.br/quem-somos/sobre-o-aquapolo/. Accessed 30 Jan 2019

ARSESP (2009) Direitos e Deveres dos Usuários de Saneamento Básico. São Paulo, Brazil. Available at: http://www.arsesp.sp.gov.br/Documentosgerais/direitos-e-deveres-saneamento.pdf. Accessed 14 May 2018

[18] A former SABESP employee estimated that around 2.3 million people in the ATB live in areas of springs. He calculated their need for housing at 50,000 housing units, for some 200,000 people: "That is for those who would need to be removed, and the housing of another 300,000 or 400,000 would need to be upgraded, which would require land use regularization" (Interview-B6).

[19] As was included in the original NWL.

ARSESP (no date) ARSESP—Institucional. Available at: http://www.arsesp.sp.gov.br/SitePages/Institucional.aspx. Accessed 4 June 2019

Bis K (2017) Enchentes em São Paulo: o que é o Plano Diretor de Drenagem? Abril, 19 Jan. Available at: https://casa.abril.com.br/ambientes/enchentes-em-sao-paulo-o-que-e-o-plano-diretor-de-drenagem/

Bógus LMM, Véras MPB (2000) A reorganização metropolitana de São Paulo: espaços sociais no contexto da globalização. Cad Metróp 3:81–98. https://doi.org/10.1590/9329

Brasil (1973) Estabelecimento das regiões metropolitanas. Available at: http://www.planalto.gov.br/ccivil_03/leis/lcp/Lcp14.htm. Accessed 23 Apr 2019

Brazil (2007) Federal Law nº 11.445 on the national guidelines of water and sanitation. Available at: http://www.planalto.gov.br/ccivil_03/_ato2007-2010/2007/lei/l11445.htm. Accessed 24 Apr 2018

Casa Civil (2015) Statute of the metropolis. Available at: http://www.planalto.gov.br/ccivil_03/_ato2015-2018/2015/lei/l13089.htm. Accessed 6 Feb 2019

Cavalcanti T (2018) Reabertos, parques lineares de SP estão sujos, com mato alto e sem segurança—02/04/2018—Cotidiano—Folha. Folha de São Paulo, 2 Apr. Available at: https://www1.folha.uol.com.br/cotidiano/2018/04/reabertos-parques-lineares-de-sp-estao-sujos-com-mato-alto-e-sem-seguranca.shtml. Accessed 12 Aug 2019

CDHU (no date) CDHU—Institucional, Companhia de Desenvolvimento Habitacional e Urbano do Estado de São Paulo. Available at: http://www.cdhu.sp.gov.br/web/guest/institucional/quem-somos. Accessed 6 June 2019

Clorosur (2015) Incidência de casos de dengue é maior em ilhas de calor | Clorosur. Clorosur, 22 Apr. Available at: http://www.clorosur.org/incidencia-de-casos-de-dengue-e-maior-em-ilhas-de-calor/. Accessed 19 Feb 2019

Constitution of Brazil (1988) Constituição da República Federativa do Brasil, 5 Oct 1988. Available at: http://www.senado.leg.br/atividade/const/con1988/con1988_atual/art_30_.asp. Accessed 24 Apr 2018

Correio Juquery (2017) 4 anos após a Lei do Saneamento coleta de esgoto em Mairiporã é mínima. Correio Juquery, 23 Feb. Available at: http://correiojuquery.com.br/noticias/2017/02/23/4-anos-apos-a-lei-do-saneamento-coleta-de-esgoto-em-mairipora-e-minima/

Costa LM, Góes E (2013) STF conclui julgamento das ADIs e decide por gestão compartilhada nas RMs entre estados e municípios. Sanear 21:20–25. Available at: http://www.stf.jus.br/arquivo/biblioteca/PastasMinistros/MarcoAurelio/Entrevistas/2013_mar_21.pdf. Accessed 24 Apr 2018

D'agostino R (2013) Ao lado de estação de tratamento, famílias usam fossa em Guarulhos. Globo, 18 Apr. Available at: http://g1.globo.com/sao-paulo/noticia/2013/04/ao-lado-de-estacao-de-tratamento-familias-usam-fossa-em-guarulhos.html. Accessed 1 July 2019

DAEE (2012) Serviços de Consultoria Especializada Para Elaboração do Terceiro Plano Diretor de Macrodrenagem da Bacia do Alto-Tietê—PDMAT 3. Available at: https://drive.google.com/file/d/0B8iXiltOrl5aMWRJTm1WRlVUTE0/view. Accessed 1 July 2019

De Sousa ACA, Costa NDR (2016) Política de saneamento básico no Brasil: discussão de uma trajetória. Hist Ciênc Saúde Manguinhos Rio de Janeiro 23(3):615–634. https://doi.org/10.1590/S0104-59702016000300002

Denizo V (2007) Os produtos da Política Estadual de Habitação na Região Metropolitana de São Paulo: Elementos para análise de uma política metropolitana de habitação. PhD thesis, Faculdade de Arquitetura e Urbanismo da Universidade de São Paulo

Denizo V (2009) O Foco do Problema Habitacional na RMSP e sua Repercussão para a Formulação da Política Metropolitana de Habitação. In: IIIENAnpur—Encontro Nacional da ANPUR—Planejamento e Gestão Territorial, Florianópolis, Brazil, p 20

dos Santos R et al (2019) Public water supply and sanitation policies and inclusive development of the urban poor in Brazil. Water Policy 21(2):351–367. https://doi.org/10.2166/wp.2019.025

EMAE (no date) Missão, Visão e Valores. Empresa Metropolitana de Águas e Energia S.A. Available at: http://www.emae.com.br/conteudo.asp?id=Missao. Accessed 2 June 2019

EMPLASA (no date) Região Metropolitana de São Paulo. Available at: https://www.emplasa.sp.gov.br/RMSP. Accessed 26 May 2019

FABHAT (2016) Relatório—I Plano de Bacia Hidrográfica do Alto Tietê, São Paulo, Brazil

Formiga Johnsson RM, Kemper KE (2005) Institutional and policy analysis of river basin management: the Alto-Tietê river basin, São Paulo, Brazil, 3650. Available at: http://econ.worldbank.org

Goldenstein S (2017) Rio Pinheiros e seu território: Conhecer para transformar, 1st edn. Águas Claras do Rio Pinheiros, São Paulo

Hordijk M, van den Brandeler F, Filippi ME (2016) Facing the floods: community responses to increased rainfall in Guarulhos, Brazil and Arequipa, Peru. In: Roy M et al (eds) Urban poverty and climate change: life in the slums of Asia, Africa and Latin America. Routledge, London

IAS (2020) SANEAMENTO 2021 Balanço e perspectivas após aprovação do novo Marco Legal—Lei 14.026/2020. Instituto Agua e Saneamento. Available at: https://www.aguaesaneamento.org.br/wp-content/uploads/2021/11/Saneamento-2021-IAS.pdf

IDS and Aliança pela Água (2017) Tarifa de Água e Esgoto: O que está por trás do valor que pagamos, São Paulo. Available at: www.aguasp.com.br. Accessed 6 May 2019

ITB (2022) Universalição. Instituto Trato Brasil. Available at: https://tratabrasil.org.br/pt/saneamento/principais-estatisticas/no-brasil/universalizacao

Jacobi P (2004) The challenges of multi-stakeholder management in the watersheds of São Paulo. Environ Urban 16(2):199–212

Jacobi P, Fracalanza AP, Silva-Sánchez S (2015) Governança da água e inovação na política de recuperação de recursos hídricos na cidade de São Paulo. Cad Metrop 17(33):61–81. https://doi.org/10.1590/2236-9996.2015-3303

Kellas H (ed) (2010) Inclusion, collaboration and urban governance: Brazilian and Canadian experiences. The University of British Columbia, Vancouver. Available at: www.pucminas.br/editora. Accessed 30 Jan 2019

KPMG (2020) Quanto custa universalizar o saneamento no Brasil? Available at: https://assets.kpmg/content/dam/kpmg/br/pdf/2020/07/kpmg-quanto-custa-universalizar-o-saneamento-no-brasil.pdf

Leite F (2014) Periferia de SP poupa dobro de água do que área nobre. Exame, 29 Aug

Lobel F, Souza F, Iory N (2015) Racionamento em São Paulo completa um ano e amplia transtornos. Folha de São Paulo, 28 Feb. Available at: https://www1.folha.uol.com.br/cotidiano/2015/02/1596237-racionamento-em-sao-paulo-completa-um-ano-e-amplia-transtornos.shtml

Machado H (2017) Parques lineares na cidade de São Paulo: Inserção na agenda pública e implementação. Fundação Getúlio Vargas. Available at: https://bibliotecadigital.fgv.br/dspace/bitstream/handle/10438/19565/MPGPP-Hannah-Arcuschin-Machado--.pdf?sequence=3&isAllowed=y. Accessed 12 Aug 2019

Marguti BO (2018) Políticas de Habitação. In: Costa MA, Magalhães MTQ, Favarão CB (eds) A nova agenda urbana e o Brasil: insumos para sua construção e desafios a sua implementação. Available at: http://www.ipea.gov.br/portal/images/stories/PDFs/livros/livros/180529_a_nova_agenda_urbana_e_o_brasil_cap08.pdf. Accessed 11 Sept 2019

Martin M (2015) Crise Hídrica: Milhares de pessoas se mobilizam contra a falta de água em São Paulo. EL PAÍS Brasil, 27 Feb. Available at: https://brasil.elpais.com/brasil/2015/02/27/politica/1424996543_317466.html. Accessed 1 July 2019

Monteiro AR et al (2017) The housing issue in Brazil. Mercator. Universidade Federal do Ceará 16(7):1–12. https://doi.org/10.4215/rm2017.e16015

Mori L (2017) Por que São Paulo ainda não conseguiu despoluir o rio Tietê? BBC Brasil. Available at: http://www.bbc.com/portuguese/brasil-42204606. Accessed 12 May 2018

Oliveira G, Marcato F et al (2014) Descarte Ilegal de Efluentes Industriais na Região Metropolitana de São Paulo. São Paulo, Brazil

Paganine J (2015) Universalização do saneamento é desafio para União, estados e municípios. Senado Notícias, 18 May

Pasternark S, D'Ottaviano C (2018) Políticas Habitacionais em Favelas: o caso de São Paulo—Observatório das Metrópoles. In: Salet W (ed) The Routledge handbook of institutions and planning in action, 1st edn. Routledge, p 398. Available at: http://observatoriodasmetropoles.net.br/wp/politicas-habitacionais-em-favelas-o-caso-de-sao-paulo/. Accessed 6 June 2019

Ribeiro J (2021) Com Cantareira em 50%, Sabesp não vê risco de falta de água em SP, 11 Mar, R7. Available at: https://noticias.r7.com/sao-paulo/com-cantareira-em-50-sabesp-nao-ve-risco-de-falta-de-agua-em-sp-11032021

Rolnik R et al (2015) O Programa Minha Casa Minha Vida nas regiões metropolitanas de São Paulo e Campinas: aspectos socioespaciais e segregação. Cad Metrop. In: Aguas Urbanas. EDUC, São Paulo, Brazil

SABESP (n.d.) Perfil. Available at: https://site.sabesp.com.br/site/interna/Default.aspx?secaoId=505

SABESP (2017) Relatório de Sustentabilidade. São Paulo. Available at: http://site.sabesp.com.br/site/uploads/file/asabesp_doctos/relatorio_sustentabilidade_2017.pdf. Accessed 23 Jan 2019

SABESP (2018) Programa Saneamento Sustentável e Inclusivo: Marco de Gestão Socioambiental do Programa. São Paulo. Available at: http://site.sabesp.com.br/site/uploads/file/asabesp_doctos/programa_saneamento_inclusivo_nov18.pdf. Accessed 6 June 2019

SABESP (2020) Companhia de Saneamento Básico do Estado de São Paulo—SABESP financial statements as of December 31, 2020 and 2019. Available at: https://api.mziq.com/mzfilemanager/v2/d/9e47ee51-f833-4a23-af98-2bac9e54e0b3/a8f48d36-4c09-911c-f720-7ece3944155a?origin=1

SABESP (2021) Projeto Tietê leva saneamento a 12,4 milhões de pessoas e contribui para reduzir poluição. SABESP, 22 Sept. Available at: https://site.sabesp.com.br/site/imprensa/noticias-detalhe.aspx?secaoId=65&id=8598

Saiani CCS, Toneto Júnior R (2010) Evolução do acesso a serviços de saneamento básico no Brasil (1970 a 2004). Econ Soc 19(1):79–106. Available at: http://www.scielo.br/pdf/ecos/v19n1/a04v19n1.pdf. Accessed 19 Jan 2019

Santos AMSP, Duarte SM (2010) Política habitacional no Brasil: uma nova abordagem para um velho problema. Os desafios atuais para a economia brasileira

São Paulo (no date) Histórico Demográfico do Município de São Paulo. Available at: http://smul.prefeitura.sp.gov.br/historico_demografico/1872.php. Accessed 5 June 2019

São Paulo Legislative Assembly (2009) Política Estadual de Mudanças Climáticas. Available at: https://www.al.sp.gov.br/repositorio/legislacao/lei/2009/lei-13798-09.11.2009.html. Accessed 12 May 2018

Schapiro MG et al (2018) Conflito de Interesses nas Empresas Estatais: Uma análise dos casos Eletrobrás e Sabesp. Rev Direito Práx Univ Estado Rio de Janeiro 9(3):1424–1461. https://doi.org/10.1590/2179-8966/2017/27922

Schmidt B, Dezem V (2014) Crise da água em SP mostra disparidade entre pobres e ricos. Exame, 22 Dec. Available at: https://exame.abril.com.br/brasil/crise-da-agua-em-sp-mostra-disparidade-entre-pobres-e-ricos/. Accessed 1 July 2019

SNIS (2016) Informações e indicadores municipais consolidados, Sistema Nacional de Informações sobre Saneamento. Available at: http://app4.cidades.gov.br/serieHistorica/#. Accessed 23 Jan 2019

SNS (2019) Saneamento, Ministério do Desenvolvimento Regional. Available at: http://www.mi.gov.br/saneamento-cidades. Accessed 11 June 2019

SOS Mata Atlântica (2017) Poluir rios custa caro. Available at: https://www.sosma.org.br/artigo/poluir-rios-custa-caro/. Accessed 6 May 2019

Souza AGD et al (2009) Política Governamental De Habitação No Brasil: Desafios E Perspectivas. Para Onde 3(2). Available at: https://seer.ufrgs.br/paraonde/article/view/22103. Accessed 17 Jan 2019

Travassos L, Momm-Schult SI (2013) Recuperação socioambiental de fundos de vale urbanos na cidade de São Paulo, entre transformações e permanências. Cad Metróp 15(29):289–312

Tucci CEM (2009) Manejo de Águas Pluviais Urbanas. In: Brasil, Ministério das Cidades (eds) Lei Nacional de Saneamento Básico: Perspectivas para as políticas e a gestão dos serviços públicos. Livro II. Conceitos, Características e Interfaces dos Serviços Públicos de Saneamento Básico. Coletânea, pp. 312–334. Accessible at: https://bvssp.icict.fiocruz.br/pdf/25711_Livro2_.pdf#page=84

van den Brandeler F (2013) Towards inclusive and resilient water governance? A case study on the implications of deliberation and participation in Guarulhos. University of Amsterdam, Brazil

Vaz JC (1997) Consórcios Intermunicipais. DICAS: Pólis-Ildesfes. Idéias para a Ação Municipal, 97. Available at: http://www.bibliotecadigital.abong.org.br/bitstream/handle/11465/1432/448.pdf?sequence=1. Accessed 4 June 2019

Chapter 7
The Implementation of IWRM/IRBM in Mexico City

7.1 Introduction

This chapter examines how drivers and institutions at multiple levels of the river basin governance regime shape water-related challenges in the Metropolitan Valley of Mexico City (MVMC). It uncovers the causal links behind water challenges and the effectiveness of existing policy instruments. It reviews the relevant context of Mexican river basin governance and its main drivers, analyses the driving forces on the river basin from local to global level (see Sect. 7.2), explores how Integrated Water Resources Management (IWRM) actors and institutions at multiple levels address water challenges at the basin scale (see Sect. 7.3), analyses the instruments of IWRM/IRBM according to their mandates, their effect on actors' behaviour given the drivers and their impacts on inclusive and sustainable water governance (see Sect. 7.4). Finally, the chapter summarizes the main empirical findings and considers how more appropriate instruments could be (re)designed for the Mexico City case study in relation to IWRM/IRBM (see Sect. 7.5).

7.2 Context and Drivers of Mexico City's River Basin Challenges

7.2.1 Context in Relation to the River Basin

Mexico is a water-stressed country with uneven water distribution. Average annual precipitation ranges from below 500 mm in the North, to more than 2000 mm in the Southeast, and around 650 mm in the central region where Mexico City is located (CONAGUA 2015). When population distribution is considered, these contrasts become even starker. More than three-quarters of the population live in regions with

F. van den Brandeler, *Scalar Mismatches in Metropolitan Water Governance*,
Water Governance - Concepts, Methods, and Practice,
https://doi.org/10.1007/978-3-031-08061-6_7

little water (OECD 2013). In addition, rainfall is unequally distributed throughout the year: 67% of rainfall occurs between June and September and droughts are frequent (OECD 2013).

Most of the MVMC lies within the lower part of the Valley of Mexico Basin (VMB), around 2200 m above sea level (masl). This is an endorheic basin, enclosed by mountains reaching 5500 masl (Martínez and Enciso 2015). Surface and groundwater resources originate from springs in these mountains. 65.5% of the VMB's surface area is urban, while 34.5% is rural, with agricultural, livestock, forest and conservation uses (World Bank 2013). Urban growth has reduced water availability in the VMB to 144 m^3/inhabitant/year.[1]

7.2.2 Main Drivers of Mexico City's River Basin Challenges

Climate

Climate change is expected to lead to warming of 1.5–5 °C from 2050 to 2100 across Mexico (Guido Aldana 2017). Precipitation is likely to be concentrated in fewer but more intense rainfalls, and to decrease overall by 5.8% in the 2020s, and by 10.4% in the 2070s (Sosa-Rodriguez 2014; Guido Aldana 2017). This may intensify of mid-Summer droughts, impact water demand and alter water quality in surface water bodies (IMTA 2007). However, vulnerability to climate change varies significantly per region. Areas in the dry North, the Centre-West, and the VMB are considered highly vulnerable (Guido Aldana 2017).

In the VMB, average annual precipitation varies between 1200 mm in the South and 600 mm in the North, mostly concentrated between May and September (Romero Lankao 2010). Heavy summer rains cause frequent floods. The basin tends to alternate between wet years and drought episodes, some of which last longer than ten years (Romero Lankao 2010). Glaciers and eternal snows in the surrounding mountains, crucial for groundwater recharge, may disappear before 2025 (Burns 2009). The resulting depletion of local aquifers risks increasing the basin's dependency on external water sources (Sosa-Rodriguez 2014). Overall, Mexico City is likely to experience more intense droughts and heat waves, more frequent floods and increased risks from waterborne diseases (Sosa-Rodriguez 2014).

Demographics

The VMB attracted settlements long before the Spaniards arrived, thanks to its temperate climate and fertile soil (Escamilla and Santos 2012). Around the Aztec city of Tenochtitlán, which would become Mexico City, peri-urban and rural communities developed highly productive floating farms (Escamilla and Santos 2012). These supplied urban areas and allowed for Tenochtitlán's growth. By the seventeenth century, the Spaniards first population census registered a total of 144,760 inhabitants (Escamilla and Santos 2012). In 2010, the basin had 20.6 million inhabitants

---

[1] A region faces absolute water scarcity if renewable water resources are below 500 m^3/inhab./year (FAO 2012).

across 85 municipalities, though mostly concentrated within the MVMC (Rodríguez Tapia and Morales Novelo 2013).

Economic Development

The 1910 revolution centralized power and led to large-scale economic development projects (Aguilar et al. 2010). In addition, technological innovations brought dams, canals, inter-basin transfers, and the electric pump (Aguilar et al. 2010). As a result, the ability to abstract and transport water resources over large distances increased significantly, and water consumption rose sharply after 1950. By the late 1980s, water infrastructure was decaying and the government, struggling to handle a severe economic crisis, was unable to address the sector's extensive needs (Wilder 2010). As of 2015, The VMB's GDP was almost a quarter of the national GDP, hosting the financial capital and important agricultural and industrial sectors (OECD 2015).

Land Use Changes

The VMB's natural hydrology has been radically altered since the Spanish conquest of the Aztec empire, when lakes covered large parts of the basin (see Map 7.1). The Aztecs developed a sophisticated system of terraces, reservoirs, canals, irrigation ditches, dikes and aqueducts to both use the hydrological conditions to their advantage and cope with its risks. From the sixteenth century onwards, the Spaniards progressively drained the lake-bed (Wilder 2010; Mazari-Hiriart et al. 2014). As urbanization accelerated, open sewage canals polluted the remaining water bodies, spreading waterborne diseases and increasing reliance on groundwater (Romero Lankao 2010). Today, only one river system, the Magdalena-Eslava, still provides water to the city, albeit heavily contaminated and with a diminished flow, while other rivers have been piped to avoid flooding and unsanitary conditions (Mazari-Hiriart et al. 2014).

The VMB's urban sprawl and population multiplied roughly 5 times between 1950 and 2000 (World Bank 2013). As urbanization spread to the mountainous areas of the basin, the capacity of these areas to infiltrate water and recharge aquifers was drastically reduced, affecting water availability and increasing flood risks through soil-sealing (Romero Lankao 2010; World Bank 2013). The growth of informal settlements without access to basic infrastructure and services also degraded the quality of local springs and aquifers.

Moreover, logging and agriculture have led to environmental degradation and land surface erosion, accelerating the desiccation of lagoons and siltation of drainage systems (Romero Lankao 2010; Pina 2011). Land use changes in neighbouring basins also affect the MVMC as these areas supply water through inter-basin transfers (see Sect. 7.4.1). Between 1980 and 2011, the population in the sub-basins of the Cutzamala System, the main inter-basin transfer, increased by almost 150%, occupying mainly informal settlements deprived of sewage collection and treatment (World Bank 2015). Agricultural activities and deforestation also increased around these reservoirs (Martínez 2018).

Table 7.1 summarizes the drivers of water-related challenges at multiple levels.

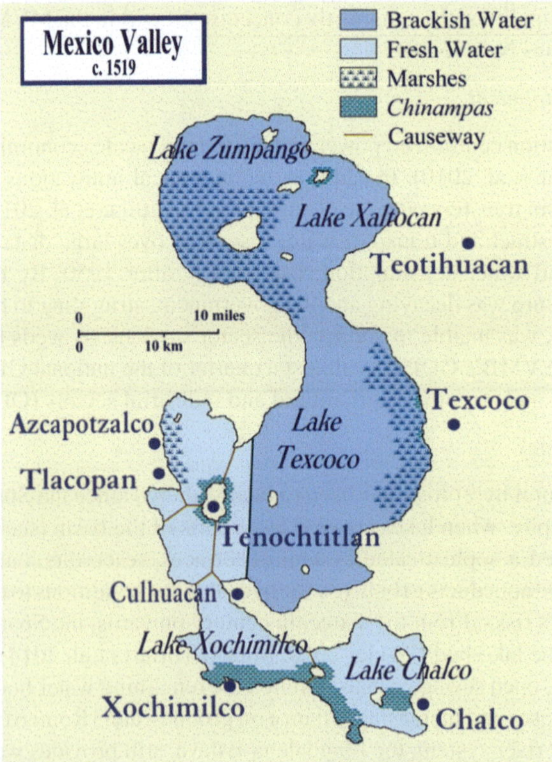

Map 7.1 Lakes in the Valley of Mexico around 1519. The lakes have been reduced to two small water bodies and a few canals—a fraction of their original size. The rest of the lake waters were drained and today the MVMC expands across the vast expanse of the dried lakebed. *Source* Madman2001/CC BY-SA 3.0

Table 7.1 Multi-level drivers of water-related challenges on the river basin

	Direct		Indirect	
	Local	Regional/global	Local	Regional/global
Land use	Urbanization, expansion into green belt	Logging and deforestation for agriculture	Transformation of original lake basin	
		Agricultural intensification		
Demography		Rapid population growth in the basin		Rapid national population growth
Climate	Heavy summer rains, frequent dry spells			Climate variability and change

Source Author

7.3 The Institutional Set Up for IWRM/IRBM in Mexico City

7.3.1 Global Level

The World Bank and IDB have provided loans to Mexican authorities for climate change adaptation, and more specifically for drought forecasting, for the protection of water reserves and for supporting the development of IWRM in the VMB (CONAGUA 2012b). The federal government and users finance the bulk of the Mexican water sector, and international actors represent only a minuscule fraction of total investments in water resources management (OECD 2013). The presence of international cooperation is stronger within UWM (see Sect. 8.3.1).

7.3.2 Transboundary Level

The MVMC lies within one river basin, the VMB, but it depends on several other basins for water supply. The VMB spreads across the entire federal entity of Mexico City, and parts of the states of Mexico, Hidalgo and Tlaxcala (CONAGUA 2012a). This has major WRM implications as these four federal entities must negotiate and agree on basin-related decisions.

7.3.3 National Level

The Federal Constitution defines all waters within Mexican territory as national property to be administered and managed by the federal government (Mexican Constitution 1917, Art. 27). Until the 1980s, water policy remained highly centralized within a limited authoritarian political system ruled by a powerful presidency and top-down regulation (Wilder 2010). It was conducted in a top-down fashion without social participation.

The emergence of political pluralism in the late 1980s, including democratic power sharing between political parties and levels of government, led to the reform of Mexico's water management institutions (Hearne 2004). The federal government introduced decentralized water management policies and adopted three WRM goals: (1) develop large infrastructure projects with combined public and private funds, (2) increase water use efficiency, and (3) control water pollution (Hearne 2004). In 1989, the National Water Commission (CONAGUA) was established to develop a new water policy to carry out these goals (Hearne 2004). This led to the 1992 National Water Law (NWL), which defines the principles and mechanisms for managing water resources, including the use of national waters, their distribution and control, as well as the preservation of their quantity and quality for 'integrated sustainable development' (NWL 2004; OECD 2013). It is centred around decentralization and marketization, with the river basin or aquifer as the management unit (Wilder 2010). River basin councils were created in 1996 to promote citizen participation and coordinate water management across three levels of government within watershed boundaries (Hearne 2004). The NWL thus indicated the possibility of a new state-citizen relationship but also increased private sector involvement in water supply and sanitation services. In 2004, the NWL was significantly rewritten, with greater emphasis on decentralization and sustainability (Wilder 2010). It led to the creation of thirteen regional Basin Agencies, based on hydrological boundaries, that act as CONAGUA's implementing agencies (Hearne 2004; OECD 2013).

CONAGUA is an autonomous agency of the SEMARNAT (Ministry for the Environment and Natural Resources). The SEMARNAT, together with CONAGUA, and state and municipal authorities, establishes official water management norms and supervises their enforcement (OECD 2013). CONAGUA receives 70% of its budget from the SEMARNAT, though it maintains significant autonomy, and is responsible for water policy, water planning, financing and strategy-setting (Spring 2011; OECD 2013). It grants water use permits, maintains the national water user registry (REPDA), constructs and operates federal water infrastructure and provides bulk water to Wat&San utilities, large industries, and irrigation districts. In addition, CONAGUA develops and manages irrigation and flood control systems (Hearne 2004).

The National Commission for Protected Areas (CONANP) and the National Forestry Commission (CONAFOR) strive to preserve ecosystems (for a detailed overview see Annex F: Main Actors in Mexico City's Metropolitan Governance). The Federal Attorney's Office for Environmental Protection (PROFEPA) is responsible for supervising the compliance with environmental regulations. The Ministry of Agriculture is responsible for addressing agricultural pollution (i.e. regulation of the use of fertilizers), whereas CONAGUA oversees water quality norms and standards issued by the Health Ministry. CONAGUA is thus limited to discharges from industries and water utilities (Spring 2014). Overall, national legislation on water quality was considered weak due to institutional fragmentation, a lack of enforcement and political will, and the PROFEPA's ineffectiveness (Interviews-M40/M42/M58).

7.3.4 State Level

At state level, the main institutions are the State Water Commissions. These autonomous entities foster coordination between the federal government and municipalities (OECD 2013). Their roles vary per state and can include WRM, irrigation, technical assistance to municipalities and the provision of Wat&San services (when municipalities choose to delegate these) (OECD 2013). States also have environmental departments that are responsible for carrying out the state environmental policy, but their capacity varies significantly. For Mexico City, this is the SEDEMA

(Environmental Secretariat of Mexico City) and its branch the CORENA (Mexico City's Commission for Natural Resources).

7.3.5 River Basin Level

The 2004 revision of the NWL introduced IRBM into Mexican water policy and transferred key functions to the river basin. The MVMC is located within the hydrological-administrative region managed by the Basin Agency 'Waters of the Valley of Mexico—Region XIII', which is responsible for formulating regional water policy, planning and maintaining water use registries (see Map 7.2). Region XIII has two river basins: The Basin of the Valley of Mexico and the Basin of the Tula River, each with its own basin council (CONAGUA 2012a). The borders of the basins do not overlap perfectly with those of Region XIII as the former are based on hydrological boundaries and the latter on municipal borders. Basin councils are multi-stakeholder consultative bodies that bring together government representatives, civil society representatives and users. Other water management entities include the basin commissions, the basin committees and the groundwater technical committees (see Table 7.2). These are voluntary and meant to support the Basin Councils (Interview-M55).

While the creation of these multi-stakeholder entities was an important step towards implementing IWRM/IRBM, they were seen as ineffective, leading

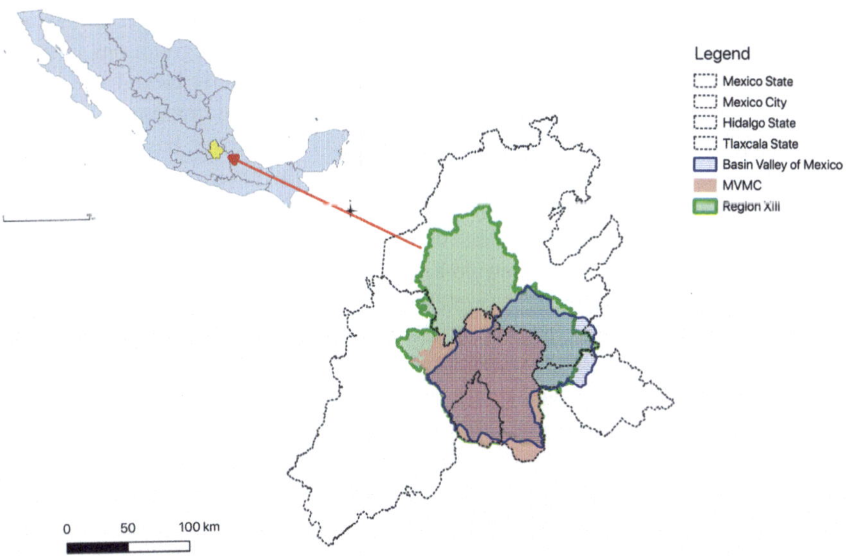

Map 7.2 Map of Region XIII and the metropolitan region of the Valley of Mexico. *Source* Author

Table 7.2 IWRM/IRBM entities

IWRM/IRBM entities	Description	Mandates
Basin agencies	13 deconcentrated agencies of CONAGUA, based on hydrographic boundaries. The MVMC is in Basin Agency Region XIII	Formulate and implement regional policy; recommend rates for water user fees and collect them; program, build, operate and maintain federal water works
Basin councils	2 basin councils in Region XIII Composed of government representatives, civil society representatives and users	Guide the work of the Basin Agencies through coordination, concertation, support, consultation and advice
Basin commissions	4 in the VMB Same composition as basin councils. Independent from CONAGUA	Provide support to basin council at sub-basin level or on a specific issue
Basin committees	2 in the VMB Micro-basin level	Address issues relevant to a specific area by bringing together residents and stakeholders
Groundwater technical committees	1 in the VMB Multi-stakeholder platform at aquifer level, independent from CONAGUA	Technical work and discussions relating to groundwater management

Source NWL (2004) and CONAGUA (2018)

many participants to drop out (Interviews-M5/M17/M38).[2] The VMB's size and complexity, and the many interests and actors involved, made negotiations virtually impossible (Interviews-M17/M55). In addition, basin councils lacked the necessary planning, regulatory, financing and enforcement power to implement decisions (OECD 2013) (Interviews-M17/M31/M33/M38/M54). Moreover, basin councils did not consider basin commission plans, resulting in incoherencies between sub-basin and basin management (Interview-M40). The groundwater committees lacked information and decision-making power, and depended on funds from CONAGUA, which restricted their autonomy (Interview-M40). The basin agency had decision-making power and funds, and ultimately implemented CONAGUA's agenda at regional level. Other spaces seemed to serve to give legitimacy to decisions already taken by CONAGUA (Interviews-M15/M40/M58).

The Basin Agency 'Regional Water Programme 2030' promotes integrated and sustainable basin and aquifer management, including strategies focused on increasing supply and reducing demand, improving water quality, attaining universal coverage of water services and reducing risks from weather events (CONAGUA 2012a).

[2] Basin council meetings are infrequent: during six months of fieldwork, the only meeting that I could hear of took place in Cancún during a parallel event. Civil society was mainly represented by (large) users, whereas representatives from CSOs and basin commissions were not given a voice or informed on how to participate (Interviews-M28/M38/M40/M48/M54/M55).

Climate change does not have a strong presence within the programme, and the NWL (Art. 13) only mentions that basin agencies should "evaluate the effects of climate change on the hydrological cycle".

7.4 Instrument Analysis

7.4.1 *Inter-basin Transfers*

Design

The MVMC has long relied on its aquifers to respond to rising water demand. As these became over-exploited, the region developed a complex system of large engineering works spanning multiple federal entities and basins to provide bulk water services to the VMB, mainly for public/urban use. These aimed to reduce the pressure on local aquifers while sustaining the MVMC's growing population and economic development. Table 7.3 shows the multiple water supply sources to this basin.

The first of these inter-basin water transfers was the Lerma System, inaugurated in 1951 and managed by the SACMEX, Mexico City's water utility (see Sect. 8.3.3) (CONAGUA 2018). It extracts and transfers groundwater from the Lerma River aquifer to Mexico City and parts of Mexico State. It measures 60 km in length, provides 5% of the VMB's water supply and has a capacity of 14 m^3/s, although actual supply is around 5 m^3/s due to the over-exploitation of the Lerma aquifers (CONAGUA 2018). The Cutzamala System supplies water to 11 districts of Mexico City and 11 municipalities of Mexico State (CONAGUA 2018). It corresponds to 17% of the volume of water used within the VMB annually (~15 m^3/s of 88 m^3/s) (CONAGUA 2018). Besides supplying water from 160 km away (one of the largest water supply systems in the world), it also pumps water up 1100 m, making it energy-intensive and expensive (Tortajada 2008; Engel et al. 2011). The Cutzamala System alone costs around USD 240 million a year (World Bank 2015). 48% of this cost is

Table 7.3 Overview of water supply in the Valley of Mexico Basin (VMB)

Source	Volume (in %)
Groundwater	*73*
Local aquifers	68
Lerma system (inter-basin transfer)	5
Surface water	*20*
Local rivers and springs	3
Cutzamala system (inter-basin transfer)	17
Water reuse	*7*

Source Adapted from Aneas (2015)

financed by water use fees, and the rest is paid with federal government resources (World Bank 2015).

Effectiveness on Actors in Terms of Mandated Goals

The MVMC's aquifers and nearby aquifers in Mexico State continue to be over-exploited and this has worsened in the last decades. Nevertheless, as external sources contribute roughly 22% to the VMB's water use, the pressure on the MVMC's aquifers would be much more drastic and water shortages more severe without centralized water resources management (Interview-M32). The MVMC's aquifers may run dry by 2060, so relying on external sources is unavoidable (EFE 2019). However, a first issue regarding inter-basin transfers concerns allocation rules, which were largely set in the 1970s (DOF 1972). Mexico City was granted around five times more water than Mexico State, under the unstated rationale that it had a much larger population. Since then, the MVMC expanded into Mexico State. Although the volumes apportioned to both entities increased over time, the decree was never adjusted and Mexico City continues to be entitled to a much larger volume of water (Mendoza 2016). This may have contributed to the sharp rise in groundwater extraction in metropolitan municipalities of Mexico State since at least the 1990s (Neri-Ramírez et al. 2013).

Moreover, the decree does not specify where imported water comes from, nor the capacity of the Cutzamala System (Mendoza 2016). This opened the door to the possibility of constant expansion of water transfers and encouraged CONAGUA and other actors to import more water rather than reduce water use (Spring 2015). Furthermore, the Federal government partially subsidizes the inter-basin transfers, so users are not incentivized to increase their water use efficiency, leading to high levels of physical and commercial water losses (i.e. leaks and theft), and excessive water consumption. Water fees for private uses and basin transfer fees paid to the Basin Agency are collected by the Federal Treasury, while bulk water supply fees (from water utilities) go to a regional trust fund for water infrastructure (Mexico 2005). The latter is mostly spent on stormwater and sanitation works in the VMB (World Bank 2015). Revenues therefore do not return to the donor basins to be reinvested in preserving regional ecosystems, managing urban and rural development or treating wastewater.

Despite the dependence on external basins, the Region XIII agency does not interact with the basin agencies where the Lerma and Cutzamala systems have their production areas, and there is no policy at that larger scale (Interviews-M28/M33). These water transfers are coordinated in a top-down manner by the central CONAGUA office (Interview-M33):

> These agencies do not meet, do not dialogue. There is no policy at that scale. [...] The federal government should guide coordination processes, create spaces for debate and joint decision-making, not tell them what to do. [...] These administrative borders are a straitjacket, as our new socio-hydrological realities no longer fit these moulds. We need to create a new entity; rethinking the scale we need to use as the policy of inter-basin transfers is unlikely to change in the short and even medium term.

The basin councils, state and municipal governments, and other regional entities [e.g. metropolitan commissions (see Sect. 8.3.4)] also do not coordinate around water transfers despite the many inter-connections between them (OECD 2013) (Interviews-M17/M28/M48).

Impacts on Inclusiveness and Sustainability

Large infrastructure works mitigate water shortages for the MVMC in the short-term and reduce pressure on the MVMC's aquifers but in the longer-term they are unsustainable, as demand will continue to rise and water will need to be imported from constantly further at increasing costs (e.g. infrastructure, energy, environmental degradation, on displaced local communities) (M6/M7/M9/M28/M40/M48/M49/M50). Moreover, climate change forecasts project an overall decrease in precipitation, which will put additional stress on the region's water systems, and this is not considered in current water supply planning (Interview-M55). The emphasis on importing water comes at the expense of efforts such as conservation in areas crucial for aquifer recharge, reusing water or reducing leaks, which are part of an integrated and long-term vision of the basin (M9/M15/M40/M44).

Large engineering works have altered the hydrological balance of the VMB basin and surrounding regions by artificially unifying not only the urban areas but also regions beyond the city (Interview-M32). Consequently, the water system has been configured into a sort of mega-basin (Perló and González 2005; Romero Lankao 2010). Around the dams, forest cover has decreased, while unsustainable farming practices and human settlements with inadequate sanitation have grown (World Bank 2015) (Interview-M51). Erosion and agricultural runoff have caused severe siltation and algae blooms (Martínez 2018) (Interview-M48). Toxicity levels in the dams reached such high levels in 2014 that CONAGUA almost suspended water imports. This was ultimately avoided by adding an additional treatment step (Martínez 2018). The degradation of the dams' water quality thus contributes to higher water treatment costs.

The power structures in place favour the urban core, where water is heavily subsidized by the central government, at the expense of water producing areas, where locals often lack access to basic water services while *their* water is exported (Interviews-M12/M28/M32/M33/M51). However, urban dwellers do not benefit equally from this system: Poorer, peri-urban areas in the East of Mexico City and the MVMC lack reliable access to water (Interviews-M28/M32/M48/M50/M52). Top-down management of basin transfers contributed to rising political opposition and socio-environmental conflicts with communities in the donor basin (Tortajada 2008; Engel et al. 2011; Pina 2011; Spring 2015). For instance, in 2004, indigenous women from the Mazahua group shut down the megacity's water supply by peacefully occupying a water treatment plant in Mexico State, after CONAGUA flooded their fields (Spring 2011). Tensions date back to the 1970s, when federal authorities started to exploit the communities' water resources without consultation or compensation. Plans for additional inter-basin transfers have stalled in part because of opposition by local communities, but tensions could escalate under drought conditions (Interview-M12). Between 2019 and early 2022, a prolonged dry spell further reduced dam levels (Garcia 2022).

Finally, the high cost of inter-basin transfers is not only borne by users but also by the general population, due to large subsidies. As the systems expand and new sources further away are connected, the costs will increase (Interviews-M15/M48). Cost–benefit analyses are not transparent, and it is not clear whether investments in alternative solutions (i.e. reducing leaks, water reuse) have been adequately considered (Interviews-M40/M48).

7.4.2 Water Use Permits

Design

Water use permits aim to control water use and provide users with rights and obligations (NWL 2004, Sec. Art. 20). Due to (relative) water scarcity, the water allocation regime is crucial to control water use, as it establishes water abstraction restrictions in certain zones through water use permits and fees (see Sect. 7.4.3). CONAGUA grants water use permits through its basin agencies or directly when appropriate (NWL 2004, Sec. Art. 20). Permits specify the maximum amount of water a user can abstract, the use's purpose (e.g. domestic, industrial, agricultural), the abstraction's location and the duration of the right (OECD 2013). They must be registered in the Public Registry of Water Rights (REPDA), created in 2004 to regulate water use and provide information and legal security to users (Art. 9, NWL). The CONAGUA also grants water discharge permits (see Annex G: Additional Instruments).

CONAGUA is responsible for verifying water availability in watersheds and aquifers for all registered water users (CONAGUA 2015). Mexico is divided in water availability zones that determine the volume of water that users can request via a permit.[3] New concessions cannot be emitted in restriction zones, such as most of Region XIII. This aims to create a ceiling for the level of abstraction within the region (OECD 2013). In addition, an environmental and economic impact assessment must be presented when requesting a water use permit (CONAGUA 2018). Water use can be temporarily restricted when it affects the minimum environmental flow[4] and during droughts, and the NWL guarantees priority for public supply in times of

[3] Surface water use permits are required if abstraction significantly alters water quality or flow (NWL 2004, Art. 17). Groundwater can be abstracted without a permit, except when the Federal Executive establishes regulatory means to limit extraction and use (NWL 2004, Art. 18). The following zoning restrictions apply: (1) In 'regulated zones', aquifers have sufficient mean annual water availability, and water can be allocated without jeopardizing the aquifer balance; (2) In 'prohibition zones', more water is leaving the aquifer (e.g. extractions, natural discharge) than entering it. No new water use permits can be issued. This concerns a large part of Region XIII; (3) 'Reserve zones' limit water use for conservation or specific uses.

[4] The term 'environmental use' or 'ecological conservation use' is used to refer to the minimum flow or volume of water needed in receiving bodies or the minimum natural discharge flow of an

scarcity. CONAGUA has the power to sanction those who violate water extraction agreements and users who grossly abuse or misuse water resources (Acevedo et al. 2013).

Domestic and public-urban uses represent about 75% of the total volume of water granted in concessions in the VMB, 18% is allocated to agricultural use and around 5% to industrial use (World Bank 2013). Permits are somewhat flexible—SACMEX can relocate its extractions points to new areas within the same aquifer and dig new, often deeper, wells (with CONAGUA's authorization) if it does not extract more water. Moreover, water users can transfer their permits to other users within a same basin or aquifer (Federal Constitution, Art. 27). CONAGUA can approve, reject or apply conditions to such transfer request, depending on whether the hydrological or environmental conditions of the concerned basins or aquifers were altered in the process (Art. 34, NWL). CONAGUA charges a small administrative fee for reviewing and authorizing the transfer and registering changes in the REPDA.

Effectiveness on Actors in Terms of Mandated Goals

Water use permits are designed to meet sustainability and inclusion criteria (i.e. preserving minimum environmental flow, guaranteeing priority for human consumption, charging fees to encourage rational use), but they are not adequately enforced. CONAGUA lacks the capacity to continuously monitor all users and rarely sanctions misuse (Acevedo et al. 2013; OECD 2013) (Interviews-M5/M6/M9/M32/M50). Bulk water use metres are also often absent. As a result, clandestine extractions (without water permits) have persisted despite regulations in place since the 1950s (prohibition zones) (Interviews-M6/M7/M9/M32). Inspections are often conducted following 'citizen complaints' that inform about irregular activities.

During the last decades, users had several opportunities to regularize illegal wells (Interviews-M9/M32). The 1992 reforms regularized a significant number of these but, as illegal wells became legal, water extractions did not decrease (Interview-M32). This meant that the groundwater allocated was greater than the rate of aquifer recharge (Interviews-M32/M48/M52). Worse, actual extraction rates were often much higher still—problems that persist due to strong resistance by permit holders to decrease consumption (see Table 7.4). 864 h m^3 are over-extracted each year from the VMB's aquifers, as shown by calculations based on government data. This corresponds to around 345,000 Olympic swimming pools of water.

Permit transfers are efficient in theory; they allow new uses without increasing extraction. However, the difficulty of obtaining a permit and the lack of monitoring have encouraged irregular water uses. In addition, permit transfers have spurred a black market where sellers and buyers agree on a (sometimes exorbitant) price (Interviews-M9/M31). Through this black market, users often buy permits for purposes other than the one on paper: Many wells for industrial or public-urban purposes in the MVMC's periphery were registered as agricultural use (Interview-M9). As the city grew, and demand for water and land increased, real estate developers

aquifer, which must be preserved to maintain environmental conditions and the system's ecological balance (Art. 3, LIV, NWL).

Table 7.4 Water availability in the Valley of Mexico Basin's aquifers (in h m^3 per year)

Aquifers	Aquifer recharge	Natural discharge	Groundwater volume allocated	Groundwater volume extracted	Over-allocation[a]	Over-extraction[b]
Cuautitlán-Pachuca	357	0	415	751	−58	−394*
Metropolitan zone of Mexico City	513	0	1104	624	−591*	−111
Tecocomulco	28	1	1	13	26	15
Apan	30	0	19	15	11	15
Chalco-Amecameca	80	0	90	128	−10	−48
Texcoco	49	0	93	465	−44	−416
Soltepec	93	42	16	18	77	75

Source Adapted from SEGOB (2016a, b, c, d, e, f)

[a] = Aquifer recharge – Natural discharge – Volume allocated to users

[b] = Aquifer recharge – Natural discharge – Volume extracted

* The Cuautitlán-Pachuca and the Metropolitan zone of Mexico City aquifers are the most important aquifers in terms of water volume used by the MVMC. The numbers in bold reveal that water allocation and water extraction are often not aligned. Much of the water extraction in the Cuautitlán-Pachuca aquifer is outside of the allocation and therefore irregular. The groundwater extracted from the Metropolitan Zone of Mexico City aquifer is lower in volume than the volume allocated, however extraction rates are still unsustainable as the aquifer has been over-allocated.

and industries bought farmland in the periphery to build housing developments, and they also often bought the farmers' water permits illegally, only regularizing this permit later with no consequences (Interviews-M6/M7/M9/M32/51/58). Developers also often infringed rules and regulations regarding land acquisition and building norms, and local governments either turned a blind eye or eventually legalized the developments (Interviews-M22/M23/M50/M51). Permit-holders who no longer need to extract water should cancel their permit, but this rarely happened (Interviews-M6/M7). Groundwater permit-holders saw themselves as 'owners' of this water and entitled to sell it as their property (Interview-M31).

Furthermore, water utilities open new, deeper wells if older ones run dry, which does not incentivize water use efficiency or savings (Interviews-M6/M7/M9/M19/M52). Permits are issued for a set amount of time, with renewals requiring an assessment, but this rarely happens due to CONAGUA's lack of capacity and will (Interview-M31). Many users had long expired permits and suffered no consequences (De Regil 2014; Valadez 2016).

The basin councils are ineffective in addressing problems related to water use and permits as they lack authority but also due to inadequate knowledge-sharing mechanisms (Interview-M31). The VMB's groundwater committees do not interact with the basin councils or other entities, impeding integration between surface and groundwater management (Interview-M32). Significant uncertainty remains regarding surface and groundwater interactions, and about aquifers themselves, due to limited available data on the location of aquifers' recharge and discharge areas, their depth, flows and more (Interviews-M31/M32). Water allocation also did not include criteria related to climate change (OECD 2013). Basin boundaries are taken as management units even though aquifer and (sub)basin boundaries do not always coincide (Interview-M32). This means that quantitative or qualitative alterations of aquifers may affect a neighbouring basin more than the one they are mostly contained in.

Impacts on Sustainability and Inclusiveness

Aquifers are recharged through summer rains and natural springs in the surrounding mountains, but extraction rates exceed recharge rates—estimates show that 45% of the MVMC's water supply comes from over-exploited aquifers (Tortajada 2008; Pina 2011; Martínez and Enciso 2015) (Interviews-M6/M7/M9). The groundwater table falls by about one metre per year, causing soil subsidence (5–40 cm per year) (Tortajada 2008; Engel et al. 2011; Pina 2011). Over-extracting groundwater can lead to the disappearance of ecosystems, wetlands and lakes, and reduced river flow, and the link is not always clear due to delayed effects (groundwater flows are slow) appearing far from their cause (Interview-M32). Mexico City could run out of water within the next decades without more sustainable water management (Spring 2015) (Interview-M9). A business-as-usual scenario for 2030 estimated that renewable water sources will only cover 53.8% of demand, with 21.1 m^3/s supplied through over-exploiting current sources and a remaining deficit of 25.1 m^3/s (World Bank 2013).

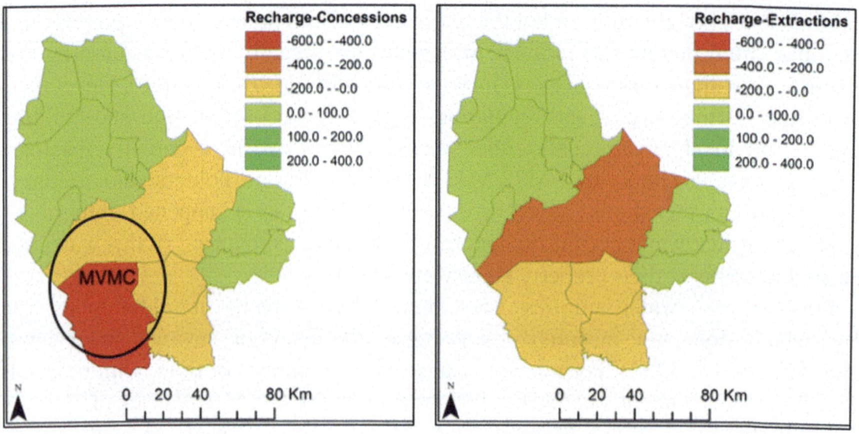

Fig. 7.1 Water balance per aquifer based on allocated water (left) and extracted water (right). *Source* Author

Some of the most stressed aquifers, such as Cuautitlán-Pachuca and Texcoco, export water to Mexico City (Escolero et al. 2016). The groundwater table of the Cuautitlán-Pachuca aquifer, just north of Mexico City, was reported to decline by two metres per year (Ramírez 2015; Escolero et al. 2016) (see Table 7.4). The inauguration of a soda drink plant that would extract large volumes of water—despite the aquifer's classification as a 'prohibition zone' and the low priority of this use—caused conflicts between civil society, CONAGUA and the soda drink company in 2017, as experts warned it threatened public water supply (Olvera 2017). Illegal abstractions are also estimated to be among the highest nationwide (Galindo et al. 2011; Ramírez 2015). The pressure on aquifers surrounding Mexico City highlights that this is a metropolitan problem requiring a regional solution with coordination between actors from the jurisdictions involved. Questions of unequal access and power relations also arise as the Cuautitlán-Pachuca and Texcoco aquifers are home to large, low-income peri-urban populations. Of every 100 residents of the VMB, only six do not suffer from water scarcity (Burns 2009). Of these six, two residents consume on average 567 l a day. As many as 77% of residents consume less than 150 l a day (Burns 2009).

The maps in Fig. 7.1 are based on Table 7.4 and show water availability based on CONAGUA's water allocation data (official calculation to determine water availability) and water extraction.[5] The MVMC spreads mainly across the Mexico City aquifer in its South and the Cuautitlán-Pachuca aquifer in its North. Comparing the two maps reveals that CONAGUA's current water availability calculations may be misleading, as water extractions can be significantly below or above the volumes allocated to users within an area. In fact, tensions between Mexico City and Mexico State (where the Cuautitlán-Pachuca aquifer, in dark orange on the right-hand map,

[5] CONAGUA calculation of water availability = Aquifer recharge − natural discharge − water allocated. Alternative calculation of water availability = Aquifer recharge − natural discharge − water extracted.

is located) are rising. This indicates that while Mexico City's aquifer may be more over-exploited, surrounding aquifers are rapidly depleting.

Water allocation remains a top-down process. The lack of data transparency creates confusion about real water abstraction and the areas that require greater support in addressing unsustainable water use practices. Ineffectiveness and corruption favour wealthy users willing and able to pay high prices for water permits, as CONAGUA lacks capacity to pursue irregular permit holders. In addition, the black market for permits is most likely facilitated by insiders. One groundwater expert claimed that CONAGUA employees attempted to sabotage the groundwater committees as monitoring and accountability-holding activities threatened entrenched (and corrupt) interests (Interview-M32). In April 2019, a fire broke out in a CONAGUA building over the weekend, and amidst rumours of an upcoming audit, within an area that held important documentation on water use permits (Martínez 2019).

7.4.3 *Water Use Fees*

Design

CONAGUA employs a 'user-pays' principle, meaning users must pay a fee for the abstraction and use of water resources (NWL 2004). Bulk water fees were introduced in Mexico in 1982 and their rates vary according to the type of use (e.g. industry, urban, hydropower, aquaculture and recreation) and geographical location. They aim to incentivize rational water use, while representing the main mechanism for water users' contribution to financing WRM (OECD 2013). Water use for agriculture, for domestic use related to agricultural activities and for rural settlements of less than 2500 inhabitants are exempt from paying water use fees (DOF 1981). Mexico's basins and aquifers are classified in four different water 'availability zones', and typically the cost of the cubic metre is higher in low availability zones (CONAGUA 2018). The bulk water rate is determined by an algorithm published in the Federal Law of Rights (DOF 1981) and adjusted annually by CONAGUA (CONAGUA 2018).

The Federal Treasury collects fees from all bulk users, except water utilities, and from inter-basin transfers via the Basin Agency (World Bank 2015). These fees fund the bulk of CONAGUA's budget and programmes such as the federal payment for ecosystem services programme (PSAH) (see Sect. 7.4.4) (World Bank 2013). Since 2004, water utilities pay bulk water fees to the regional Fideicomiso 1928 trust fund, which invests in large-scale storm and wastewater infrastructure in the VMB (World Bank 2015). The governments of Mexico State and Mexico City are the fund's settlors and CONAGUA acts as president and technical coordinator (World Bank 2015). The SACMEX also includes this trust fund as part of its financing plan (OECD 2013). Users also pay a fee, set at Federal level, for wastewater discharge, based on the 'polluter-pays' principle (NWL 2004). However, the fee is only charged if the discharge does not meet minimum quality standards (Interviews-M5/M6/M7). The fee aims to cover investments in wastewater treatment (DOF 1981).

Effectiveness on Actors in Terms of Mandated Goals

The delimitation of 'availability zones' means that water use fees in the VMB should reflect the gap between water availability and water demand (Interview-M32). Indeed, as water availability decreased in the VMB, bulk water use fees increased significantly (World Bank 2013), even if they remained overall low and restricted to industrial uses (Interviews-M5/M32). Agricultural users were exempted if they remained within their licensed quotas and the rate for water use exceeding their licensed quota was only between 0.7 and 8% of the general rate (depending on the availability zone) (OECD 2013) (Interviews-M32/M50).

Most water users self-report their volume used and pay fees accordingly. There are incentives to regularize water use and pay some fees, such as energy subsidies for groundwater pumping, as users must present their water use permit to qualify. Audits of industries also often start with a review of water use permits and bills from water use fees (Interview-M32). Nonetheless, Basin Agencies cannot afford to inspect most users, so these self-reported values were rarely verified (Interview-M58). The lack of metering also encouraged under-reporting (Interview-M19). Lack of monitoring, enforcement and exemptions also affected wastewater discharge fees (Interview-M40). The total value of the collected bulk water use fees in Mexico increased slightly between 2008 and 2017, reaching around USD 860 million in 2017, and Region XIII contributes over a third of this amount (CONAGUA 2018). In 2017, the value of wastewater discharge fees collected in Mexico was only USD 80 million. The fees go to the Federal Treasury and are reinvested in Federal level programmes, which are unlikely to be related to preventing contamination (Interview-M58).

Impacts on Sustainability and Inclusiveness

The limited effectiveness of bulk water use and wastewater discharge fees aggravates that of water use and wastewater discharge permits, reinforcing the same impacts (see Impacts—Sect. 7.4.2). Users lacked incentives to reduce water use, invest in water saving technologies or prevent water contamination (Interviews-M5/M9). Fees paid by water utilities in the VMB were partly reinvested in water supply and sanitation infrastructure through the Fideicomiso 1928 trust fund (see Sect. 7.4.1), which supported the cash-strapped utilities (OECD 2013) (Interview-M19). However, a large volume of water used by utilities in the MVMC comes from other basins, and these fees benefit infrastructure within the VMB. This does not compensate donor basins nor benefit local residents. Fees from other users are absorbed at national level and do not benefit the MVMC or its donor basins.

7.4.4 Payment for Ecosystem Services (PES)

Design

Several water-related PES programmes have been implemented in Mexico. At federal level, the CONAFOR (National Forestry Commission) manages the Payments for

Hydrological Ecosystem Services Programme (PSAH). It was implemented in 2003 to protect and restore ecosystems, and preserve ecosystem services through direct economic compensation to their providers (OECD 2013; Perevochtchikova and Torruco Colorado 2014). It is one of the world's largest PES programmes focused specifically on watershed services, and it was mainly funded by stable, long-term revenues from bulk water use fees (OECD 2013). Landowners were eligible for different payment levels depending on the type of ecosystem and their score on a deforestation risk index. Payments were issued yearly after verification of the forest cover (through satellite image analysis or ground visits) (OECD 2013). Areas that lost forest cover were removed from the programme and payments were reduced proportionally (OECD 2013).

In Mexico City, the CORENA (Mexico City's Natural Resources Commission) manages the PROFACE (Fund for Ecosystem Conservation and Restauration) programme, which covers 13,000 ha. It financially compensates landowners or designated groups for conservation efforts in important sites, including through fire prevention, productive reconversion, the preservation of agroforestry systems and silvopastoral systems. Each participating area trains Brigades for environmental monitoring, including through GIS (Interview-M30). Payments are concentrated within the Conservation Land (see Sect. 8.4.4). Mexico State also has a PES programme, similar to the PSAH, financed by a portion of the state's Wat&San tariff.

Effectiveness on Actors in Terms of Mandated Goals

Results were mixed. To be effective, PES programmes must compensate for the opportunity cost of preserving rather than developing land. This opportunity cost is high close to Mexico City and lower in isolated, distant areas (Interview-M48). The PSAH's financial compensation was much lower than the opportunity cost of refraining from exploiting or selling the land, as payments only average USD 15–20 per ha per year (Interviews-M2/M9/M26/M48/M50/M54). Landowners worry about informal occupations and illegal loggers, opting to develop or sell their land (Interviews-M10/M26/M48). As a result, CONAFOR encourages productive activities compatible with conservation, such as honey production or ecotourism (Interview-M10). Those involved with the programme compared the payments to seed money supporting local projects in their initial stages (Interview-M55).

There were virtually no assessments of PROFACE, although respondents familiar with the programme claimed it had modest positive results, despite a low budget (Interviews-M15/M17/M29/M37). Supporters argued that it is more permissive but also more pragmatic than the PSAH. The lack of understanding of the programme's impact prevented expansion (Interview-M30). The PES programme in Mexico State ensures a stable, reliable inflow of funds, although these remain low due to subsidized tariffs and consumers who did not pay their water bills (Interview-M15). Nevertheless, this financial stability gave recipients security. Furthermore, despite the existence of multiple PES programmes, there were no coordination mechanisms between these or with similar policy instruments.

Impacts on Sustainability and Inclusiveness

PES programmes have potential to preserve areas important for springs and aquifer recharge, and thus preserve water supply sources for the MVMC. However, near large urban centres the opportunity cost of protecting land rather than selling it or using it productively is too high for many landowners (Interviews-M46/M48). Recent studies show that areas of Mexico City included in the PSAH programme sometimes had higher deforestation rates than other areas (Saavedra Díaz et al. 2017). As those areas were under pressure from real estate and agricultural interests, it was unclear whether deforestation would have been worse without the programme. Although there were no assessments of PROFACE, respondents claimed that biodiversity and reforestation had increased in the areas involved (Interview-M30). In addition, water-focused PES programmes were rarely designed based on basin or aquifer boundaries, hindering the ability to measure effects on water systems (Interview-M55). Another critique was that PES programmes put an arbitrary value on water and reinforce its commodification (Interview-M58). It was also politically difficult to charge fees or taxes in one region to reinvest revenues elsewhere for results only apparent in the medium to long term (Interview-M28).

7.5 Instrument Assessment and Redesign

The instruments employed in the MVMC's river basin governance showed promise, but often failed to effectively change actors' behaviour and enhance sustainability and inclusiveness. The linear approach still dominated key institutions and policy frameworks. Water resources from multiple states were transported into the MVMC and out (as wastewater) through a complex infrastructural network of inter-basin transfers, deep drainage tunnels and canals. This artificially connected four basins and multiple jurisdictions, lowering incentives for rational water use, reducing contamination and preserving vital ecosystems (see Table 7.5).

Redesign

Based on the instrument evaluation above, the following redesign options are promising.

Inter-basin transfers: As aquifers are increasingly over-exploited, it is unrealistic to support the dismantling of inter-basin transfers in the short or medium-term, despite their many negative externalities. In the longer-term, a combination of measures could alleviate the need for external resources but, in the meantime, decision-making on water imports could be more transparent and inclusive of marginalized interests, as decisions regarding inter-basin transfers are centralized and made behind closed doors. While basin councils do not have mandates to influence infrastructure decisions, their involvement would improve social inclusion. Members should at least have access to financial information to ensure accountability. Investing more resources in land use management and environmental conservation in the donor

Table 7.5 Assessment of IWRM/IRBM policy instruments in the MVMC

	Instrument design	Effects on actors	Impact
Inter-basin transfers	[0] Necessary to meet growing demand and reduce pressure on aquifers Costly and energy intensive	[0/+] Ability to respond to rising water demand (for now), though water supply is unequal across the MVMC Subsidies reduce incentives to decrease water use and losses	Env: [−−] Reduced pressure on aquifers. Externalities are shifted to donor basins Soc: [−−] Peri-urban and rural dwellers receive less water Econ: [−−] Subsidies hide real costs Rel: [−−] Conflicts with donor basin. Decisions are centralized and lack transparency
Water use permits	[++] Based on water availability and type of use Requires environmental impact assessment, preservation of minimum flow. Can be transferred	[−] Lack of capacity and political will to enforce regulations. Transfer system enabled users to bypass rules No coordination on regional water use, despite interlinked systems	Env: [−−] Groundwater table has fallen, and local sources may dry up in next decades Soc: [−−] Peri-urban poor lack adequate access to drinking water Econ: [−] Industry and real estate actors obtain permits, but low enforcement leads to (unsustainable) economic/urban growth Rel: [−−] Industrial users receive permits despite scarcity. Lack of transparency
Water use and wastewater discharge fees	[+] Rates vary per use and area and went to storm/wastewater works in the VMB or Federal Treasury	[−] Prohibition zone and other regulations exist, but lack of metres, cost of monitoring and impunity hindered enforcement. Low fees and energy subsidies promoted high use. Only part of the fees was reinvested regionally	Env: [−−] Same as water/wastewater permits Soc: [−−] Same as water/wastewater permits Econ: [−−] Not all fees were reinvested locally or even in water management Rel: [−−] Donor basins are not compensated for their ESSs

(continued)

Table 7.5 (continued)

	Instrument design	Effects on actors	Impact
Payments for ecosystem services programmes	[+] Aim to compensate ESS providers Funded by water use fees, water tariffs, or city budget	[0] Does not always compensate for opportunity cost, but demand grew, and funds are insufficient. Stable funds are important	Env: [−] Deforestation and degradation continue on urban fringe. Difficult to link programmes to local impact Soc: [0] Restricting urbanization does not address the affordable housing deficit Econ: [+] Programmes are costly, but degradation is much more so in long-term Rel: [+] Alternative for landowners. Recognition of the value of ESSs

Relative assessment scores ++ Very positive; + Positive; 0 Neutral; − Negative; −− Very negative (see Sect. 2.4)

basins, as well as basic infrastructure for communities living near water supply dams, would reduce water losses, increase the productivity of water systems and include currently marginalized groups. This could be through federal funds, PES programmes or revenues from bulk water fees. This will sustain supply dams by ensuring that springs continue to replenish them, reducing the need for new sources further away at higher costs. Indeed, the strategy of supplying the MVMC with inter-basin transfers is reaching the limits of economic viability. Investments in donor basins could also come from the water use fees for uses other than public supply and from basin transfer fees. Instead of going to the Federal Treasury, fees from basin transfers could be reinvested in donor basins. Changes to water tariffs for domestic consumption (see Sect. 8.4.1) could also increase water conservation within the MVMC and reduce the need for water imports.

Water use permits: To increase their effectiveness, water use permits need to be better enforced. More monitoring could reduce irregular groundwater use, though this is costly. Applying adequate fines that reflect the severity of the transgression would address the current impunity for violations (Interview-M9). This would make non-compliance more costly, and reduce irregular water use. Fines would also help fund monitoring costs. These efforts can focus on large water users, such as industries, as their use significantly affects other (potential) users, and their development and investment plans are generally less affected by the payment of water use fees. Smaller users need technical assistance through CONAGUA, as the process of obtaining a permit can be burdensome and expensive (Sanz 2015).

Another challenge relates to the regulation of permit transfers. Large water users, such as real estate developers and industries, sometimes obtain water use permits irregularly, aggravating unsustainable urban and economic growth. Requiring environmental impact assessments could mitigate this by addressing impacts on aquifers, access to water for nearby residents and potential compensation. Strengthening the basin councils and groundwater committees could also better include diverse, local interests into water allocation decisions.

Moreover, data collection and calculations on water availability and abstraction require more transparency. CONAGUA calculates water availability based on water allocation rather than on water abstraction, but these are far from similar. If water availability calculations are not adjusted, at least its methods could be made more transparent.

In addition, to enhance sustainability and inclusiveness, the enforcement of priority uses is crucial. Water-intensive industries could undergo environmental impact assessments that emphasise their potential impacts on water availability for other users. If it is determined that these industries would jeopardize the sustainability of the aquifer and surrounding communities' drinking water access, licensors would be able to withhold permit approvals.

Water use fees: Expanding mandatory and tamper free metering as well as water use fees could promote more rational water use (Interviews-M13/M15/M46). This requires sanctions for transgressions. While it was difficult to implement fees for agricultural users, incentives can help farmers transition to less wasteful irrigation techniques. Such incentives could include subsidies for irrigation technologies and the strengthening of irrigation districts to seek collaborative solutions to growing water scarcity.

Another issue is that not all water use fees return to the basin where they were charged, as is the case in the MRSP. This creates a disconnect between water users and the areas that produce water resources. By applying a system similar to that of the MRSP, the willingness to pay of users may increase and it could strengthen participatory basin management by creating a source of revenue for basin councils.

Payment for ecosystem services: As most land in Mexico is privately owned, implementing strict land use restrictions with no compensation would amount to expropriation and lead to time-consuming and costly litigation and uncertainty. Considering this, PES programmes have potential in areas under pressure from urban and agricultural expansion. However, funding sources for PES programmes remain scarce. Integrating a small fee within Wat&San tariffs for such programmes, as is the case of Mexico State, is viable as these are currently very low (see Sect. 8.4.1). Low-income households should be exempted from paying this additional fee. The fee could increase proportionally with consumption. This would not only ensure stable funding and incentivize rational water use, but it could also raise awareness among consumers by emphasizing the link between water sources outside the city and their taps.

Furthermore, returning the collected fees to the areas where they were charged can enhance sustainability and inclusiveness. More environmentally sustainable management of the rivers that drain into the Cutzamala system could reduce water imports. However, while there is some recognition (with still limited action) of the ecosystem services provided by areas within the metropolitan area, there is virtually none for the ecosystem services provided in these other basins. This requires integrating PES programmes with basin and aquifer management and could strengthen the link between IWRM/IRBM and urban water services. It could also strengthen basin councils' roles (Interviews-M40/M55). PES programmes would then account for interconnections and spillovers across different basins. Ultimately, a mix of regulatory and economic instruments are necessary to preserve environmentally significant areas and landowners require greater support to contribute to these measures. Coordinating PES programmes and environmental agencies with urban planning and land use actors could also enhance environmental preservation through solutions for affordable housing in other areas or by promoting densification rather than sprawl.

Missing Instruments

Water use should be compatible with the needs of current and future consumers but also with those of the basin(s) from which water is supplied. Reinvesting bulk water use fees and water tariffs in programmes to preserve ecosystem services are potential mechanisms for this purpose. Reinvesting water tariffs requires Wat&San planning that looks beyond the boundaries of the utility's jurisdiction, at a larger regional, scale.

Some alternative approaches have been gaining support, including facilitating groundwater recharge in the city through green infrastructure and artificial aquifer injections (where rainwater or treated wastewater is injected into the ground), or environmental protection in areas of aquifer recharge (particularly in the mountains surrounding the basin). These both mitigate excess surface water runoff and replenish water supply sources. The effectiveness of these measures is difficult to determine due to the many unknowns surrounding groundwater flows (e.g. velocity, impact on water quality from contaminants). Understanding groundwater flow dynamics is important for prioritizing preservation efforts and for installing water injection technology in optimal locations (Interview-M32). More groundwater studies are needed to evaluate the potential impacts and costs of such measures, and thereby the extent to which it could reduce the dependence on external water resources.

References

Acevedo B et al (2013) The institutional structure of water policy in the Mexico City metropolitan area. Available at: http://www.hardrainproject.com

Aguilar LA, Gardida DB, Trejo JAG (2010) El manejo de las aguas mexicanas en el siglo XX. In: Cisneros BJ, Armentia MLTY Aguilar LA (eds) El agua en México: cauces y encauces. Academia Mexicana de Ciencias, Mexico City, D.F., pp 21–50

Burns E (ed) (2009) Repensar la Cuenca: La Gestión de los Ciclos del Agua en el Valle de Mexico. UAM, Mexico City, D.F.

CONAGUA (2012a) Programa Hídrico Regional Visión 2030: Región Hidrológico-Administrativa XIII Aguas del Valle de México. Mexico City

CONAGUA (2012b) Recuento de la Cooperación Internacional de la Conagua 2009–2012. Mexico City. Available at: www.conagua.gob.mx. Accessed 13 May 2019

CONAGUA (2015) Estadísticas del Agua en México. México, D.F.

CONAGUA (2018) Estadisticas del Agua en México. Mexico City, D.F.

De Regil M (2014) Conagua dará facilidades a usuarios de aguas nacionales. El Financiero, 22 Apr

DOF (1972) DECRETO por medio del cual se crea un organismo técnico administrativo dependiente de la Secretaría de Recursos Hidráulicos, que se denominará Comisión de Aguas del Valle de México

DOF (1981) Ley Federal de Derechos. Available at: http://www.diputados.gob.mx/LeyesBiblio/pdf/107_281218.pdf. Accessed 16 July 2019

EFE (2019) El Valle de México agotará la capacidad de extraer agua del acuífero en 40 años: expertos. Sin Embargo, 12 May

Engel K et al (2011) Big cities. Big water. Big challenges. Water in an urbanizing world. Berlin. Available at: https://www.wwf.se/source.php/1390895/Big_Cities_Big_Water_Big_Challenges_2011.pdf. Accessed 27 Dec 2018

Escamilla IH, Santos CC (2012) La Zona Metropolitana del Valle de México: Transformación Urbano-Rural en la Región Centro de México. In: XII Coloquio Internacional de Geocrítica, Bogotá, p 15

Escolero O et al (2016) Diagnóstico y análisis de los factores que influyen en la vulnerabilidad de las fuentes de abastecimiento de agua potable a la Ciudad de México, México. Bol Soc Geol Mex 68(3):409–427. Available at: http://www.scielo.org.mx/scielo.php?script=sci_abstract&pid=S1405-33222016000300409&lng=es&nrm=iso. Accessed 18 May 2019

FAO (2012) Coping with water scarcity: an action framework for agriculture and food security. Rome. Available at: http://www.fao.org. Accessed 11 Dec 2018

Galindo E et al (2011) Sobreexplotación del acuífero Cuautitlán-Pachuca: balance hídrico 1999–2020. In: González EAR (ed) Avances en Impacto, Tecnología y Toxicología Ambiental. Universidad Autónoma del Estado de Hidalgo, pp 9–32

Garcia G (2022) Sequía acecha en el Edomex; Cutzamala se reduce casi un 5%. El Heraldo, 14 Feb. https://heraldodemexico.com.mx/nacional/2022/2/14/sequia-acecha-en-el-edomex-cutzamala-se-reduce-casi-un-5-378497.html

Guido Aldana PA (2017) Cambio climático: selección, clasificación y diseño de medidas de adaptación. Jiutepec, Mor. Available at: www.imta.gob.mx

Hearne RR (2004) Evolving water management institutions in Mexico. Water Resour Res 40(12). https://doi.org/10.1029/2003WR002745

IMTA (2007) Efectos del cambio climático en los recursos hídricos de México. Gaceta del IMTA 7

Martínez P (2018) El reto del agua para el Valle de México será cada vez mayor. Animal Político, 28 Oct. Available at: https://www.animalpolitico.com/2018/10/reto-abasto-agua-valle-mexico/. Accessed 3 Nov 2018

Martínez MA (2019) Incendio en Conagua levanta sospechas; se quemaron documentos de fiscalización. La Silla Rota, 25 Mar

Martínez ÁC, Enciso A (2015) Logros parciales de México en los objetivos del milenio. La Jornada, 5 Jan

Mazari-Hiriart M et al (2014) Final opportunity to rehabilitate an urban river as a water source for Mexico City. PLoS ONE 9(7). https://doi.org/10.1371/journal.pone.0102081

Mendoza HC (2016) The crisis of potable water in Mexico City: institutional factors and water property rights as conditions for creating adequate metropolitan water governance. University of Southern California. https://doi.org/10.25549/USCTHESES-C40-217168

Mexican Constitution (1917) Constitution of the United Mexican States. Available at: https://www.juridicas.unam.mx/legislacion/ordenamiento/constitucion-politica-de-los-estados-unidos-mexicanos. Accessed 18 May 2019

Mexico (2005) Fideicomiso 1928. Available at: https://dof.gob.mx/nota_detalle_popup.php?codigo=2102731

Neri-Ramírez E et al (2013) Evaluación de la Sustentabilidad del Acuífero Cuautitlán-Pachuca Mediante el Uso de la Metodología Mesmis. Rev Chapingo Ser Cienc For Ambiente. Universidad Autónoma Chapingo, Coordinación de Revistas Institucionales 19(2):273–285. Available at: http://www.scielo.org.mx/scielo.php?script=sci_arttext&pid=S2007-40182013000200009. Accessed 15 May 2019

NWL (2004) LEY DE AGUAS NACIONALES. Available at: http://www.diputados.gob.mx/LeyesBiblio/pdf/16_240316.pdf. Accessed 24 Sept 2017

OECD (2013) Making water reform happen in Mexico. In: OECD studies on water. Paris. https://doi.org/10.1787/9789264187894-en

OECD (2015) OECD territorial reviews: Valle de México, Mexico. OECD Publishing, Paris. Available at: https://books.google.com/books?id=cEO8CgAAQBAJ&printsec=frontcover#v=onepage&q&f=false. Accessed 20 Aug 2019

Olvera D (2017) Peñafiel, "el refresco de Peña", le saldrá caro a Edomex: chupará el agua de millones de ciudadanos. Sin Embargo, 14 Apr

Perevochtchikova M, Torruco Colorado VM (2014) Análisis comparativo de dos instrumentos de conservación ambiental aplicados en el Suelo de Conservación del Distrito [Federal comparative analysis of two environmental conservation instruments used on conservation land in the federal district]. Soc Ambiente 1(3):3–25

Perló M, González AE (2005) ¿Guerra por el agua en el Valle de México? Estudio sobre las relaciones hidráulicas entre el Distrito Federal y el Estado de México. UNAM–PUEC, Mexico City, D.F.

Pina P (2011) Addressing water crisis in Mexico City: a political and institutional perspective

Ramírez É (2015) Sobreexplotación del "oro líquido", alerta de seguridad nacional. Contralinea, 4 Mar. Available at: www.contralinea.com.mx/archivo-revista/2015/03/04/sobreexplotacion-del-oro-liquido-alerta-de-seguridad-nacional/

Rodríguez Tapia L, Morales Novelo JA (2013) Integración de un sistema de cuentas económicas e hídricas en la Cuenca del Valle de México. Rev Int Estad Geogr 4(1):73–89. Available at: https://www.inegi.org.mx/RDE/RDE_08/Doctos/RDE_08_Art5.pdf. Accessed 8 June 2019

Romero Lankao P (2010) Water in Mexico City: what will climate change bring to its history of water-related hazards and vulnerabilities? Environ Urban 22(1):157–178. https://doi.org/10.1177/0956247809362636

Saavedra Díaz ZM, Perevochtchikova M, Saavedra Díaz ZM (2017) Evaluación ambiental integrada de áreas inscritas en el programa federal de Pago por Servicios Ambientales Hidrológicos. Caso de estudio: Ajusco, México. Investig Geogr. Instituto de Geografía, UNAM 93. https://doi.org/10.14350/rig.56437

Sanz M (2015) Micro and small enterprises, water and developing countries: a challenge for sustainability in Colombia. University of Amsterdam. Available at: www.taylorandfrancis.com. Accessed 17 July 2019

SEGOB (2016a) ACUERDO por el que se da a conocer el resultado de los estudios técnicos de las aguas nacionales subterráneas del acuífero Zona Metropolitana de la Cd. de México, clave 0901, en la Ciudad de México, Región Hidrológico-Administrativa Aguas del Valle de México. Diario Ofícial de la Federación. Available at: https://www.dof.gob.mx/nota_detalle.php?codigo=5453759&fecha=26/09/2016#gsc.tab=0

SEGOB (2016b) ACUERDO por el que se da a conocer el resultado de los estudios técnicos de las aguas nacionales subterráneas del acuífero Cuautitlán-Pachuca, clave 1508, en el Estado de México, Región Hidrológico-Administrativa Aguas del Valle de México. Diario Ofícial de la Federación. Available at: https://cofemersimir.gob.mx/expedientes/19151

SEGOB (2016c) ACUERDO por el que se da a conocer el resultado de los estudios técnicos de las aguas nacionales subterráneas del acuífero Tecocomulco, clave 1319, en el Estado de

Hidalgo, Región Hidrológico-Administrativa Aguas del Valle de México. Diario Oficial de la Federación. Available at: https://www.dof.gob.mx/nota_detalle.php?codigo=5447503&fecha=11/08/2016#gsc.tab=0

SEGOB (2016d) ACUERDO por el que se da a conocer el resultado de los estudios técnicos de las aguas nacionales subterráneas del acuífero Apan, clave 1320, en el Estado de Hidalgo, Región Hidrológico-Administrativa Aguas del Valle de México. Diario Oficial de la Federación. Available at: https://www.dof.gob.mx/nota_detalle.php?codigo=5447504&fecha=11/08/2016#gsc.tab=0

SEGOB (2016e) ACUERDO por el que se da a conocer el resultado de los estudios técnicos de las aguas nacionales subterráneas del acuífero Chalco-Amecameca, clave 1506, en el Estado de México, Región Hidrológico-Administrativa Aguas del Valle de México. Diario Oficial de la Federación. Available at: https://sidof.segob.gob.mx/notas/docFuente/5572647

SEGOB (2016f) ACUERDO por el que se da a conocer el resultado de los estudios técnicos de las aguas nacionales subterráneas del acuífero Texcoco, clave 1507, en el Estado de México, Región Hidrológico-Administrativa Aguas del Valle de México. Diario Oficial de la Federación. Available at: https://www.dof.gob.mx/nota_detalle.php?codigo=5572270&fecha=13/09/2019#gsc.tab=0

Sosa-Rodriguez FS (2014) From federal to city mitigation and adaptation: climate change policy in Mexico City. Mitig Adapt Strateg Glob Change 19(7):969–996. https://doi.org/10.1007/s11027-013-9455-1

Spring ÚO (2011) Aquatic systems and water security in the Metropolitan Valley of Mexico City. Curr Opin Environ Sustain 3:497–505. https://doi.org/10.1016/j.cosust.2011.11.002

Spring ÚO (2014) Water security and national water law in Mexico. Earth Perspect 1(7):1–15. https://doi.org/10.1186/2194-6434-1-7

Spring ÚO (2015) Managing water resources in Mexico in the context of climate change. In: Shrestha S et al (ed) Managing water resources under climate uncertainty. Springer. https://doi.org/10.1007/978-3-319-10467-6_18

Tortajada C (2008) Challenges and realities of water management of megacities: the case of Mexico City metropolitan area. J Int Aff 61(2):147–166

Valadez S (2016) Conagua: deben usuarios renovar sus concesiones | Expreso—Expreso. Expreso, 12 Oct. Available at: https://expreso.press/2016/10/12/conagua-deben-usuarios-renovar-sus-concesiones/. Accessed 16 July 2019

Wilder M (2010) Water governance in Mexico: political and economic aperatures and a shifting state-citizen relationship. Ecol Soc 15(2):22. https://doi.org/10.5751/ES-03469-150222

World Bank (2013) Agua Urbana en el Valle de México ¿un camino verde para mañana? Washington, DC. Available at: www.bancomundial.org.mx. Accessed 8 May 2019

World Bank (2015) Cutzamala: Diagnóstico integral. Washington, DC

Chapter 8
The Implementation of UWM in Mexico City

8.1 Introduction

This chapter examines how different drivers and institutions at multiple levels of the urban water governance regime shape water-related challenges in the Metropolitan Valley of Mexico City (MVMC). It uncovers the causal links behind these water challenges and the effectiveness of existing policy instruments. It reviews the relevant context of Mexican urban water governance and its main drivers (see Sect. 8.2). It analyses the driving forces on the city from local to global level. Section 8.3 then explores which formal actors and institutions shape Urban Water Management (UWM) in the MVMC. Section 8.4 analyses the instruments of UWM according to their stated mandates, their effect on actors' behaviour and their impacts on inclusive and sustainable water governance. Finally, the chapter summarizes the main empirical findings and considers how more appropriate instruments could be (re)designed for the Mexico City case study in relation to UWM (see Sect. 8.5).

8.2 Context and Drivers of Mexico City's Urban Water Challenges

8.2.1 Context in Relation to the Metropolitan Region

Mexico City was founded by the Spaniards in 1521 on top of the ruins of Tenochtitlán, which had approximately 70,000 inhabitants at the time of the Spanish conquest (Escamilla and Santos 2012).[1] As the capital expanded, it formed the Metropolitan

[1] Population estimates for Tenochtitlán vary from 50,000 to 175,000 and even 300,000 inhabitants (Evans 2016; Matos Moctezuma 1988; Hardoy 1964; Vaillant 1941). Recent estimates are more conservative.

F. van den Brandeler, *Scalar Mismatches in Metropolitan Water Governance*, Water Governance - Concepts, Methods, and Practice,
https://doi.org/10.1007/978-3-031-08061-6_8

Valley of Mexico City (MVMC), which includes Mexico City and 60 municipalities located primarily in Mexico State, but also in the states of Hidalgo and Tlaxcala (Spring 2011). Mexico City itself is a federal entity, composed of 16 districts that are similar to municipalities.

8.2.2 Main Drivers of Mexico City's Urban Water Challenges

Urbanization

The MVMC grew rapidly during the twentieth century, especially after 1950, through urban–rural migration. By 2010 it had a fifth of Mexico's population, with 20 million inhabitants (INEGI 2014). While Mexico City's growth rate declined after 1990, it increased in surrounding municipalities, intensifying pressure on water resources through higher water use (Spring 2011; Escamilla and Santos 2012). Moreover, the city could no longer absorb all newcomers into the labour market, leading to high levels of unemployment and underemployment and the spread of informal settlements in peri-urban areas (Romero Lankao 2010; Spring 2015). The speed of urbanization, the lack of planning and local governments' weak financial and institutional capacities led to neighbourhoods without basic infrastructure (World Bank 2013).

Climate

The MVMC is likely to experience the greatest impacts of climate change in Mexico due to ecological degradation and population density (Sosa-Rodriguez 2014). Temperatures have increased by almost 4 °C due to the heat-island effect and climate variability and change (Sosa-Rodriguez 2014). Current models predict an increase in the mean temperature by 2–4 °C and a decrease in mean precipitation of up to 20% by 2080, which could further disrupt Mexico City's aquifers (Sosa-Rodriguez 2014). The metropolis is highly vulnerable to floods due to soil subsidence and its location within an endorheic basin (Spring 2011).[2]

Economic Development

National economic policies since the 1980s have transformed Mexico City into a hub of financial and service activities, stimulating, along with the increasing lack of economic prospects in rural areas, population flows to the capital (Romero Lankao 2010). The Mexican capital has the largest concentration of wealth in Mexico, generating a third of the national GDP (Romero Lankao 2010). Poverty and socioeconomic inequalities across the country continues to foster migration to the MVMC (Table 8.1).

---

[2] Soil subsidence has inverted the land slope, so expelling wastewater requires more engineering and energy and represents a greater flood risk (Martínez and Enciso 2015).

Table 8.1 Multi-level drivers of water-related challenges on the city

	Direct		Indirect	
	Local	Regional/global	Local	Regional/global
Urbanization	Urban growth Informal expansion at MVMC's margins	Urban growth in surrounding regions	Soil subsidence	
Economic			Economic centre of the country	Poverty and inequality across Mexico
Climate	Heavy summer rains		Heat island effect	Climate variability and change

Source Author

8.3 The Institutional Framework for UWM in Mexico City

8.3.1 Global Level

International institutions in the MVMC influence UWM by setting standards and providing financing. For instance, in 2010, the World Bank approved a USD 450 million loan to support the design of climate change adaptation policies for the water sector (Score 2010). As Mexico continues to develop, the role of development partners in financing Mexico's water sector is expected to diminish (OECD 2013). By 2010, the supply of previously disinfected piped drinkable water in Mexico reached 91% of the population, slightly above the 2015 target of 89% set by the Millennium Development Goals (UN 2013). Access to sewerage reached 88% in 2010, above the 79% target defined for 2015. However, data about access to piped water includes households with taps that do not receive water every day (Martínez and Enciso 2015).

8.3.2 National Level

During the twentieth century, Wat&San functions were gradually transferred to states and municipalities. However, Wat&San has no overarching legal framework at national level, which has hindered efforts to expand these services (OECD 2013). This vacuum was brought to light in 2012, when Wat&San services were recognized as a human right within the Constitution (OECD 2013).[3] Guaranteeing this right across the three levels of government required the publication of a new National Water Law. However, as of March 2022 no new bill was approved. In 2014, CONAGUA sent

[3] Article 4 states: "Every person has the right to water access, disposal and sanitation for personal and domestic consumption in a sufficient, clean, acceptable and affordable manner".

a bill to Congress but it was met with strong rebuttal and was removed from the parliamentary agenda by March 2015 (Godoy 2015).[4]

Despite the transfer of UWM responsibilities to lower levels, CONAGUA intervenes within the MVMC through its management of deep drainage infrastructure and the canal that discharges wastewater to the Tula Basin (Spring 2011) (see Sect. 8.4.3). The SEMARNAT, CONAGUA, and the Ministries of Health and Economy have issued UWM standards, such as pollutant limits in wastewater discharges, guidance on how to evaluate Wat&San tariffs and quality standards for drinking water (CONAGUA 2018).

The General Law for Climate Change was adopted in 2012 and led to the creation of the National System for Climate Change (Mexico (Cámara de Diputados) 2012). CONAGUA states that water-related climate measures that diminish the vulnerability of the population are a priority. Various federal programmes have implemented such measures, including the Special Programme for Climate Change (2014–2018). These programmes focus on mitigation, whereas national level adaptation measures are mostly limited to monitoring and forecasting of extreme weather events (Interviews-M6/M7/M12). In addition, although several laws have been established, implementation has been slower, especially at lower levels of government, in part due to a lack of stable funding sources (Interviews-M6/M7/M15/M46).

8.3.3 State Level

The lack of an overarching framework for Wat&San services means that the role of state governments varies significantly. State governments are responsible for planning, regulating and developing infrastructure for water resources. State congresses approve tariffs and are involved in approving state water plans and allocating financial resources for water infrastructure (OECD 2013). Municipalities can choose to delegate Wat&San services to the State Water Commissions. Besides sometimes acting as Wat&San operators, the latter may be responsible for standard setting, monitoring service provision, improving efficiency and promoting public participation, but they often lack capacity to adequate regulate Wat&San services (OECD 2013). Within the MVMC, the state water commissions also coordinate with the basin agency and the civil defence and firefighters in case of flood risks (Interview-M19). However, information sharing between these commissions, basin institutions and academics could be improved through the integration of water information systems (OECD 2013). Overall, state governments have few mandates in relation to water, creating a gap between the federal government and the almost 2500 municipalities (Interviews-M4/M15).

[4] A coalition of CSOs and academics argued that the bill commodified water, banned water quality research and favoured dam construction and river diversion (Godoy 2015). Members of the ruling coalition retorted that public investment was insufficient to meet water demand and only the private sector could fill this gap (Adler 2015).

Mexico City's water commission is the SACMEX (Water System of Mexico City) and operates as a water utility. It also operates several large wells within the city and the Lerma System (see Sect. 7.4.1). It is part of the SEDEMA and has limited technical, budgetary and decision-making autonomy (Interview-M20). Mexico State's Water Commission (CAEM) manages large volumes of water (e.g. water abstraction from the Lerma-Cutzamala systems) and provides bulk water to 16 municipalities of the MVMC, treats wastewater and provides Wat&San services in some municipalities. The mandates of states and state water commissions in UWM lead to fragmentation as the metropolis spreads over several states.

8.3.4 Metropolitan Level

The Federal Constitution states that municipalities within a metropolitan area must coordinate their actions (Mexican Constitution 1917, Sec. Art. 122). It establishes a normative framework that enables the creation of metropolitan committees with representatives from municipalities, the Mexico City and Mexico State governments, and sectoral entities. Several committees were created in the MVMC, such as the Metropolitan Committee for Water and Drainage (CADAM) (1994), Metropolitan Committee for Human Settlements (1995) and Metropolitan Committee for Civil Defence (2000). The Environmental Commission of the Megalopolis, created in 2013, focuses on air pollution in the MVMC (Interviews-M14/M15/M16/M17/M48/M57). There is no policy for conservation at metropolitan or regional scale.

Most metropolitan commissions did not coordinate in terms of shared planning and decision-making (Interviews-M9/M14/M46). Metropolitan-level coordination was mainly limited to day-to-day technical operation of large-scale infrastructure, as the hydraulic systems were completely interconnected (Interview-M14). The Metropolitan Fund financed metropolitan-scale hydraulic infrastructure but also roads and transportation in Mexico's metropolitan regions (SAF 2019) (Interview-M15). Municipalities were not directly involved, and there was no strategic regional vision for water supply management (Interview-M14).

Metropolitan coordination was a divisive topic, with respondents defending decentralization to local levels, while others argued that metropolitan management was necessary to address shared water-related challenges (Interviews-M9/M17/M19/M40/M52). Coordination is also particularly challenging in the MVMC, as it involves three different federal entities (Mexico City, Mexico State and Hidalgo State). It was more effective between governments from the same political party (Interviews-M6/M7). Party politics reinforced the fragmentation as they created tensions between levels of government and between state, municipalities and districts across the MVMC (Interviews-M5/M9/M14/M15/M17/M33/

M47/M49/M54). Metropolitan entities did not coordinate with basin entities (Interview-M48).

8.3.5 Local Level

Democratization reforms allocated many UWM responsibilities to local governments. The Constitution gives municipalities the mandate to provide water services (Mexican Constitution 1917, Sec. 115), including drinking water, stormwater, and sewage collection, treatment and disposal. State laws specify the form in which these services will be provided. Municipalities are responsible for developing a municipal water policy.

Across Mexico, the provision of these services is highly heterogeneous due to the lack of an overarching regulatory framework and the lack of financial and human capacity in most municipalities, which depend on federal and state funds (OECD 2013) (Interviews-M4/M6/M7/M15/M17/M33). Municipalities can delegate Wat&San services to the state or a private company. Within the MVMC, Mexico City has one public water utility, the SACMEX, for its sixteen districts (although it outsources many activities to third parties), some municipalities of Mexico and Hidalgo States delegate this service to their respective state water commissions and others have their own utility, causing a regional coordination challenge (Interviews-M2/M6/M7/M28). Some regions created inter-municipal Wat&San utilities.[5]

Water utilities cannot provide regular services in informal settlements, as their legal status prohibits the installation of basic infrastructure. Consequently, these areas are supplied through water trucks or alternative systems (e.g. rainwater harvesting) (Interviews-M24/M26/M49). NGOs and companies mainly filled this gap at the time of field research, with little support from local or state authorities (Interviews-M41/M42/49). However, in 2020, SEDEMA started collaborating with NGOs to support rainwater harvesting initiatives, promising to install over 100,000 systems (Breña 2020). The municipal civil defence addresses water-related risks through monitoring, implementing security measures and elaborating civil defence programmes (SEGOB 2014). The Municipal Civil Protection Programme is based on a risk diagnosis and outlines goals, strategies and actions, and the necessary resources to implement this (SEGOB 2014). It requires municipalities to develop a Risk Atlas that identifies the type and severity of local risks (SEGOB 2014). Municipalities are also expected to develop, implement and evaluate a municipal climate change policy framework and instruments in accordance with the national and state climate change frameworks (Mexico 2012).

Municipalities are responsible for land use management and zoning, and can create ecological reserves within their borders (Mexico 2009). Although a federal

[5] One challenge for this form of water provision is that, when municipalities have wells, they often consider themselves owners of the water resources and are reluctant to share these (Interview-M4).

entity, Mexico City's central government is responsible for its planning, delegating certain implementation tasks to its districts (Connolly and Wigle 2017). Mexico City's districts coordinate certain actions (e.g. on environmentally protected areas, through the SEDEMA), but coordination is absent between districts and surrounding municipalities (Interview-M26). Water and other urban policies lack continuity as local administrations change every three years (Interviews-M14/M17/M21/M55). Moreover, officials are often political appointees, as there is no professionalization of the careers within these departments (Interviews-M17/M55/M57) (Fig. 8.1).

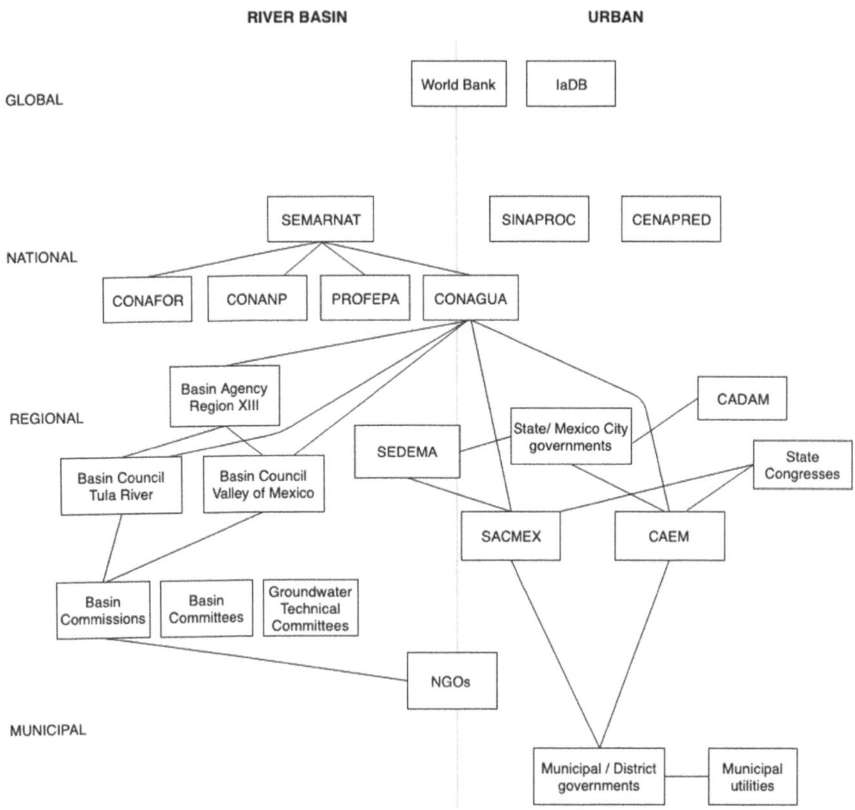

Fig. 8.1 Basin and urban water stakeholders in the MVMC. *Source* Author

8.4 Instrument Analysis

8.4.1 *Water Tariffs*

Design

With no overarching legal or policy framework for Wat&San, utilities largely defined their own objectives. However, the Constitution defined Wat&San as a human right, implying that all utilities should universalize these services within their jurisdiction. Broadly, utilities in the MVMC, such as SACMEX, CAEM and municipal utilities, established goals such as the provision of clean and accessible drinking water, including to marginalized communities that comply with land tenure laws, and tariff systems that allow for autonomy, financial self-sufficiency, and rational water use (CAEM, no date; SEDEMA 2012; ODAPAS 2016). Residents in informal settlements are therefore excluded from such services.

Multiple factors fostered wide inter-municipal disparities in Wat&San tariffs: The absence of a national framework, highly unequal distribution of water resources across Mexico, different levels of costs, and differences in cost-recovery efforts (OECD 2013; CONAGUA 2018). However, even within regions there were stark disparities. In Region XIII, water tariffs, set by different utilities, range from USD 0.95/m^3 (in Pachuca de Soto, Hidalgo State) to USD 0.28/m^3 in several municipalities of Mexico State (Hernández 2014). A price variability analysis revealed that municipalities with similar water availability sometimes had significant disparities in tariffs (Hernández 2014). While the State Congress generally approved tariffs, this was sometimes done by the board of directors of municipal operators, and municipal councils have the option to grant subsidies, surcharges or cancel fines (OECD 2013).

Tariffs were typically composed of a fixed charge, a volumetric charge (sometimes divided in increasing blocks) and a surcharge for wastewater treatment (OECD 2013). The latter was not applied by all utilities. For instance, the SACMEX charged a water tariff with a fixed rate up to 15 m^3 with increasing block rates above this volume, but no sewage tariff. Tariffs depended on the quantity of water consumed, with rates increasing steeply for higher consumption (Interview-M9). The SACMEX applied a development index for subsidized rates based on socio-territorial characteristics, including social marginalisation, property values and income, allowing for cross-subsidies between consumers paying the full rate and those receiving subsidies. Each street block had a specific tariff based on these characteristics (Interview-M24). As a result, tariff rates varied across the city (Interview-M2). Comparing water tariffs across the MVMC is difficult due to differences such as the inclusion of fixed rates or sewage fees, but overall tariffs in Mexico City were much lower than in other metropolitan municipalities (OECD 2015).

Effectiveness on Actors in Terms of Mandated Goals

SACMEX's tariff structure aimed for equity, by charging increasing block rates and lower tariffs for low-income households, and other utilities in the MVMC applied similar policies. In fact, many municipalities heavily subsidized water tariffs, including in Mexico City (Interviews-M6/M19/M28/M38/M42/M46/M50/M56). Subsidized tariffs were often a tool used by local politicians in their campaigns (Interviews-M4/M19/M38). Mexico City residents without subsidies paid a flat rate of approximately USD 25 per month for water consumption up to 10 m^3 (see Annex H: Water Tariffs). However, it was estimated that at least 75% of Mexico City's consumers received subsidies of up to 91% (Roa 2018). Therefore, many residents with the financial means to pay higher tariffs pay subsidized water tariffs. There are four categories of subsidies for domestic consumers: Popular, Low, Middle and High. Their respective water tariffs for consumption up to 10 m^3 were USD 2.22, 2.52, 8.32 and 9.99 per month. Water tariffs in all groups increased from 2015 to 2022. While this better reflected the region's water scarcity, tariffs in the popular group increased relatively more than tariff groups for higher income consumers. Residents purchasing water from trucks spent around 14 times more than those with access to the piped network, for much lower quality water (Mendoza 2016). Subsidized tariffs thus do not reach the most marginalized citizens. In addition, many residents paid fixed tariffs, regardless of their consumption, because they lacked water metres (Interviews-M6/M7/M15). In Mexico State, only 43% of CAEM's consumers had water metres. The rest paid fixed rates, which also varied according to subsidies (Hidalgo 2018). Although Mexico City has the same tariffs across its 16 districts, tariff rates in the rest of the MVMC vary per municipality.

Utilities struggled with cost-recovery and could not operate optimally, as water tariffs remained low and many households still lacked water metres or did not pay their water bills (Interviews-M5/M50/M56). Tariffs collected within the MVMC barely covered 51% of the costs of service provision (World Bank 2013; SEGOB 2020). In Mexico City proper, the annual revenue collected from water tariffs is USD 116 million, whereas service provision costs reach USD 405 million (Ciencias 2018). SACMEX was therefore in complete deficit and was subsidized by the federal and Mexico City governments (OECD 2015) (Interviews-M7/M9/M46/M51/ M56). In 2018, it was estimated that 45% of SACMEX's consumers did not pay their water bills regularly (Navarro and Gómez 2018). SACMEX can cut off the water to households that recurrently do not pay as a last resort, but usually attempts to seek agreements with consumers and it sometimes reduces the water pressure in the network as a deterrent (Interview-M19).

Tariffs are largely determined by congressmen, who are reluctant to increase rates due to the political cost (IMCO 2014; Flores et al. 2016) (Interviews-M4/M6/M15/M32/M46). Water tariffs increased above inflation rates in many cities, but remained far from allowing for cost-recovery (OECD 2013) (Interview-M6). For many municipal utilities in the MVMC, their small scale was an obstacle for cost-recovery (IMCO 2014).

Impacts on Sustainability and Inclusiveness

Approximately 97% of the MVMC's population had access to piped water infrastructure (Romero Lankao 2010). However, access—in both quantitative and qualitative terms—was not equally distributed across the city (Hackett and Eakins 2015; Spring 2015). Drinking water services were irregular and unreliable in certain areas, particularly for peri-urban residents (Romero Lankao 2010; Acevedo et al. 2013). While Mexico City's central districts received about 300 l/person/day, the Eastern districts of Tláhuac, Xochimilco and Iztapalapa received 177, 214 and 238 l respectively (Rodea 2016). Even these numbers hide stark disparities, as wealthy neighbourhoods of Mexico City consume up to 800 l/person/day, while certain marginalized areas receive only 14 l/person/day (Spring 2015; Rodea 2016). Some marginalized areas are not connected to the public supply network, leading to precarious access to drinking water. Survey evidence across the MVMC indicates that drinking water access is even more precarious outside Mexico City (OECD 2015). In many parts of the MVMC, residents had water a few hours a day every other day, or less frequently, but at a higher cost as water was brought by trucks (Burns 2009) (Interviews-M14/M24;M46). Water delivered by trucks was generally free in regularized neighbourhoods, but in informal settlements residents organized and paid for this delivery themselves.[6] Subsidized water tariffs therefore did not benefit those who most needed these.

Residents sometimes resorted to clandestine connections, which caused water losses as well as health risks from contamination. In one informal settlement visited in July 2016 in the Xochimilco District, in the South of Mexico City, flimsy drinking water hoses criss-crossed the polluted canals that were the vestiges of the ancient lake city (see Fig. 8.2). Data on contaminants is not provided, and although data on water quality in the Cutzamala is available, there is no information on water quality after water enters local delivery networks (OECD 2015). The South and Southeast of Mexico City were among the most affected by poor drinking water quality (Sosa-Rodriguez 2010). Although the CONAGUA stated that 88% of water in the MVMC was chlorinated, independent studies showed that only 60% of tap water was suitable for human consumption (Spring 2011). Sources of contamination included deteriorated pipes, insufficient treatment processes and high levels of dissolved minerals in groundwater (Spring 2011). Although SACMEX frequently states that the piped water meets official quality standards, there is no transparency around this data (Durán and Gómez 2018).

Unequal access to water services reflects and reinforces broader patterns of inequality and power imbalances between residents across the MVMC (Interviews-M47/M50), with impoverished communities in the East suffering much more than richer inhabitants in the West (Interview-M14). Women carry the burden of ensuring that their families have water and they are vulnerable to harassment by water truck delivers (Interviews-M27/M59/M60). Furthermore, the lack of effective communication to the population about drinking water reliability has resulted in a general distrust of tap water (Interviws-M47/M50). In combination with inconstant supply,

[6] Land tenure was the main obstacle, but the physical layout and topography also prevented access to the piped water network.

Fig. 8.2 Improvised water supply in Amalacachico, Mexico City. *Source* Author

this has led Mexicans to become the world's largest consumers of bottled water per capita, at 234 l/person/year in 2015 (Martínez and Enciso 2015).

Although residents are wary about drinking water from public supply, highly subsidized tariffs encourage high water consumption for other purposes—e.g. washing, cleaning, filling swimming pools (Interviews-M6/M15/M46). This is compounded by the many households without metres or with fixed tariffs (Interviews-M9/M15/M19/M56). Consumers with access to the (subsidized) piped network have few incentives to switch to alternative systems, such as rainwater harvesting or water reuse (Interviews-M32/M46/M49). In addition, up to 40% of water was lost in leakages due to decaying infrastructure and soil subsidence (Martinez et al. 2015; Ciencias 2018) (Interview-M46). Besides contributing to water shortages, leaks reduce revenues and investment capacity (Interview-M51). Under current conditions, the deficit between demand and supply may reach between 25 and 46 m^3/s by 2030 (World Bank 2013).

Furthermore, as utilities were unable to recover costs, they depended significantly on federal, state and municipal funds (IMCO 2014). As a result, taxpayers across Mexico subsidized utilities by paying an annual economic cost of around USD 1.5 billion, or almost a triple of the tariff revenue (World Bank 2013). With access to external funds, cheap bulk water and low tariff revenues, utilities lacked incentives to improve services (Interviews-M19/M32).

8.4.2 Metropolitan Drainage System

Design

Managing stormwater is challenging due to the MVMC's location in a closed basin and much of the soil's impermeability. While the region has always been flood-prone, soil subsidence aggravated this risk as parts of the city sank by over 12 m. The macro-drainage system aims to mitigate flood risks by rapidly removing excess water from the basin and depositing it in the Tula basin. Four artificial exits (canals and large tunnels) interconnected across the metropolis were built over the years to discharge stormwater, with the first inaugurated in 1789 and the last in 1975 (SACMEX 2013). In late 2019, CONAGUA inaugurated an additional mega-tunnel to evacuate up to 150 m^3/s from the basin (CDMX 2019).

CONAGUA, Mexico City and Mexico State jointly manage this infrastructure through the CADAM and shared operation protocols (e.g. infrastructure maintenance, warning systems, emergency strategies). The Fideicomiso 1928 funds the approximately USD 160 million annual in costs (Burns 2009) (Interviews-M15/M19). The Valley of Mexico Metropolitan Fund also finances infrastructure in the MVMC. However, the fund was used mainly for transportation and urbanism, and just USD 1.3 million were spent between 2014 and 2017 on stormwater management. It is managed by the governments of Mexico City, Mexico State and Hidalgo State, to reduce vulnerability to extreme weather events through studies, plans, evaluations, and hydraulic infrastructure (DOF 2011). In 2019, the budget was USD 170 million.

Effectiveness on Actors in Terms of Mandated Goals

There was an agreement between the federal government, states and sectoral entities that managing stormwater and floods requires a metropolitan approach, as the metropolitan drainage infrastructure was entirely interconnected (Interviews-M14/M19). This coordination was efficient in controlling large infrastructure works for daily operations, with the three governments in constant communication (Interviews-M7/M14/M19). While respondents recognized the impressive engineering feat of managing stormwater in a highly challenging location, many criticized the lack of an integrated, long-term vision for water management and metropolitan planning (Interviews-M9/M14/M15/M46). SACMEX was "in the day-to-day, in constant battle", focusing exclusively on flushing water out of the city with pumps and pipes (Interview-M14). The CADAM, which should plan stormwater management at metropolitan scale, only addressed hydraulic infrastructure (SACMEX 2013) (Interview-M46).

Furthermore, the metropolitan scale was not always considered ideal to address stormwater challenges. Municipalities were responsible for drainage systems within their borders, but these were often clogged with waste, which aggravated flooding (SACMEX 2013) (Interviews-M5/M19). There were arguments for focusing stormwater efforts at local or sub-basin levels rather than exporting these problems. This would still require metropolitan coordination, but to a lesser degree. Supporters of such an approach argue for capturing surface runoff, treating it, reusing it when possible and returning it to the sub-basin or aquifer, thereby reducing dependence on external sources and limiting discharges downstream (Interviews-M40/M52).

Impacts on Sustainability and Inclusiveness

Rapidly expelling water from the city has prevented floods in recent decades, but tragedies such as a flood in September 2021 that killed 17 people downstream of Mexico City and affected over 70,000 indicate deep flaws in the stormwater management system. The mega-tunnel opened in 2019 is believed to have accelerated the floods (BNAmericas 2021). The metropolitan stormwater system presents risks across the MVMC, as infrastructure failures expose communities to highly contaminated floodwaters. This often affects informal settlements in the East of the MVMC, disadvantaged by historical power relations and their peripheral location (Interviews-M27/M57). With limited financial resources, these municipalities struggle with their stormwater responsibilities (Interviews-M6/M15).

While many industries moved out of the MVMC, domestic wastewater was still mostly discharged untreated (Interview-M40). As stormwater mixes with sewage and diffused pollution, water contamination has worsened in the metropolis and mega-infrastructure exports this problem to the neighbouring basin (Interview-M52). These waters were then reused for irrigation in the Tula basin, contaminating surface and groundwater (Interview-M58). Flood risks have also increased with soil subsidence, which inverted the slope of canals that discharge water out of the basin (Interviews-M50/M57). Diesel pumps push these waters out, raising costs and emissions (Interview-M40). Siltation adds further risks and costs.

Although the system was meant to foster a regional approach, in practice the governors and finance departments of each federal entity determined the destination of funds, and these often went to projects of local and personal interest, rather than metropolitan scale projects (Interviews-M15/M17/M33/M57). Negotiations on metropolitan measures between governments at state and federal levels were even more complicated between different political parties (Interview-M7). While the CADAM supposedly managed this coordination, it was not transparent and did not involve other regional entities (e.g. basin councils, other metropolitan commissions) or public participation (Interviews-M15/M17/M28/M46/M48/M52).

8.4.3 Metropolitan Wastewater Infrastructure

Design

In 2017, only 10% of Mexico City's wastewater was treated, despite dozens of treatment plants spread across the MVMC (Ambiental 2017). To address the lack of sewage treatment in the MVMC, the Atotonilco wastewater treatment plant was inaugurated in 2017 at the receiving end of the enormous stormwater infrastructure and the Tula River (Ambiental 2017). This strategic location allowed for easily capturing wastewater and stormwater from the entire metropolis. The treatment plant was part of CONAGUA's planning for the VMB and received funding from the Fideicomiso 1928 (SACMEX 2013). The plant discharges treated waters into the Tula basin's irrigation canals. Its treatment capacity of 35 m^3/s make it the largest in Latin America and the treated discharge can be more safely reused by farmers to irrigate 80,000 ha than the previously untreated flows (SACMEX 2013; Tamargo 2016; Ambiental 2017).

Effectiveness on Actors in Terms of Mandated Goals

Most treatment plants in the MVMC were underused or even abandoned due to under-investments. The Atotonilco plant has significantly increased the volume of wastewater that was treated before being discharged *in natura*. Respondents claimed that building large sewage treatment plants was more attractive to local politicians than rehabilitating existing ones, as they often bring in substantial federal funds (Interviews-M4/M15/M40/ M42). However, federal funds do not cover operational costs. Moreover, despite the Atotonilco plant's direct effect on basin management, sanitation was not discussed within or between basin councils, nor between basin councils and state and municipal governments, and the metropolitan commissions (Interviews-M17/M28/M48).

Furthermore, critics argued that this mega-plant reproduced the linear approach of taking, using and expelling water. It perpetuates the dependence on water imports, as large volumes of sewage and stormwater leave the VMB. These critics argued that sewage treatment capacity should have been distributed across the VMB to allow for wastewater recycling and reuse and groundwater infiltration (Enciso 2018) (Interviews-M9/M32M40/M52). Recycled wastewater was mainly used to irrigate parks and fill the Xochimilco canals, although it could be used more extensively (e.g. in industrial processes, to infiltrate into aquifers or for toilet flushing) (Interview-M40). Demand for treated wastewater must be created for wastewater reuse initiatives and policies to be successful (Interviews-M27/M44). However, the authorities seem hesitant to support such initiatives (Interviews-M6/M40/M44).

Impacts on Sustainability and Inclusiveness

In 2017, Mexico City treated only around 3 m^3/s, or 15% of total wastewater, significantly below the national average of 57.7% for municipal wastewater, and 38.3% for industrial and other wastewater (CONAGUA 2018). These numbers increased with the opening of the Atotonilco plant, but as of October 2021, the latter was operating at only 20% capacity. Although this increases the capital's wastewater treatment to roughly 10%, with positive effects on the Tula basin's surface and groundwater quality, the Atotonilco plant was far from a silver bullet. Building it was expensive, at approximately USD 630 million, several times the initially projected amount (*Acciona Agua* 2018). Maintaining operations costs around USD 5.5 million per month (Interview-M9). These numbers do not include the additional costs of building sewage pipes within the city and pumping this sewage out. The construction of the plant was also clouded by corruption allegations (Interview-M9). Moreover, residents living near the treatment plant in the Tula Basin had much higher rates of skin and gastrointestinal diseases (Peña et al. 2013; González 2018). Earlier studies revealed that irrigated lands in the Tula Basin received 44,000 tons of nitrogen and 17,000 tons of phosphorus yearly, causing severe risks of waterborne diseases and contaminated crops (Burns 2009).

8.4.4 The Conservation Land

Design

Within Mexico City proper, an area known as the 'Conservation Land' occupies 59% of the city's surface area, across nine southern districts (Escamilla and Santos 2012). This area was defined in 1976 through a Federal District law, modified in 1982, which classified the territory in either urban or conservation land (Federal District Legislative Assembly 1982). This zoning category established a zero-growth land use policy (Aguilar and Santos 2011). The Conservation Land holds high ecological value and its preservation is considered crucial for the city's sustainable development (SEDEMA 2013). Half of the Conservation Land is covered in forests, a third by farmlands, 12% by grasslands and 11% by urban uses (Escamilla and Santos 2012). The Conservation Land hosts over 1800 flora and fauna species and provides important ecosystem services for the MVMC, including climate regulation, water provision (57% of the water consumed by the city came from aquifers in this area), soil and water retention and contaminant reduction (Escamilla and Santos 2012; SEDEMA 2013). The Conservation Land is part of the green belt around the MVMC and extends across state borders.

The Conservation Land requires collaboration between sectoral entities and districts and a common land use management vision. Local authorities implemented conservation measures, such as monitoring activities, forest fire prevention, land use planning and PROFACE (see Sect. 7.4.4). The state and federal government also established small-scale Environmentally Protected Areas across the MVMC, including within the Conservation Land (SEDEMA 2016). These areas were particularly valuable in terms of biodiversity and had individual plans and strategies. About 30% of the Conservation Land was within environmentally protected areas, with their own advisory councils (SEDEMA 2013).

Effectiveness on Actors in Terms of Mandated Goals

This instrument recognizes the value of the area's ecosystem services. However, the region is under pressure from continued urban growth. The district governments are responsible for enforcing zoning regulations, but lack capacity and political will (Connolly and Wigle 2017) (Interview-M17). Residents pressured local politicians to legalize informal settlements, aggravating corruption and clientelism (Interviews-M17/M26). For instance, district authorities changed zoning regulations without informing the Mexico City government (Interviews-M6/M7). This prevented alternatives that addressed the affordable housing gap while preserving the Conservation

Land, such as housing programmes with higher densities and better access to basic infrastructure and services in less ecologically valuable areas. It also meant that official land use maps were often inaccurate (Escamilla and Santos 2012).

Disjointed environmental and urban policies within Mexico City lead to incoherent planning, as sectoral departments for conservation, housing and agriculture have conflicting interests (Interview-M15). This ultimately hinders the city's ability to contain urban sprawl in the Conservation Land. Moreover, there is also no regional or metropolitan land use planning, and basin entities and metropolitan commissions are not involved, even though the green belt extends beyond Mexico City and the VMB (Interviews-M9/M17). This coordination is the role of the Environmental Commission of the Megalopolis. Additionally, 70% of the Conservation Land is communally held by traditional agrarian communities. Any intervention in these areas involves negotiations with (all) the landowners (Interview-M48).

Impacts on Sustainability and Inclusiveness

Despite land use restrictions and multiple conservation measures, informal settlements have expanded, causing environmental degradation with severe impacts for the city, including in terms of aquifer recharge (Escamilla and Santos 2012) (Interviews-M15/M46). Informal urbanization of 2700 ha in the Conservation Land may have prevented the infiltration of around seven billion litres of water per year (Macdonald 2016). Forest cover could entirely disappear by 2030, further jeopardizing the city's groundwater resources (Hernández 2016) (Interview-M29). Although more investments were planned for 2021/2022, funds had to be partially diverted for responses to the Covid-19 pandemic (Santiago 2021). Moreover, weak land use regulation and monitoring enabled the expansion of illegal logging and intensive agriculture, causing groundwater pollution from pesticides (Interviews-M48/M50). In some cases, conflicts erupted between communities in the Conservation Land and local authorities (Interview-M33). Although multiple agencies at Mexico City and district levels were involved, non-governmental voices were not included in planning.

8.5 Instrument Assessment and Redesign

UWM in the MVMC was disconnected from environmental management and the real cost of water services (see Table 8.2). Although the basin agency, Mexico City and Mexico State closely coordinated mega-infrastructure that expelled stormwater and wastewater from the basin, their approach was technical, top-down and linear (Interviews-M6/M7/M19/M40).

Redesign

Water tariffs: Subsidies for low-income households are crucial to ensure affordability, as the population has unequal abilities to pay. However, subsidies do not benefit the MVMC's most marginalized residents, as utilities are not allowed to provide services in informal settlements. In the MRSP, special programmes to provide

Table 8.2 Assessment of UWM policy instruments in the MVMC

	Instrument design	Effects on actors	Impact
Water tariffs	[0] No overarching framework leads to large disparities Many utilities use subsidies for low-income residents	[0] Tariff design promoted inclusion of low-income residents through block rates and subsidies Lack of metres and cost-recovery reduced investments Informal settlements were excluded	Env: [−−] Excessive consumption and high leaks Soc: [0] High access to drinking water except in informal settlements. Low quality Econ: [−] Utilities rely on external funds, transferring costs to taxpayers Rel: [−] The most marginalized rely on water trucks at higher prices
Metropolitan drainage system	[−] Coordination between state governments for infrastructure funding and operation	[−] Metropolitan management is needed as infrastructure is interlinked. Coordination does not involve regional planning. Disconnect with local level hindered efforts	Env: [−] Combined sewers contaminated potentially reusable rainwater Soc: [−] Residents exposed to flood risks are also at risk from contaminated waters Econ: [−−] Pumping water out is costly Rel: [−] Investments depend on political negotiations
Metropolitan sewage infrastructure	[0] Mega- sewage plant built down-stream of MVMC, centrally managed Treated water can be reused for irrigation	[0] Sewage treatment increased with new mega-plant, but its location prevented reuse of treated wastewater within the basin	Env: [+] Positive effects were expected on environmental quality in Tula basin Soc: [0] Farmers and local residents are negatively affected Econ: [−] Expensive and unclear if operation costs are sustainable. Potentially more efficient alternatives exist Rel: [−] Top-down management with no voice for farmers, residents

(continued)

Table 8.2 (continued)

	Instrument design	Effects on actors	Impact
Conservation land	[+] 60% of Mexico City under special zoning and land use restrictions Cross-sectoral planning Limited to Mexico City	[0] Weak enforcement of regulations due to lack of capacity and political will. Fragmented policies further weaken efforts	Env: [−] Degradation from urbanization, deforestation and agriculture continued Soc: [0] Lack of tenure of residents led to precarious living conditions Econ: [−] Local politicians and landowners saw opportunities in developing land, but degradation had severe costs in longer-term Rel: [0] Cross-sectoral planning limited to government agencies

Relative assessment scores ++ Very positive; + Positive; 0 Neutral; − Negative; −− Very negative (see Sect. 2.4)

drinking water, and when possible sanitation, have expanded access in established informal settlements, through cooperation between utilities and local governments. This could be considered in the MVMC. Besides benefits to residents, it can reduce commercial water losses (i.e. clandestine connections). Furthermore, as subsidies extended to around 75% of Mexico City's population, a review of criteria to qualify for these is necessary, as many households may have the capacity to pay. This process requires the creation of an independent regulatory agency to set criteria and approve tariff rates, transferring this mandate from the State Congress. The design of an overarching national policy framework for Wat&San could establish minimum standards and general goals.

In addition, the tariff system should reflect the social, economic, relational and environmental impacts of importing water from other basins and over-exploiting aquifers (see Sect. 7.5). Criteria for designing tariffs are based on local socio-economic characteristics to ensure affordability (in formal settlements) but do not consider water availability within the basin and the effect of tariffs on water demand. A more independent process of setting tariffs could set rates that would improve cost-recovery for utilities and reduce excessive consumption. This could incentivize repairs of leaks and the installation of water metres where there are none and water saving equipment. The reduction or elimination of subsidies for households with the capacity to pay could lead enhance investments. A portion of the added revenue could also fund PES programmes or conservation measures in donor basins, as is done in Mexico State.

Macro-drainage: It is necessary to shift from linear to more circular stormwater management, while remaining realistic about the basin's propensity to flood. Efforts could be made to retain surface runoff within the Conservation Land through preservation, and in the rest of the MVMC through a combination of decentralized grey and green infrastructure (e.g. bioswales, green roofs) adapted to the local context (i.e. available space, permeability of soil, slope). Small-scale initiatives to capture, filter and inject surface runoff into the ground have increased in recent years to simultaneously address groundwater over-exploitation, soil subsidence and recurrent floods. Retaining water within the basin could also reduce the costs of pumping storm and wastewater out of the basin. These could be scaled up within a regulatory framework with strict norms for the infiltrated water quality. This decentralized infrastructure could be managed locally or at sub-basin level to adjust to local needs and to be treated and reused or returned to streams or aquifers rather than to be funnelled downstream. Decentralized stormwater management should be part of a regional plan that considers basin and aquifer dynamics, with coordination between the MVMC's municipalities and state governments and shared financing. Nevertheless, centrally managed artificial drainage exits will continue to play an important role because of infrastructure lock-in. In addition, the concentration of rainfall in a short timeframe makes it more difficult to store water (Interview-M28).

Mega-sewage: Storm and wastewater are currently combined and expelled together, and as with stormwater, efforts are needed to retain, treat and reuse wastewater within the MVMC. Separate sewer and stormwater systems could be installed in new developments or when replacing old pipes, but the total replacement of combined systems is prohibitively costly and unrealistic in the short-term. Initiatives for decentralized storm and wastewater treatment, recycling and reuse already exist, and have great potential for non-drinking uses such as school toilets and industrial processes (CCRAC 2011) (Interviews-M4/M52). Similarly, such measures require a clear normative framework. Users, such as industries, were often reluctant to shift to water reuse because this meant giving up their water extraction permits and thereby the security of stable access to water. Introducing flexibility within water use permits could facilitate transitions to water reuse while maintaining a water allocation guarantee, thereby enhancing environmental sustainability and avoiding the alienation of large users.

A mega-sewage plant was recently built downstream of the MVMC, despite most sewage treatment plants within the metropolis operating significantly below capacity. The new mega-plant is also underused. Investing in the existing, smaller plants rather than the mega-plant would have reduced the need for transporting sewage across the entire MVMC. The mega-plant disincentivizes measures to retain and treat storm and wastewater within the basin. Nevertheless, smaller plants within the basin could still play a role as not all wastewater is collected and treated. Expanding the network has to do with the financial capacity of utilities, but also issues of land tenure, as many unconnected households are in informal settlements. This justifies a greater involvement of local governments within regional wastewater planning.

Conservation land: Efforts are fragmented as the Conservation Land is limited to Mexico City, and there was no regional approach for preserving the MVMC's green belt. However, coordinated actions between Mexico City and surrounding states are crucial for preserving ecosystem services. More effective coordination could be achieved by harmonizing the states' land use management plans and conservation policies. Funding for a regional approach could be linked with PES programmes or compensation mechanisms. Such an idea has emerged from civil society actors, through the Water Forest Conservation Strategy, which promotes a regional approach to conservation and sustainable land use management. Part of this strategy is to raise awareness about the interdependencies between the megacity's ecosystems and the water it uses. Whether through local or regional efforts, conservation of these ecosystems would require engaging local rural dwellers, including indigenous communities, whose livelihoods are threatened by urban encroachment. As land is largely privately owned, the government may need to buy land from landowners in areas identified as significant in terms of ecosystem services and at risk of development (Intervew-M9/M15). In most cases, cooperation with landowners to ensure sustainable land use practices is the more pragmatic approach.

Ultimately, political leadership and political will are necessary. In 2018, Mexico City's newly elected governor committed to a fivefold increase in the Conservation Land's budget and to double resources for PES and compensation programmes. This presented an opportunity for Mexico City's leaders to engage surrounding regions for regional conservation efforts. Although the Covid-19 pandemic resulted in a partial diversion of these resources, investments in the Conservation Land have increased and led to the rehabilitation of at least 200 ha of land (Arguelles 2021).

Missing Instruments

UWM is currently excluded from urban policy and planning. To achieve greater social inclusion, assessments could be conducted to identify informal settlements that can be legalized based on criteria such as the settlement's age, viability of installing basic infrastructure and environmental impacts. Residents in areas considered at risk or highly valuable in terms of ecological functions should be relocated as part of a just regional housing strategy. Urban upgrading programmes coordinated between water utilities, sectoral departments and local authorities could lead to joint efforts to expand sewage collection. In addition, in upgraded and newly built neighbourhoods, storm and wastewater systems could be separated and managed more locally. As climate change will bring more intense precipitation, this can mitigate floods of contaminated waters. Funds could partially come from construction companies.

Moreover, rainwater harvesting initiatives have multiplied and complement water supply for non-drinking uses or for aquifer recharge. It was increasingly common in marginalized peri-urban areas without piped water but with space for rainwater tanks and filtering systems. This could significantly reduce dependence on water trucks, at least during the rainy season, and lower household expenses. While there were legitimate concerns about water quality, rainwater could be used for less noble uses such as toilets and watering gardens. The popularity and success of such initiatives indicated they had potential to be scaled up. Although it is estimated that the volume of rainwater that enters the MVMC's drainage systems is greater than the volume of water imported from the Cutzamala System, only a fraction of rainwater could realistically be captured by rainwater harvesting systems (Sosa-Rodriguez 2010) (Interview-M40). This is due to the surface area needed to capture rain and the irregularity of rainfall. While rainwater harvesting systems may make only a dent in reducing groundwater over-exploitation and water imports, it could be significant for households and users with precarious access to water, in particular those currently depending on water trucks.

References

Acciona Agua (2018) La EDAR de Atotonilco (México), la mayor planta de tratamiento de aguas residuales del mundo, cumple un año desde su puesta en marcha, 25 July. Available at: https://www.acciona-agua.com/es/salaprensa/a-fondo/2018/julio/la-edar-de-atotonilco-méxico-la-mayor-planta-de-tratamiento-de-aguas-residuales-del-mundo-cumple-un-año-desde-su-puesta-en-marcha/. Accessed 21 May 2019

Acevedo B et al (2013) The institutional structure of water policy in the Mexico City metropolitan area. Available at: http://www.hardrainproject.com

Adler D (2015) The war for Mexico's water. Foreign Policy, July

Aguilar AG, Santos CC (2011) Aguilar, Adrian Guillermo Escamilla, Irma Herrera. In: Aguilar AG, Escamilla IH (eds) Periurbanización y Sustentabilidad en Grandes Ciudades. Estudios Urbanos, Serie, Mexico City, D.F., pp 277–317

Ambiental T (2017) Detectan irregularidades en los recursos invertidos en programa de la CONAGUA. Tecnología Ambiental, 28 July

Arguelles E (2021) Recuperan 200 hectareas de reserva ecologica em CDMX. Excelsior, 3 Mar. Available at: https://www.excelsior.com.mx/comunidad/recuperan-200-hectareas-de-reserva-ecologica-en-cdmx/1435816

BNAmericas (2021) Acusan que tubería de drenaje de Ciudad de México aceleró inundaciones en Hidalgo. BNAmericans, 14 June. Available at: https://www.bnamericas.com/es/noticias/acusan-que-tuberia-de-drenaje-de-ciudad-de-mexico-acelero-inundaciones-en-hidalgo

Breña CM (2020) La dificultad de cosechar la lluvia en Ciudad de México. El Pais, 23 June. Available at: https://elpais.com/sociedad/2020-06-22/la-dificultad-de-cosechar-la-lluvia-en-ciudad-de-mexico.html

Burns E (ed) (2009) Repensar la Cuenca: La Gestión de los Ciclos del Agua en el Valle de Mexico. UAM, Mexico City, D.F.

CAEM (no date) Misión, Visión y Objetivos—Comisión del Agua del Estado de México. Available at: http://caem.edomex.gob.mx/mision_vision_objetivos. Accessed 19 May 2019

CCRAC (2011) Plan Hídrico de las Subcuencas Amecameca, La Compañía y Tláhuac-Xico. Comisión de Cuenca de los Ríos Amecameca y La Compañía. Available at: https://comisiondelacuenca.mx/wp-content/uploads/2016/02/PlanHidricoSubcuencasAmecamecaCompaniaTlahuac-1.pdf

CDMX (2019) Inauguran Túnel Emisor Oriente que permitirá evitar inundaciones en la Ciudad de México y Edomex. Available at: https://jefaturadegobierno.cdmx.gob.mx/comunicacion/nota/inauguran-tunel-emisor-oriente-que-permitira-evitar-inundaciones-en-la-ciudad-de-mexico-y-edomex

Ciencias AMD (2018) Sobreexplotación, mal uso y elevado subsidio, entre los problemas del agua en la CDMX. Vanguardia, 23 Oct

CONAGUA (2018) Estadisticas del Agua en México. Mexico City, D.F.

Connolly P, Wigle J (2017) (Re)constructing informality and "doing regularization" in the conservation zone of Mexico City. Plan Theory Pract 18(2):183–201. https://doi.org/10.1080/14649357.2017.1279678

DOF (2011) Acuerdo por el que se emiten las Reglas de Operación del Fondo Metropolitano. Diario Oficial de la Federación. Mexico City, D.F. Available at: http://dof.gob.mx/nota_detalle_popup.php?codigo=5250215. Accessed 21 May 2019

Durán M, Gómez L (2018) Autoridades de la CDMX examinan la calidad del agua. Heraldo de México, 9 Nov. Available at: https://heraldodemexico.com.mx/cdmx/autoridades-de-la-cdmx-examinan-la-calidad-del-agua/. Accessed 20 Aug 2019

Enciso AL (2018) La obra de Atotonilco es una aberración, de acuerdo con experto. Jornada, 3 Sept. Available at: https://www.jornada.com.mx/2018/09/03/sociedad/036n1soc. Accessed 21 May 2019

Escamilla IH, Santos CC (2012) La Zona Metropolitana del Valle de México: Transformación Urbano-Rural en la Región Centro de México. In: XII Coloquio Internacional de Geocrítica, Bogotá, p 15

Evans ST (2016) Location and orientation of Teotihuacan, Mexico: water worship and processional space. Processions in the Ancient Americas, Penn State University Occasional Papers in Anthropology 33:52. Available at: https://journals.psu.edu/opa/article/download/60110/59995

Federal District Legislative Assembly (1982) Reglamento de Zonificacion para el Distrito Federal. Available at: http://www.paot.org.mx/centro/reglamentos/df/otrasdispjuridicas/zonificaciondf.pdf. Accessed 20 Aug 2019

Flores CC, Vikolainen V, Bressers H (2016) Water governance decentralisation and river basin management reforms in hierarchical systems: do they work for water treatment policy in Mexico's Tlaxcala Atoyac sub-basin? Water 8(5). https://doi.org/10.3390/w8050210

Godoy E (2015) Water politics polarised in Mexico. Inter Press Service, 22 Apr

González J (2018) Exigen cierre de planta tratadora en Atotonilco de Tula. Criterio Hidalgo, 29 Nov

Hackett E, Eakins H (2015) ISSC "transformations to sustainability" programme concept note water governance challenges, Mexico City and Phoenix introduction and sustainability challenge

Hardoy JE (1964/1973). Pre-Columbian cities. George Allen & Unwin Ltd., London. ISBN 0-04-722001-5

Hernández S (2014) Impera desorden en cobro de agua. El Universal, 23 Apr. Available at: http://archivo.eluniversal.com.mx/periodismo-datos/2014/-articulos-87397html-87397.html

Hernández L (2016) Mancha urbana devorará al sur; amenaza suelo de conservación | Excélsior. Excelsior, 22 Nov. Available at: https://www.excelsior.com.mx/comunidad/2016/11/22/1129619. Accessed 22 May 2019

Hidalgo C (2018) Legisladores locales revisan tarifas de agua para 2019. Milenio, 27 Dec

IMCO (2014) Guía para la creación de organismos metropolitanos de agua potable y saneamiento en México. Available at: http://imco.org.mx/wp-content/uploads/2014/03/AguaPotable.pdf. Accessed 19 May 2019

INEGI (2014) Cuaderno estadístico y geográfico de la zona metropolitana del Valle de México 2014. México, D.F. Available at: http://internet.contenidos.inegi.org.mx/contenidos/productos/prod_serv/contenidos/espanol/bvinegi/productos/nueva_estruc/valle_mex/702825068318.pdf. Accessed 10 May 2018

Macdonald LF (2016) IMPORTANCIA AMBIENTAL DE LOS SUELOS DE CONSERVACIÓN EN LA CDMX. Mexico City, D.F. Available at: http://www.paot.org.mx/micrositios/FORO_CONS_RN/pdf/mesa_2/Luis_Fueyo.pdf. Accessed 22 May 2019

Martínez ÁC, Enciso A (2015) Logros parciales de México en los objetivos del milenio. La Jornada, 5 Jan

Martinez S, Escolero O, Perevochtchikova M (2015) A comprehensive approach for the assessment of shared aquifers: the case of Mexico City. Sustain Water Resour Manag 1(2):111–123. https://doi.org/10.1007/s40899-015-0010-y

Mendoza HC (2016) The crisis of potable water in Mexico City: institutional factors and water property rights as conditions for creating adequate metropolitan water governance. University of Southern California. https://doi.org/10.25549/USCTHESES-C40-217168
Mexican Constitution (1917) Constitution of the United Mexican States. Available at: https://www.juridicas.unam.mx/legislacion/ordenamiento/constitucion-politica-de-los-estados-unidos-mexicanos. Accessed 18 May 2019
Mexico (Cámara de Diputados) (2012) Ley General de Cambio Climático. Available at: http://www.diputados.gob.mx/LeyesBiblio/pdf/LGCC_190118.pdf. Accessed 13 May 2018
Mexico (Camara de Diputados de la Nación) (2009) Ley de Uso del Suelo y Ordenamiento. Available at: http://www.cafedelasciudades.com.ar/imagenes96/Proyecto_AUGSBURGER_Ley_de_Ord_Territorial_Presentaci%F3n_2009.pdf. Accessed 13 May 2018
Moctezuma EM (1988) The great temple of the Aztecs: treasures of Tenochtitlan. 192 pages, 23 colour, 135 black-and-white illustrations. Thames & Hudson, London & New York
Navarro C, Gómez L (2018) Morosos, la mitad de los capitalinos. Heraldo de México, 31 Oct. Available at: https://heraldodemexico.com.mx/cdmx/morosos-la-mitad-de-los-capitalinos/. Accessed 20 Aug 2019
ODAPAS (2016) Reglamento Interno del Organismo Público Descentralizado Para la Prestación de los Servicios de Agua Potable, Alcantarillado y Saneamiento del Municipio de Ixtapaluca 2016–2018. Ixtapaluca, Mexico, p 35
OECD (2013) Making water reform happen in Mexico. In: OECD studies on water. Paris. https://doi.org/10.1787/9789264187894-en
OECD (2015) OECD territorial reviews: Valle de México, Mexico. OECD Publishing, Paris. Available at: https://books.google.com/books?id=cEO8CgAAQBAJ&printsec=frontcover#v=onepage&q&f=false. Accessed 20 Aug 2019
Peña MEDL, Ducci J, Zamora V (2013) Tratamiento de aguas residuales en México
Roa W (2018) Así se factura el agua en la CDMX. El Excelsior, 24 Aug. Available at: https://www.excelsior.com.mx/comunidad/asi-se-factura-el-agua-en-la-cdmx/1260692. Accessed 19 July 2019
Rodea F (2016) Hay familias en la capital que vive con 14 litros diarios de agua: UAM. El Financiero, 26 June. Available at: https://www.elfinanciero.com.mx/nacional/hay-familias-en-la-capital-que-vive-con-14-litros-diarios-de-agua-uam.html. Accessed 24 May 2019
Romero Lankao P (2010) Water in Mexico City: what will climate change bring to its history of water-related hazards and vulnerabilities? Environ Urban 22(1):157–178. https://doi.org/10.1177/0956247809362636
SACMEX (2013) EL GRAN RETO DEL AGUA EN LA CIUDAD DE MÉXICO PASADO: PASADO, PRESENTE Y PROSPECTIVAS DE SOLUCIÓN PARA UNA DE LAS CIUDADES MÁS COMPLEJAS DEL MUNDO. Mexico City, D.F.
SAF (2019) Fondo Metropolitano del Valle de México. Secretaría de Finanzas del Distrito Federal. Available at: http://procesos.finanzas.cdmx.gob.mx/Metropolitano/. Accessed 21 July 2019
Santiago D (2021) Sheinbaum anuncia 1,000 mdp para áreas naturales protegidas la CDMX. Expansion, 24 July. Available at: Sheinbaum anuncia 1,000 mdp para áreas naturales protegidas la CDMX (expansion.mx)
Score A (2010) Climate change adaptation policy for the water sector in Mexico. Climate adaptation knowledge exchange. Available at: https://www.cakex.org/case-studies/climate-change-adaptation-policy-water-sector-mexico. Accessed 10 June 2019
SEDEMA (2012) Programa de Gestion Integral de los Recursos Hídricos 2012–2018. SACMEX, Mexico City, D.F.
SEDEMA (2013) Primer Informe de Trabajo. Mexico City, D.F. Available at: http://data.sedema.cdmx.gob.mx/sedema/images/archivos/noticias/primer-informe-sedema/informe-completo.pdf. Accessed 22 May 2019
SEDEMA (2016) Suelo de Conservación. SEDEMA. Available at: https://www.sedema.cdmx.gob.mx/storage/app/media/Libro_Suelo_de_Conservacion.pdf

SEGOB (2014) Boletín Especial: Protección Civil. Mexico City, D.F. Available at: https://www.gob.mx/cms/uploads/attachment/file/138375/proteccion_civil_boletin.pdf. Accessed: 13 May 2018

SEGOB (2020) PROGRAMA Nacional Hídrico 2020–2024. Mexico City, D.F. Available at: https://www.dof.gob.mx/nota_detalle.php?codigo=5609188&fecha=30/12/2020

Sosa-Rodriguez FS (2010) Impacts of water-management decisions on the survival of a city: from ancient Tenochtitlan to modern Mexico City. Int J Water Resour Dev 26(4):675–687. https://doi.org/10.1080/07900627.2010.519503

Sosa-Rodriguez FS (2014) From federal to city mitigation and adaptation: climate change policy in Mexico City. Mitig Adapt Strateg Glob Change 19(7):969–996. https://doi.org/10.1007/s11027-013-9455-1

Spring ÚO (2011) Aquatic systems and water security in the Metropolitan Valley of Mexico City. Curr Opin Environ Sustain 3:497–505. https://doi.org/10.1016/j.cosust.2011.11.002

Spring ÚO (2015) Managing water resources in Mexico in the context of climate change. In: Shrestha S et al (eds) Managing water resources under climate uncertainty. Springer International. https://doi.org/10.1007/978-3-319-10467-6_18

Tamargo JLL (2016) Corrupción escandalosa detiene la Planta de Atotonilco. El Universal, 27 June. Available at: https://www.eluniversal.com.mx/entrada-de-opinion/columna/jose-luis-luege-tamargo/metropoli/2016/06/27/corrupcion-escandalosa

UN (2013) The millennium development goals report 2013. New York. Available at: https://www.un.org/millenniumgoals/pdf/report-2013/mdg-report-2013-english.pdf. Accessed 10 June 2019

Vaillant GC (1941) Aztecs of Mexico: origin, rise and fall of the Aztec nation. Doran, Garden City: Doubleday

World Bank (2013) Agua Urbana en el Valle de México ¿un camino verde para mañana? Washington, DC. Available at: www.bancomundial.org.mx. Accessed 8 May 2019

Chapter 9
Comparing São Paulo and Mexico City: Evidence of Scalar Mismatches

9.1 Introduction

This chapter compares the case studies of São Paulo and Mexico City to provide insights on how to best respond to water-related challenges in terms of quantity, quality and climate change, and through which policy instruments. First, it examines both cases in terms of the drivers of their water challenges (see Sect. 9.2). It then analyses the organizational set up of the institutions that respond to these challenges in each case (see Sect. 9.3). Subsequently, it describes the policy instruments employed in both case studies (see Sect. 9.4), and compares their effects on actors (see Sect. 9.5). This is followed by an evaluation of the instruments' effectiveness in achieving their stated mandates (see Sect. 9.6). It then compares recommendations for redesign and considers lessons that can be learned from each case study (see Sect. 9.7). Finally, the chapter summarizes the conclusions and introduces the concept of 'Scalar Mismatches', which, if unaddressed, impede sustainable and inclusive water governance in large metropolises (see Sect. 9.8).

9.2 Comparing Drivers of Metropolitan Water Challenges

Overall, similar drivers shaped water-related challenges in both São Paulo and Mexico City: Urbanization (associated with population growth), economic development and climate change.

Urbanization was the most influential driver in both case studies. Urbanization was characterized by nation-wide rapid although slowing rural to urban migration accompanying population growth especially concentrated in a few cities, and by the inability of cities to account for the daily influx of migrants. Unplanned urbanization has exacerbated inequalities, as informal, precarious settlements have mushroomed in peri-urban areas, leaving local authorities unable to cope with the high

F. van den Brandeler, *Scalar Mismatches in Metropolitan Water Governance*, Water Governance - Concepts, Methods, and Practice,
https://doi.org/10.1007/978-3-031-08061-6_9

demand for affordable housing and related services. At regional level, urban expansion encroached on surrounding areas of springs or aquifer recharge. Repercussions include erosion and increased floods and mudslides, water degradation and increased water demand. At local levels, soil-sealing and the occupation of floodplains and steep hill sides, as well as inadequate sanitation and solid waste management (i.e. clogged drains, water-borne diseases), exacerbate flood risks. The type of housing and land tenure also affect residents' access to water services and exposure to risks. Moreover, water use was often unsustainable and inefficient, particularly in the MVMC, with heavy losses and excessive consumption in wealthier parts of the city.

Economic development is an important driver for both Mexico City and São Paulo as they are their respective countries' economic centres. At national and regional levels, policies have prioritized economic growth over environmental preservation, leading industrial, mining and agricultural development at local and regional levels. These sectors are important water users, thereby driving tensions around the allocation of limited water resources. Contamination remains a challenge and diffused pollution is particularly hard to address.

Climate variability and change, including global and regional processes (e.g. deforestation of the Amazon, El Niño) may have significant influence on extreme weather events experienced in Mexico City and São Paulo. Both cases record increases in extreme weather events and in temperatures in line with climate change forecasts. The size of the two megacities also contribute to strong heat island effects.

The combination of rapid, unplanned urbanization and a focus on economic growth have led Mexico City and São Paulo to grow into wealthy mega-cities with rampant inequalities—including in access to safe drinking water, safety from contaminated water, and protection from water-related risks. Climate change could act as a risk multiplier to the existing challenges.

9.3 Comparing Actors and Institutions

9.3.1 Institutional Changes in UWM and IWRM/IRBM

Both Mexico and Brazil have made efforts to develop IWRM/IRBM over the last three decades, implementing water laws inspired by the Dublin principles. These legal frameworks recognize the multiple uses of water, introduce a multilevel governance system and create new policy instruments. IWRM is more advanced in São Paulo, where basin committees have broad representation, meet regularly, develop basin plans in a participatory manner and have budgets for projects within the basin. Even then, the ATB (Alto-Tietê Basin) Committee is often bypassed in decision-making processes, as happened during the 2013–2015 water crisis. In Mexico, basin agencies oversee WRM at regional level, but these are regional CONAGUA offices and are thus government agencies that implement decisions in a top-down manner. Participatory basin councils exist on paper but lack representation, resources and influence.

Regarding UWM, Wat&San responsibilities have been decentralized in both cases. Municipalities have been granted a wide range of mandates after Brazil and Mexico's democratic transitions, including land use, local environmental issues, stormwater, urban planning and civil defence. At higher levels, however there is often a vacuum. In Mexico, there is no national framework for Wat&San policy, which means states and municipal governments implement vastly different standards and goals. Brazil introduced a national Wat&San Law in 2007, but the role of states remains limited in the legislation, despite the crucial role of state water companies. Existing frameworks in both cases also do not specifically address metropolitan regions. A 2013 Supreme Court decision in Brazil stated that, within metropolitan regions, functions of common interest, including Wat&San services, should be a shared responsibility between municipalities and the State. In the MVMC, stormwater and flood management are, to some extent, coordinated at metropolitan level by national and state actors.

9.3.2 The Influence of Federalist Structures

While both are in federal regimes, Mexico concentrated more power at federal level, and Brazil at state level. Despite formal decentralization, local governments lack financial and human capacity to implement adequate measures. Higher levels decisions often bypass metropolitan/basin platforms and local actors, while local level decisions are often not coordinated between municipalities. This results in fragmented policies at regional level. In the MRSP, municipalities develop Wat&San plans separately despite interlinkages with neighbouring municipalities. This fragmentation has led state-level actors to take the lead. For instance, SABESP developed its own planning and investment priorities as there was no Wat&San state policy and it operates in a majority of São Paulo State's municipalities. The dependence on external water resources strengthens higher levels of government in both cases. As the MVMC spreads across three states and imports water resources from even further, the involvement of federal actors is inevitable. State level actors are more prominent in the MRSP, as most water resources are contained within the state, but dependence on other states increased as water supply systems expanded following the 2013–2015 water crisis.

Furthermore, the three-layered structure of federal regimes (i.e. federal, state and municipal) made it more difficult to legitimize any intermediary governance level (e.g. metropolitan, basin) (see Sects. 6.3.4 and 8.3.4). Institutions at such levels are mainly voluntary and rely on political will, consensus and alternative funding sources. Moreover, the MRSP is composed of 39 municipalities, and the MVMC of 60 municipalities and 16 districts in three federal entities, with vast differences in financial and human capacity. Differences in political party affiliations further aggravate regional fragmentation, as governments often prioritize politics and the interests of their jurisdiction over regional needs. Party politics hinder horizontal coordination (i.e. between municipalities), and vertically (i.e. between local and state government).

Nevertheless, the two megacities have developed metropolitan initiatives related to water and more. In the MVMC, this took a highly top-down form (e.g. metropolitan funds for mega-infrastructure), whereas the MRSP's initiatives were voluntary and often failed to materialize. Either way, developing a shared, metropolitan vision is challenging both in terms of structural design (how to create decision-making bodies) and resources to implement projects.

9.4 Comparing Instruments

9.4.1 Instruments per Country and Level of Governance

Table 9.1 lists the policy instruments selected in each case and the levels at which they are implemented. Brazil and Mexico have designed similar instruments, but sometimes at different levels. Unlike the MVMC, the MRSP has no (substantial) PES programmes. Suasive instruments, such as awareness campaigns for rational water use, exist in both cases but are considered of minor influence as they are generally

Table 9.1 Instruments at multiple levels of governance in Brazil and Mexico

	Brazil	Mexico
National	Water use permits Wastewater discharge permits[a]	Water use permits Water use and wastewater discharge fees Wastewater discharge permits[a] Classification of water bodies[a] Metropolitan water supply system Metropolitan sewage and drainage system
State	Water use permits Wastewater discharge permits[a] Areas of protection and rehabilitation Macro-drainage plan Integrated metropolitan system	Climate change plan and fund[a] Conservation land
River basin	Water use and wastewater discharge fees Classification of water bodies[a]	Water use permits
Municipal	Water tariffs: local governments are involved if the Wat&San services are provided by a municipal utility	Water tariffs
Other	Water tariffs: determined by the utility and the state regulatory agency	PES programmes: multiple: federal and state

Source Author
[a]See Annex G: Additional Instruments

implemented at a small scale and in a piece-meal manner. Relevant instruments that are absent in both cases include environmental taxes[1] and flood insurance.[2]

9.4.2 Evaluating and Comparing Instruments in Terms of Design

Water Use Permits

In both countries, water belongs to the nation or state, which grants access to users under similar specific conditions. In the VMB, the federal government issues permits through CONAGUA. In the ATB they are issued by the state government, through the DAEE, except for water import permits from basins that cross state borders (e.g. permit for imports from the Cantareira System), which are issued by the ANA. Brazilian and Mexican water permits both aim to control water use. In the VMB, the granting of new permits was restricted due to the Basin's low 'water availability zone', meaning that a user could only obtain a permit through a transfer. The ATB does not apply such restrictions. On paper, the MVMC's water use permits system is therefore more adequate for water use management in a water scarce basin.

Water Use Fees

Water use fees aim to incentivise rational water use while financing WRM. In Mexico, fees are set at national level and differ according to the basin or aquifer's 'water availability zone', increasing as availability decreases; the MVMC's basin has the highest fees. In principle, this may encourage certain water users, such as industries, to move to areas where water is more affordable. In São Paulo, the river basin committee sets water use fees through a participatory and deliberative process, in consultation with its technical boards. The two cities reinvested revenues in different ways. The MVMC only collected and reinvested water utility fees for storm and wastewater infrastructure in the VMB's. Investments were not returned to donor basins to compensate users or invest in ecosystem preservation. The Federal Treasury collected fees from other uses and reinvested these in various national programmes. In the ATB, all collected water use fees were reinvested in basin-scale projects defined in the basin plan.

Water Tariffs

In Brazil, national policy requires water tariffs to support the universalisation of water services while allowing for financial sustainability. There is no overarching framework for water and sanitation in Mexico, but the MVMC's water operators share comparable goals. In São Paulo, an independent state regulator approves water

[1] Brazil's National Water Law originally included compensation for environmental damage, but this was later vetoed. In 2018, a federal law created a fund for compensation from environmentally damaging activities (MMA 2018). This fund aims to support the management of Protected Areas.

[2] Flood insurance exists in both cases although it was not widely used. This research did not examine measures adopted at individual or household level.

tariffs,[3] unlike in the MVMC, where this is generally done by the State congress, making tariff-setting a political decision. In both cases, low-income households receive subsidies and different rates are charged for commercial and industrial consumers connected to the water utility. Tariffs rise exponentially as consumption increases, except for industrial and commercial users in the case of São Paulo. While some metropolitan municipalities had local utilities, most were serviced by a state-level utility.

Inter-basin Transfers/Metropolitan Water Supply Systems

Inter-basin transfers were created to meet water demand in the MRSP and the MVMC as local water supply sources became insufficient or inadequate. For both cases, these supply-oriented approaches are core strategies for responding to water demand. CONAGUA builds and manages these systems for the MVMC, with SACMEX's help. Both national and state entities are involved in the ATB, depending on whether the imported water comes from a basin that is at least partly located in another state. In Brazil, basin committees are involved in negotiations, but ultimately have limited influence. In the MRSP, water is mainly supplied from large reservoirs through inter-basin transfers within and beyond the ATB. These systems are interconnected, forming the Integrated Metropolitan System, so that, if one system is dry, the area supplied by it can receive water from another system. Mexico City also relies heavily on inter-basin transfers, which import both surface and groundwater from other basins.

Metropolitan Wastewater Infrastructure

In both cases, metropolitan wastewater infrastructure centralizes the efforts to collect, treat and discharge wastewater, based on the premise of greater efficiency through economies of scale. It represents a conventional approach to UWM. Again, CONAGUA (national level) is directly involved in decisions and operations, whereas SABESP (State level) oversees this infrastructure in the MRSP. SABESP has developed an integrated, metropolitan wastewater system, based on a few treatment plants, including one mega-plant. It takes an integrated view of the basin and metropolitan sanitation infrastructure and relies on municipalities and industries to connect to the sewage network. In the MVMC, a giant sewage treatment plant opened in 2017, addressing the significant backlog in sewage treatment. It pumps sewage water out of the VMB, and discharges treated effluents into the neighbouring Tula basin.

Macro-drainage Plan of Alto-Tietê

The Alto-Tietê Macro-drainage plan, led by the DAEE, aims to mitigate flood risks through an IRBM approach and broad coordination. It is integrated with basin planning and the basin committee funds certain projects. As the Alto-Tietê Basin and MRSP almost overlap, this scale facilitates a regional approach to urban/metropolitan macro-drainage and flood management. Drainage districts are further delimited according to sub-basin boundaries. The plan combines hard and soft measures,

[3] The design of tariff rates is nevertheless complex and lacks transparency.

including environmental rehabilitation. In addition, it emphasizes that flood management requires coordinating the urban stormwater plans of all the basin's municipalities, acknowledging upstream/downstream interlinkages. Although the CONAGUA in the MVMC coordinates metropolitan stormwater infrastructure, this is not based on basin management principles or a regional plan but focuses on rapidly expelling excess stormwater.

Environmental Protection Measures

In the MRSP, state legislation for APRMs regulates land use and development in areas crucial for water supply. The areas are delimited by sub-basin and aim to protect springs while improving living conditions. Municipal governments are expected to control land use within their borders and therefore bear significant responsibility for these laws. Within the MVMC, part of Mexico City proper, including the mountainous South that recharges the city's aquifers, is in 'Conservation Land', which restricts land use and development. The Mexico City government played a key role but depended on cooperation from district authorities. A limitation is that the ecologically valuable green region extends beyond the borders of Mexico City, but there is no regional or metropolitan land use planning for environmental conservation. While there are Protected Areas in Mexico City and surrounding states, these do not foster a regional approach to preserving water-relevant ecosystems. Mexico City's environmental and urban policies are also not coordinated, leading to inconsistencies between sectoral plans.

PES Programmes

Multiple PES programmes have been implemented in the MVMC, by different levels of government and with different strategies (e.g. strict rules on land use change or ability to develop sustainable activities, technical assistance, different payment frequency). They aim to protect ecosystem services, especially hydrological services, by supporting or compensating the providers of these services. As much of the land is communal, authorities must collaborate with a large number of individuals. No PES programmes are implemented within the MRSP.

9.4.3 Inferences

Most instruments were evaluated positively or as neutral in terms of their objectives. They emphasize environmental sustainability (e.g. preservation of minimum environmental flows, protection of green areas around urban areas) and social equity (e.g. subsidized tariffs for low-income households, compensation to landowners in the green belt). Although many instruments linked to climate change adaptation, this was generally not their primary goal. Moreover, infrastructural and planning instruments were mainly implemented at the metropolitan or basin scale, although they were managed at higher levels. Instruments focused on ecosystem protection were implemented at multiple levels, including through funding mechanisms, although their scope was local or regional.

Overall, responsibilities spread across multiple levels (e.g. legislation or policy framework at national or state level, land use management is a local responsibility, implementation and coordination takes place at regional level) and between actors with different interests. This has implications for how policy instruments are coordinated. Although the goals of UWM and IWRM/IRBM significantly overlap in both cases, they were managed separately. This is clearly illustrated by the absence of linkages between water use or discharge instruments (i.e. permits, fees) and water services instruments (i.e. tariffs), as IWRM/IRBM and UWM mandates and goals are designed at different spatial scales. Interactions between IRBM and UWM entities mainly focus on large-scale water supply infrastructure and stormwater/flood control.

9.5 Comparing the Effect of Instruments on Actors' Behaviour

This step involved evaluating how effective the instruments were in achieving their stated mandates.

9.5.1 Regulatory Instruments

Water Use Permits

In both cases, water availability within basins and aquifers determined water allocation through water use permits, but decisions were made at state or national levels with little transparency. The possibility of transferring permits between users in Mexico meant that these changed hands over time from farmers to industries and real estate companies, as the city expanded. The value of these permits spurred a black market. These permit transfers, combined with weak enforcement of regulations regarding abstractions, failed to reduce or even stabilize water use levels in the region. In fact, they enabled unsustainable urban and industrial growth.

In the ATB, there were no restrictions for allocating additional water volumes and permits were granted through lax criteria, even though large volumes were imported from other basins. In both cases, no effective measures limited who could obtain permits. Interlinkages with surrounding basins were ignored, and reliable groundwater extraction data was lacking. Moreover, water imports from other basins simply diverted water resources from rural to urban areas. Through water use permits, these cities gained legal control over water resources and thereby over areas far from their borders.

Environmental Protection Measures

The enforcement of Protected Areas, with restrictions on land use and land occupation, relied disproportionately on local governments in the periphery of the metropolises and was ineffective due to a lack of monitoring. These municipalities lacked capacity for monitoring and were under pressure from the local population to provide housing and urban infrastructure and services. In fact, informal settlements multiplied around these areas, with reports of clientelism by local politicians.

9.5.2 Economic Instruments

Water Use Fees

Bulk water use fees were low and lacking in both cases (e.g. they excluded agricultural users, despite their heavy water use), discouraging actors from internalizing the real cost of bulk water supply—as discussed in the water tariffs section below. Low fees reduced the effectiveness of the fees in both cases. However, water use fees in the MRSP were reinvested within the basin where water was extracted, and according to the priorities defined by the basin committee. This gave users a sense of shared ownership and better acceptance of fees, meaning that a progressive rise in the fees' value and its expansion to agricultural users was more likely to be accepted than in the MVMC. Industrial users in both cases were more likely to reduce use, due to the cost and the possibility of switching to recycled water and other water saving practices.

Water Tariffs

Water tariffs for water supply services were considered relatively affordable in both cases, with subsidized rates for low-income households, and even for most residents in the case of Mexico City. However, the lack of an independent regulator in the MVMC politicized the design of water tariffs. Heavily subsidized water tariffs and fixed tariffs due to a lack of water metres have led to very high consumption rates. As a result, cost-recovery was low, and Mexican utilities depended on external funds. This hindered their ability to invest in the sector, mainly affecting low-income peri-urban residents who received subpar services or none at all. Informal settlements in the MVMC did not benefit from subsidized tariffs as they were not connected to the public network, and they often depended on water trucks that charged exorbitant prices. In the MRSP, utilities such as SABESP had special programmes to provide drinking water services in informal settlements, although not all such neighbourhoods were included.

In the MRSP, significant funds were invested in expanding water supply, and profits were redistributed to shareholders. Critics argued that these profits were insufficiently reinvested in expanding wastewater services or reducing leaks. Nevertheless, SABESP's state-wide cross-subsidies have allowed it to expand services to peri-urban and rural areas, while applying the same tariffs (including subsidies for low-income households) regardless of the cost of service provision. All in all, the lack of transparency prevents an in-depth evaluation, but this indirect evidence, combined with high levels of water losses and the constant need to increase water supply, indicates that there is significant room for improvement in balancing the water use fee/water tariffs within SABESP.

Payment for Ecosystem Services

The value of the PES programmes in the MVMC is low compared to the opportunity cost for landowners to develop their land or log the wood. Nevertheless, at least around 2.5 million ha of forests have benefitted from the federal PES programme across Mexico. The Mexico State PES programme has stable funds as it receives part of the revenue from water tariffs, and this also created a direct link between water consumption and WRM. Although there is no PES programme in the ATB, the neighbouring PCJ basin has led to the reforestation of over 500 ha out of 35,000 ha needed. A challenge in both cases is cooperation with private landowners, as the economic benefits are minimal.

9.5.3 Infrastructural Instruments

Inter-basin Transfers

Overall, residents have high levels of water access in both metropolitan regions. Inter-basin transfers and the interconnection of supply systems have ensured relatively stable water flow to both metropolitan regions, although marginalized communities still have precarious access, especially in the MVMC. The integration of systems may lead to greater resilience in the face of droughts. However, in both the MRSP and MVMC, inter-basin transfers were managed by higher-level actors in a top-down manner. The lack of transparency was more pronounced in the MVMC, as basin and local actors were not involved in discussions. The focus on increasing water supply through inter-basin transfers (rather than reducing demand, rehabilitating local water sources or investing in alternative water supplies[4]) has increased dependence on bulk water suppliers, thereby further empowering these actors.

Metropolitan Wastewater Infrastructure

Despite investments in large-scale sewage infrastructure in both megacities, sewage treatment was low. Utilities could not collect and treat sewage in a conventional

[4] In both cases, industrial users seemed more likely to invest in wastewater reuse technologies if there were sufficient incentives to do so. While cost through fees and environmental norms can push them in this direction, they also needed a guarantee of stable access to water. They were otherwise reluctant to give up water use permits. In Mexico, respondents mentioned cases of industries or buildings switching to reuse/rainwater harvesting, but still being charged fees as they still had a water permit, and this can dissuade them from making these switches.

manner in informal settlements, as pipes could not be installed due to the physical layout and lack of land tenure.

In São Paulo, wastewater treatment plants often ran under capacity, due to the cost and practical challenge of transporting sewage through extensive piping networks. Municipal treatment plants performed worse in both cities because they lacked financial resources, as well as natural economies of scope and scale usually associated with these services. Neither very large nor very small sewage treatment plants were optimal in the more urbanized areas, but small-scale solutions were effective in the sparsely populated periphery of the MRSP.

However, the downstream location of both mega-plants prevented the reuse of treated wastewater within the metropolitan region or the replenishment of local waterways and aquifers. While the mega-plants expanded sewage treatment, this linear approach justified the continued import of water from other basins and was not an ecologically sustainable path.

Macro-drainage Plan (São Paulo)

Although this plan aims for a basin approach, the interactions between municipalities and the State (DAEE) were stronger than horizontal interactions (between municipalities). Tensions occurred between municipalities, especially those with rival political parties (e.g. construction of dykes that aggravate flooding for neighbouring municipality). Local mandates to address stormwater and flood challenges and differences in financial and human capacity further prevented inter-municipal coordination. The plan mainly reinforced the DAEE's power and its vision of flood prevention through large, hard infrastructure, and did not substantially help foster changes in urbanization or land use. Nevertheless, it fostered a common vision of flood risks and priority areas in the basin. While the MVMC coordinated the discharge of sewage and stormwater between CONAGUA and the three state governments, this was limited to financial decisions and the operation of infrastructure and did not involve a multi-stakeholder basin-oriented planning process.

9.5.4 Inferences

Large-scale infrastructure remains the preferred choice for state and national governments to address water quantity, water quality and climate change adaptation-related challenges. It requires significant coordination. While this allows them to (partly) fulfil their main goals (e.g. supplying drinking water, treating sewage), it also consolidates their monopolistic power and reproduces the linear approach of taking, using and expelling water, with little incentive for long-term sustainable water use.

In terms of economic instruments, water use fees are overall too low to reduce water use. When implemented, they were more effective when they were reinvested in the region where they were collected through an inclusive decision-making process. Even if the funds were dwarfed by those of other entities, it helped foster a basin-oriented approach. Regarding water tariffs, subsidized rates did not always increase access to services for marginalized residents. When low rates were applied to the majority of the population, they affected cost-recovery and thus the quality of services.

Although regulatory instruments were crucial for better oversight of water resources and crucial ecosystems in and around the two metropolises, their enforcement was limited. In addition, they did not address drivers of urban and economic growth. Suasive instruments played a minimal role. After the 2013–2015 water crisis in São Paulo, there was greater awareness of the region's relative water scarcity and water use remained below pre-crisis levels. The experience of the crisis and the use of bonuses and fines for reductions or increases in water consumption were more effective than awareness campaigns for water saving.

9.6 Comparing the Impact on Sustainable and Inclusive Water Governance

In this step, I assessed the impact of instruments in terms of the four dimensions of sustainable and inclusive development, in the context of existing driving forces.

9.6.1 Ecological Impacts

Environmental criteria were often unambitious or weak. For instance, water use permits in São Paulo mainly relied on quantitative standards for surface water (i.e. the Q7,10), disregarding water contamination and groundwater extraction. In Mexico City, there were restrictions on granting new water use permits, but water was already over-allocated and monitoring was lacking. The disconnect between water use fees and tariffs from water availability levels, especially in the MVMC, hindered more rational water use. Industrial plants still obtained permits through transfers, to extract large volumes of water, for low fees, from over-exploited aquifers. Furthermore, as utilities were often not allowed or able to provide Wat&San services in informal settlements, free flowing sewage entered nearby streams. Industrial contamination decreased as industries moved to nearby regions in both cases, but this shifted contamination to other basins, some of which supply the megacities through inter-basin transfers.

The instruments address the symptoms of metropolitan water challenges, but not their drivers, in particular uncontrolled urbanization and a focus on economic growth at the expense of environmental sustainability. Both cases studies revealed that WRM was disconnected from land use management, environmental protection and urban planning, in part because these were the mandates of actors at different governance levels. Ultimately, the instruments reflect a mismatch between ecosystems and where their services are used, which led to importing water from increasingly further away, and ignoring high water losses, contamination and encroachment on surrounding green belts.

9.6.2 *Social Impacts*

Water for human consumption was prioritized over other uses in both cases. Water tariffs were subsidized for low-income households. Cross-subsidies in the MRSP support poorer, peri-urban municipalities. Nevertheless, marginalized residents in informal settlements still have limited access to reliable and affordable drinking water. During São Paulo's water crisis, poorer households in the periphery experienced severe water shortages, while residents in central areas more rarely experience dry taps. This issue was more severe in Mexico City, where residents in peri-urban areas continuously struggled to access water and relied on water trucks, often paying much higher prices. Further from the metropolitan area, indigenous communities were cut off from water sources after these were incorporated into inter-basin transfers. Meanwhile, industries still expanded their water use, despite protests from surrounding communities that depended on the same aquifers. Similarly, instruments addressing water-related risks mainly focused on technical fixes and did not address residents' socio-economic vulnerability or uncontrolled urbanization that lead to the occupation of floodplains and hillsides.

9.6.3 *Economic Impacts*

Water use permits grant users security and enable economic development. However, unsustainable water allocation harms the economic prospects of future generations and of the areas where more water is imported from. In Mexico City, land subsidence from over-drafted aquifers also damaged buildings and infrastructure. Water availability and the risk of water depletion were not effectively incorporated within a regional, long-term economic strategy.

Water use fees can redistribute gains across the basins and recognize the value of ecosystem services. In São Paulo, fees were reinvested within the basin where water was abstracted, in projects selected by the basin committees. In the MVMC, these fees disappeared within a federal budget or were invested in large-scale infrastructure through a top-down process. The ineffectiveness of wastewater discharge fees led to heavy economic costs due to reduced water availability and the potential waterborne diseases it can cause.

Regarding water tariffs, some respondents in São Paulo argued that profits were redistributed to shareholders or invested in increasing water supply, rather than invested in reducing water losses or contamination. In Mexico City, tariffs were too low for cost-recovery. The cost of mega-infrastructure for water supply was borne by taxpayers nationwide. The MRSP, on the other hand, de facto subsidizes many of São Paulo State's municipalities.

Large infrastructure (e.g. mega-sewage plants, inter-basin transfers, flood management works) was often very costly. No comparative analyses were carried out (or disseminated publicly) to evaluate if these measures were the most economically efficient.

9.6.4 Relational Impacts

Decision-making on UWM and WRM lacked transparency, prioritized short-term political and economic interests and excluded marginalized communities through policies and actions focused on importing water, exporting storm and wastewater and ignoring issues of environmental preservation or land tenure. It led to conflicts with indigenous communities and users in donor basins, who had no say in the process. Besides water tariffs, most instruments did not aim to put the needs of the most vulnerable first, but even subsidized tariffs often failed to benefit marginalized residents. Clientelism was common between local politicians and residents in the MVMC and between municipal utilities and residents in the MRSP. In Mexico City, CONAGUA's control over water use and wastewater discharge permits and fees hindered more local level involvement. Participatory river basin planning was virtually absent. In São Paulo, basin and local actors had more influence, but ultimately, state-level actors dominated decision-making processes. Partly, more advanced implementation of IWRM/IRBM may be due to the near overlap of the ATB and MRSP's territories and being contained within one state, creating fewer coordination challenges. In addition, water use fees increased the ATB committee's budget and thereby its influence.

Mega-infrastructure played prominent roles in both cases and reinforced the power of state and national actors. Some respondents argued that mandates and funds should be transferred to local actors, but besides the lack of technical capacity, it was not clear how they could address their regions' collective water-related challenges.

9.6.5 Inferences

Overall, instruments applied in the MRSP performed better in terms of design, effect on actors given their mandates and impacts on inclusive and sustainable development than those in the MVMC, although there was room for improvement in both cases. Social inclusiveness was greater in the MRSP, which had high levels of access to drinking water and cross-subsidies that supported access in poorer municipalities. However, in both cases, the patterns of economic growth that reproduced socio-economic inequality and unequal exposure to water-related risks remained unaddressed. Negative ecological impacts were clear in the MRSP through water contamination, and in the MVMC through land subsidence from groundwater over-exploitation. Even instruments specifically designed to address this, such as PES programmes or environmental protection areas were mostly ineffective. In particular, such instruments did not address the root causes of environmental degradation, namely informal urban growth that encroaches on green areas and a lack of basic infrastructure and services. In terms of economic impacts, many instruments favoured short-term outcomes and supply-focused approaches disregarding (not-so-distant) future economic consequences. They were often economically inefficient, preventing cost-recovery, or with a budget too low compared to the needs (i.e. PES programmes). Finally, power largely remained in the hands of state or national level actors. Table 9.2 provides scores based on the evaluation of Mexico City and São Paulo's instruments in terms of their design (see Sect. 9.4), their effect on actors' behaviour (see Sect. 9.5) and their impacts in terms of sustainability and inclusiveness (see Sect. 9.6).

9.7 Comparing Redesign

Finally, based on the instruments' effects on actors and impacts in terms of sustainability and inclusiveness, I conclude about the lessons that each city can learn from the other in terms the (re)design of policy instruments within urban water and river basin governance regimes.

9.7.1 Macro-Drainage

A participatory, basin-scale plan such as the MRSP's Macro-drainage Plan, which strengthened metropolitan-wide collaboration and more context-appropriate responses in the MRSP, could benefit the MVMC as the current focus on expelling storm and wastewater ignored basin and local considerations. However, the MRSP's plan mainly focused on hard infrastructure, and more emphasis could be shifted to drivers of vulnerability such as land use, land tenure, affordable housing and climate change. Green infrastructure was rare despite its potential for flood mitigation and other co-benefits (e.g. leisure, climate regulation, groundwater recharge). In the MVMC, small-scale infrastructure to infiltrate or harvest rainwater were more common. These measures mitigate floods and retain water within the basin for reuse or aquifer recharge, alleviating the pressure on over-drafted aquifers and external water resources. While aquifers were not under similar pressure in the MRSP, retaining stormwater through green infrastructure would alleviate over-burdened grey stormwater infrastructure during heavy rains and reduce diffused pollution in waterways. Scaling up these measures in both cities could involve incentives and the revision of norms, combined with risk assessments.

Table 9.2 Evaluation of the effectiveness of instruments in the MRSP and MVMC

	Brazil			Mexico		
	Design	Effect on actors	Impact on Sust&Incl	Design	Effect on actors	Impact on Sust&Incl
Water use permits	+	–	Ecol: – Soc: – Econ: – Rel: –	++	–	Ecol: –– Soc: –– Econ: – Rel: ––
Water use and wastewater discharge fees	++	+	Ecol: + Soc: ++ Econ: + Rel: ++	0/+	–	Ecol: –– Soc: –– Econ: –– Rel: ––
Water tariffs	+	++	Ecol: –– Soc: ++ Econ: – Rel: –	0	0	Ecol: –– Soc: 0 Econ: – Rel: –
Metropolitan water supply		+	Ecol: –– Soc: + Econ: – Rel: ––	0	0/+	Ecol: –– Soc: –– Econ: –– Rel: ––
Metropolitan wastewater infrastructure	+	0/+	Ecol: – Soc: – Econ: – Rel: –	0/–	0	Ecol: – Soc: – Econ: –– Rel: –
Macro-drainage	++	0/+	Ecol: – Soc: – Econ: –– Rel: –	–	–	Ecol: – Soc: – Econ: –– Rel: –
Environmental protection areas	+	0	Ecol: + Soc: + Econ: + Rel: +	+	0	Ecol: – Soc: 0 Econ: – Rel: 0
PES programmes				+/++	0/+	Ecol: – Soc: 0 Econ: + Rel: +

Relative assessment scores ++ Very positive; + Positive; 0 Neutral; – Negative; –– Very negative (see Sect. 2.4)

Moreover, the MRSP's Macro-drainage Plan did not increase coordination between municipalities. This could be addressed if local governments adjust their stormwater plans to be coherent with the macro-drainage plan. As an incentive, this can be a pre-requisite for local governments to apply for funding from the basin committee for stormwater and flood management-related projects. In the MVMC, three state governments and the federal government coordinate infrastructure across the metropolis, but municipalities also did not coordinate with each other. Regional, long-term strategic planning that includes local authorities, metropolitan infrastructure and basin hydrology could better address underlying vulnerabilities. In the Brazilian experience, this was more effective if local governments were incentivized or required to make local stormwater and flood mitigation plans coherent with regional planning. Measures can be implemented at sub-basin level (as in MRSP) to adjust to local needs and return surface runoff to streams rather than funnelling it downstream.

9.7.2 Metropolitan Wastewater Systems

Despite relative water scarcity within the ATB, there were few efforts to promote recycling and wastewater reuse. Treated wastewater was reused more often in the MVMC, for urban and rural agriculture, industrial use and the replenishment of canals. This can reduce pressure on (blue) water systems. However, the downstream locations of mega-sewage plants in both cities discouraged measures to retain treated wastewater within the basin for reuse or replenishing streams (or aquifers through artificial groundwater infiltration). (Treated) wastewater reuse could be promoted through larger and smaller treatment plants distributed according to population density and managed at sub-basin level. Smaller plants could especially expand sewage collection and treatment in informal settlements through recycling and reuse.

Many sewage plants in the MRSP operated under capacity due to legal and physical limitations to connect households in informal settlements to the sewage network. This was particularly true for the mega-sewage plant, surrounded by informal settlements. Expanding sewage treatment services was less an issue of treatment capacity but of connecting households in informal settlements and installing sewage mains. This requires utilities to coordinate with local governments, responsible for land use and housing. The latter need greater support in enforcing land use restrictions and in upgrading informal settlements. Within the APRMs, municipalities could receive technical and financial support for the upgrading of settlements from the basin committee, conditional on their harmonization of local plans with the basin or regional Wat&San plans. Expanding decentralized wastewater treatment—and therefore reuse—in the MVMC was also more likely to succeed if utilities and local governments coordinated within a regional wastewater management plan.

In addition, water users were often reluctant to take the risk of switching their water supply system to treated wastewater. Incentives to large users, such as industries, could help, including lower water use fees, easier access to water use permits or a water allocation guarantee (e.g. if the alternative system was inadequate for a user's needs, they can reclaim their permit for bulk water supply within a set deadline).

9.7.3 Water Tariffs

In both cities, water tariffs were overall affordable and included of low-income residents through block rates and subsidized tariffs. In the MRSP, water tariffs allowed for cost-recovery. However, subsidized tariff rates in the MVMC were granted to most households, including those with the capacity to pay full rates, which crippled utilities' revenues and ability to invest in the services. It also led utilities to depend on federal funding, meaning taxpayers subsidized water services in the country's wealthiest region. Adjusting eligibility criteria for such tariffs would allow utilities to better recover costs and encourage more rational water use and the installation of water metres. Additional revenue could further support leak repairs or service expansion to areas currently relying on water trucks. The MVMC could learn from the MRSP, which has legal frameworks at national and state levels to promote minimum standards and goals for access to drinking water and an independent regulator. In addition, programmes for informal settlements in the MRSP have expanded access to drinking water and reduced commercial water losses. This could also be implemented in the MVMC through collaboration between local authorities and utilities. In addition, coordination between utilities and local governments regarding land use and land tenure could facilitate the identification of settlements to be legalized, upgraded or serviced through alternative means.

Furthermore, SABESP applied the same tariff rates and subsidies in all the municipalities it operated in through cross-subsidies, enabling service provision in unprofitable areas. However, the tariff structure did not consider water availability within the basin or aquifer. The exception was when bonuses and fines were applied during the water crisis of 2013–2015 to incentivize water saving. In addition, local water utilities that obtained bulk water from SABESP did not always pay for this and likely disregarded water availability in the broader basin when setting their tariffs. This disconnect by SABESP and local utilities from the water resources that they depend on could be addressed by incorporating environmental impact criteria and compensation mechanisms into tariff-setting.

The tariff design in the MVMC also ignored water availability within the basin, the cost of importing water and the effect of low tariffs on water demand. The State Congress was not accountable to the basin and was incentivized to apply widespread subsidies as a political tool. A more independent process of setting tariffs (e.g. through a regulator) could improve cost-recovery for utilities and increase investments, but also incentivize rational water use and better match the reality of the basin.

9.7.4 Water Permits

Water permits in both metropolises prioritized water for domestic consumption and preserved minimum environmental flows. In the MRSP, they were a pre-requisite for obtaining an environmental license and in the MVMC they required an environmental impact assessment. In the MVMC, there were stricter regulations than in the MRSP for obtaining a water permit as a basin-wide restriction zone imposed a moratorium on issuing new permits. Nevertheless, permits in the MVMC could be transferred between users. The introduction of such a moratorium could be considered for the MRSP, but the case of the MVMC highlighted the importance of an accountable regulatory system. Unregulated permit transfers and weak land use and building regulations fostered urban expansion in peri-urban areas, which increased water demand. This showed a disconnect between WRM at basin or aquifer level and urban planning at local level. In some cases of over-allocation, it may be preferable for the relevant authority to terminate a permit rather than to allow its transfer to another user, especially if it is not for a high priority use (e.g. drinking water).

In addition, water-centric Strategic Environmental Assessments could help decision-makers identify where new projects are viable without causing environmental or social impacts. This would go beyond current Environmental Impact Assessments as it would allow region-wide impacts in a continuous manner and support sustainable planning processes. Such regional planning that includes stakeholders from the multiple basins inter-connected in metropolitan water use, the water allocation authority, users and local governments could lead to concerted action to ensure that urban and economic growth considers resource limitations and benefits local residents. More groundwater data and the integration of groundwater into participatory basin management could also lead to more sustainable groundwater use.

9.7.5 Water Use and Wastewater Discharge Fees

Although they promoted rational water use and considered water availability, the value of water use and wastewater discharge fees in the MRSP remained low. Increasing these fees, at least for certain larger users, would help reduce water demand and increase the basin committee's budget for basin-wide projects. In Mexico, these fees were set at national level and varied per 'availability zone'. The value of fees in the MVMC was relatively low, despite its location in a low availability zone. Requiring the use of water metres for bulk water users could increase rational water use and the revenue collected. Applying sanctions to users who lack metres or use water illegally would also incentivize compliance (as non-compliance is now met with virtual impunity) and finance intermittent inspections. Although the largest water use was public supply, incentives for farmers to switch to more efficient irrigation systems could reduce water use in the MVMC's rural hinterlands and reduce pressure on aquifers.

In the MRSP, bulk water use and wastewater discharge fees were returned to the committee of the basin where they were charged. Thanks to these fees, the ATB committee's budget increased and renewed stakeholder interest and involvement due partly to a greater capacity to invest in projects of basin relevance. This system of keeping fees within the river basin and involving users and other stakeholders has potential for the MVMC. The MRSP's system increased willingness to pay of users and stakeholder engagement, leading to more sustainable and inclusive outcomes at basin level.

9.7.6 *Inter-basin Transfers*

For both the MRSP and the MVMC, inter-basin transfers were key to responding to rapidly growing water demand, and it was unrealistic that either city could stop importing water in the short to medium term. Nevertheless, the focus on inter-basin transfers as the main strategy to achieving water security disincentivized water demand management. As both cities were reaching the limits of the economic viability of importing water and as climate change forecasts indicate an overall decrease in precipitation, better water demand management was urgent.

In both cases, water imports transferred externalities to donor basins. In the MRSP, this led to tensions with the PCJ committee and its reluctance to renew transfer agreements. In the MVMC, there were conflicts with indigenous communities who lost access to water resources without prior consultation or any form of compensation. Basins exporting water to the MRSP receive compensation through water use fees. However, these fees do not fully compensate for externalities. Cancelling the discount on inter-basin transfer fees from the Cantareira System to the MRSP would more fairly reflect the costs transferred onto the donor basin. This could be done progressively to allow time for users to adjust (e.g. by investing in water saving technology). The additional funds could be reinvested in preserving the Cantareira System and sustainable activities around it, thereby economically supporting the donor basin and ensuring the preservation of ecosystem services that benefit the MRSP. In addition, actors in the PCJ basin, could be given a greater role in basin transfer negotiations through a regional platform, such as the Integrated Urban Development Plan.

Returning water use fees to donor basins, eliminating subsidized water tariffs for those with the capacity to pay, improving billing by installing water metres where they were absent and incentivizing investments into water saving and recycling technologies could reduce water demand and the pressure on donor basins. Returning water use fees to basins could also strengthen basin councils and give stakeholders in donor basins more voice in negotiations.

9.7.7 *Environmental Protection Measures*

In both cases, environmental protection measures were characterized by sectoral fragmentation and a lack of coordination between local governments and across levels of governments. In the MRSP, APRMs were managed at sub-basin level and struggled to contain informal urbanization, especially as land use management and housing were primarily municipal mandates. It was therefore essential that the sub-basin committee, municipalities, state entities and Wat&San utilities coordinate their actions and goals and develop coherent planning. Municipalities that update their Wat&San plans to align with the basin plan could receive points in the FEHIDRO system, increasing their qualification for funds from the basin committee. Coherent plans would facilitate coordination between utilities and local governments regarding land use limitations for service provision, enabling solutions for informal settlements.

As in the MRSP, the preservation of Mexico City's Conservation Land involved multiple local governments (i.e. districts), but the green belt expanded far beyond the Conservation Land and there was no inter-state cooperation. A regional vision and plan that englobed the MVMC's green belt was designed by NGOs but failed to attain broad government support. However, such a plan could guide investments for PES programmes and from revenue generated by water use fees, as well as facilitate coordination between state conservation programmes. As in the MRSP, managing funds and coordination activities could be the responsibility of a regional body (e.g. a basin organization or an inter-state conservation body). Although PES Programmes still have relatively small budgets, they can support over-burdened local governments and strengthen collaboration across administrative boundaries.

9.8 Inferences

In both case studies, rapid and unplanned urbanization, economic development prioritizing the interests of industry and the economic elite, and extreme weather events aggravated by climate change were the major drivers of metropolitan water challenges. Developing coherent UWM and IWRM/IRBM that leads to changes in actors' behaviour and that affects inclusive and sustainable development was challenging in both the MRSP and MVMC. In part, this was due to the many jurisdictions involved and the need to align mandates at different levels, spatial scales and sectors. Instruments were sometimes disconnected from the water cycle (e.g. water tariffs focusing on affordability but ignoring water availability) or failed to include marginalized communities within or outside the metropolis. Many suggestions for redesign therefore relate to the spatial scale of instruments and the challenge of overcoming mismatches between the scale of their design and the scale of their impacts. Harmonizing plans at different levels and between sectors is crucial in that regard. Importing water from other basins remains necessary in the short and medium term but externalities could be accounted for through regional planning that incorporates inter-linked basins and gives a voice to all stakeholders, and through compensation mechanisms, which would also incentivize more water demand measures. Drainage and wastewater management require a metropolitan plan for interlinked infrastructure, but stormwater and wastewater can often be managed at sub-basin level, retaining water closer to its source. Although the MRSP's river basin and urban water governance regimes are overall more effective in terms of inclusiveness and sustainability, policy redesigns were recommended for both cases and each had relevant lessons for the other.

Reference

MMA (2018) Publicada lei da compensação ambiental. Ministério do Meio Ambiente, 29 May

Chapter 10
Towards a Theory of Metropolitan Water Governance

10.1 Introduction

Large metropolitan regions are multiplying and increasingly facing challenges in terms of water quantity, quality and adaptation to risk from extreme weather events. Although they generally have more resources to address these challenges, their complexity increases exponentially with their size, as the number of actors and institutions involved directly or indirectly in metropolitan water governance also multiplies. The failure to reconcile metropolitan water governance across urban and river basin scales can translate into water policies that lead to a 'scalar mismatch'. This can create redundancies that enhance resilience within water governance, as different actors and instruments work towards similar goals through different means. This can help cope with the impacts of climate variability and change. However, overlapping and disconnected policies and a lack of coherence and coordination between levels of government and between urban and basin scales can render measures ineffective or lead to externalities and conflicts over limited resources.

This research investigates river basin and urban water governance in relation to metropolitan water challenges through the following overarching question: **How do interactions between drivers and institutions at different spatial and institutional scale levels shape metropolitan water challenges, and how can policy instruments from river basin and urban water governance frameworks be (re)designed to foster more sustainable and inclusive metropolitan water governance?**

Section 10.2 reviews the lessons learned from the case studies in terms of the scalar mismatches that influence sustainable and inclusive metropolitan water governance. Section 10.3 highlights important considerations for metropolitan regions in federal states around the world. Following this, Sect. 10.4 identifies five elements that need to be considered individually and jointly to address scalar mismatches in metropolitan water governance. Section 10.5 then assesses the implications of this research for global policy and the SDGs.

F. van den Brandeler, *Scalar Mismatches in Metropolitan Water Governance*,
Water Governance - Concepts, Methods, and Practice,
https://doi.org/10.1007/978-3-031-08061-6_10

10.2 Taking Stock from the Experiences of São Paulo and Mexico City: Scalar Mismatches in Metropolitan Water Governance

The case studies illustrate the challenge of developing sustainable and inclusive metropolitan water governance as multiple levels of government and spatial scales are involved. This challenge translates into mismatches across scales.

10.2.1 Managing Bulk Water Supply and Water Contamination at the Multi-basin Scale

Applying the spatial unit of the river basin to managing bulk water supply and water contamination does not internalize the two megacities' externalities on neighbouring basins. Both the MVMC and MRSP are contained within one river basin (although the VMB is much larger than the MVMC, whereas the MRSP and ATB roughly overlap). However, the impacts of the two metropolitan regions in terms of water use and water contamination extend far beyond their basins and artificially interlink several basins. Water availability is insufficient: Between 39.4 and 47.9 m^3/s are imported into the ATB from other basins out of the 81.25 m^3/s that are used; 20 m^3/s are imported to the VMB out of 88 m^3/s used and over 27 m^3/s are estimated to be over-extracted within the VMB's aquifers (see Sects. 5.4.1 and 7.4.1). In both cases, higher levels of government intervene to guarantee water to the megacities from other basins, thereby transferring the cities' water scarcity to donor basins, with impacts on the latter's communities, economies and ecosystems. Similarly, contamination from the MRSP flows down the Tietê River far beyond the borders of the ATB, while contaminated waters were reused in irrigation in the VMB's neighbouring basin until a large treatment plant opened in 2017. In fact, the impact of large cities extends much further as they rely on 'virtual water' through the import of food and other goods, and local changes in the hydrological cycle can have repercussions far away. This regional approach may not fully account for these impacts.

A regional scale that includes all concerned basins can better account for these externalities. It is unrealistic for cities such as Mexico City and São Paulo to reduce their demands and impacts on surrounding basins in the short or medium-term. Centralized management of large-scale infrastructure for water supply (e.g. inter-basin transfers, metropolitan-scale integration of supply systems) at the multi-basin scale will remain in the short- and medium-term due to infrastructure lock-in, but also as it allows for economies of scale and integrates water scarce areas and areas without the means (or scale) to develop local infrastructure. The MRSP can learn from Mexico's restrictions on issuing new permits in low water availability zones, which represents an effort to stabilize overall water use despite remaining concerns around irregular water use and regulatory enforcement. The MVMC could adopt a water use fees system similar to the MRSP's, where funds are collected by the basin entity and reinvested in projects that benefit the population and ecosystems within that basin. Wastewater discharge fees could also partly be used to compensate areas downstream of the metropolitan regions that receive contaminated water.

Nevertheless, the two cases show that **measures that reduce water demand, increase water use efficiency, integrate alternative sources of water and treat wastewater closer to the source can wean metropolises from their dependency on distant water resources and reduce contamination far beyond their borders**. Although they require long-term planning and regional coordination, they can often be managed locally. Decentralized management generally ensures more context-relevant responses and better user compliance (Xiao et al. 2018). These initiatives exist but receive little support from governments or utilities (e.g. rainwater harvesting for non-drinking water purposes, bioswales or artificial injections to infiltrate surface runoff). While certain initiatives had clear benefits, utilities lack incentives to reduce the volume of water they supply or of wastewater they treat, as this affects their revenue. It also requires actors to radically change their approach and is therefore met with scepticism.

In the longer-term, with these measures, the spatial scale of managing bulk and water supply and water contamination can be reduced. However, parts of these metropolises may never become self-sufficient in terms of water supply, and some degree of intervention through inter-basin transfers might remain necessary in the long-term (see Sect. 7.4.1). The MVMC is expected to face a deficit of 25.1 m^3/s of water by 2030, even after importing water from the Cutzamala and Lerma systems and over-drafting the local aquifers (World Bank 2013). These calculations do not account for the effects of climate change on precipitation, evaporation and water use. The over-exploitation of aquifers and the deficit would correspond to 51% of the water demand (of approximately 90 m^3/s). Reducing the pressure on local aquifers and addressing the remaining gap of 25 m^3/s could partially be done through reducing leaks (projected to be 28.1 m^3/s in 2030 in a business-as-usual scenario) and commercial losses (projected at 6.8 m^3/s in 2030) and other measures mentioned above. Nevertheless, these require significant investments by the government and households. Halting water imports from other basins would require supplying another 20 m^3/s through water demand management and alternative sources, which is unrealistic to be achieved by 2030 and even beyond.

10.2.2 Coordinating Surface and Groundwater Management

The river basin scale is also too restrictive for bulk water management and water contamination as it mainly focuses on surface water dynamics. The MVMC and MRSP cases highlight that **groundwater is not adequately integrated in basin management**. Significant uncertainty remains around groundwater dynamics. In the MVMC, groundwater use is very high, but little is known about recharge rates, groundwater flows and quality levels. In the MRSP, how much groundwater is used is unclear, but it is estimated to be much higher than official rates. Groundwater use could be expanded in certain areas, thereby decreasing water imports, but this would require more studies. The result is an absence of groundwater management in the MRSP and the absence of effective groundwater management within basin management in the MVMC. Although the MVMC has rules and restrictions to reduce groundwater over-exploitation, the relevant authorities lack capacity and incentives to enforce these (see Sect. 7.4.2).

The aquifer's invisibility hinders the sense of a shared resource among users across a larger spatial scale. It is harder to monitor, and users tend to feel that the water beneath their feet belongs to them. Water use permits and fees do not consider interlinkages between surface and groundwater. However, the effects of inadequate groundwater management can spread far beyond water extraction wells and lead to land subsidence and dried up springs and wetlands. In the MVMC, efforts to recharge local aquifers exist (e.g. artificial groundwater recharge, retention reservoirs in the mountains around the city). Their effects on the groundwater table are uncertain and depend partly on the speed of groundwater flows. However, as they also mitigate flood risks it is worthwhile to further assess such options and scale them up.

10.2.3 Retaining and Reusing Greywater: Semi-decentralized Stormwater and Wastewater Management

The approach to storm and wastewater has focused on metropolitan-scale infrastructure to rapidly evacuate these effluents, ignoring downstream effects. This centrally managed linear approach is disconnected from local mandates and basin dynamics. Local responsibilities are not made coherent with a basin vision and plan. This has led to flood control infrastructure in one municipality that aggravates flood risks for a neighbouring municipality. Erosion or inadequate sewage treatment upstream affects downstream municipalities. In addition, this linear approach **consolidates a dependence on water imports**.

The ATB's macro-drainage plan defines a shared vision with **differentiated measures at sub-basin levels** according to local characteristics, and vertical coordination between the state and municipalities to ensure that local stormwater measures are coherent with the regional plan. This combines local and metropolitan-scale measures on soil-sealing, sewers and the occupation of floodplains, as well as basin-scale measures to reduce erosion and siltation upstream and maintain or rehabilitate wetlands and floodplains and retain stormwater upstream. Smaller-scale measures (e.g. warning systems, civil defence emergency measures, green infrastructure) can be more effectively implemented locally within a regional plan. To avoid a lack of coordination between municipalities—resulting, for instance, in stormwater being pushed to neighbours—it is crucial that local stormwater plans are made coherent with the ATB's plan. Meanwhile, government agencies at higher levels can more efficiently coordinate large-scale infrastructure, such as deep drainage systems, and monitor large rivers.

While CONAGUA and state actors in the MVMC coordinate large stormwater infrastructure, this is not part of a regional plan that includes local governments or basin management principles. Land subsidence from groundwater depletion made certain areas more vulnerable to floods. Measures such as artificial groundwater infiltration, which mitigate floods and recharge aquifers, can be implemented in densely urbanized areas, while green infrastructure and environmental conservation retain stormwater in the surrounding mountains, recharge aquifers and reduce floods. Scaling up rainwater harvesting could also reduce surface runoff and provide water for non-drinking uses. The MVMC could therefore benefit from a regional plan such as the ATB's macro-drainage plan, which considers local needs, implements measures within sub-basin or aquifer units and promotes regional collaboration.

Wastewater management can also be better addressed through a multilevel framework. Decentralized wastewater management allows for discharging treated effluents closer to the source, recharging aquifers or replenishing local wetlands, or it can supply local industries and other users. Designing wastewater treatment measures at sub-basin level allows for better consideration of local characteristics to design context-appropriate measures (e.g. population density, local hydrology, potential demand for recycled wastewater, lack of land tenure and physical obstacles for connecting informal settlements to sewage pipes). Meanwhile, regional coordination remains needed as mega-treatment plants and large sewage mains will remain in place for decades, and the failure to treat sewage has a regional impact.

In the longer-term, retaining stormwater and (treated) wastewater within cities' sub-basins can restore ecosystems and reduce reliance on external water resources. As climate change will bring more irregular and heavy rains, such measures will enhance urban resilience.

10.2.4 Water Services Are Fragmented Across the Metropolis and Disconnected from Water Resources and Environmental Management

Whether the services are provided by a local or a state utility, metropolitan municipalities generally depend on shared water resources and inter-linked infrastructure. However, different utilities do not coordinate in terms of planning, which leads to fragmented infrastructure, inefficiencies and downstream impacts from inadequate sewage treatment. In the MRSP, municipal governments are responsible for developing Wat&San plans, even when they have delegated this service to SABESP. However, these plans are often not (adequately) designed and/or implemented, and SABESP de facto elaborates state planning for Wat&San. In addition, there is no requirement for neighbouring municipalities to harmonize their plans or seek synergies. However, as SABESP has control of planning it could—in theory—ensure greater coherence across municipalities. In the MVMC, the total vacuum in terms of Wat&San policy at national level meant there are no shared standards and goals between local and state governments. This hinders the ability of local actors to coordinate actions on shared challenges.

While municipal provision may enable greater proximity to consumers and knowledge of the local context, some degree of **regional planning and management can better address the reliance on shared water resources and infrastructure and the regional impact from a lack of wastewater collection and treatment**. This would allow for economies of scale (e.g. water purification and wastewater treatment plants can be shared between multiple municipalities), greater technical capacity (local utilities often struggle to hire and retain skilled professionals, especially in poorer municipalities) and cross-subsidies between municipalities. Moreover, in Brazil, the Supreme Court decided in 2013 that services of common interest (e.g. Wat&San services) should be co-managed by the state and municipalities within metropolitan areas. This is a recognition of the need for regional cooperation in Wat&San services, due to shared benefits and externalities. However, the ruling fails to address the reliance of Wat&San services on distant water resources and the externalities transferred beyond the metropolis.

Even with regional planning, utilities cannot control local factors such as land use and land tenure, which often restrict where they can provide Wat&San services. Utilities struggle to expand Wat&San services to informal settlements, and the effects are felt at regional level (e.g. leaks from clandestine connections, sewage discharge into local streams and reservoirs). This was even more problematic in protected areas, but environmental actors rarely collaborated with utilities and municipalities to find common ground. Overcoming this requires inter-sectoral coordination, sometimes between government agencies at different levels.

Furthermore, **water tariffs can be a tool for facilitating more equal access to Wat&San services at regional level**. Affordable tariffs in metropolitan regions are crucial to ensure social inclusion, but still need to reflect regional water availability and enable cost-recovery on average—where some consumers pay more to compensate for the lower costs to others. Cross-subsidies across São Paulo State enable SABESP to apply the same tariffs (including subsidized rates) to all its consumers. In the MVMC, tariffs are highly subsidized for most consumers, despite water shortages, which leads to irrational water use and transfers costs to future generation and donor basins, as well as to taxpayers nation-wide. The poorest did not benefit as they live in informal settlements and spend much more buying water from water trucks. The spatial scale for designing water tariffs therefore involves a solidarity component that considers who benefits and who pays within the utility's jurisdiction but also beyond it. The adequate scale for water tariffs also has a water availability component (i.e. how is water consumption affected, where is water taken from, at what cost). People and water resources beyond a utility's jurisdiction are often affected by its water tariffs but excluded in their design. The presence of an independent regulator at regional or higher level was therefore found to be crucial for sustainable and inclusive tariffs (see Sect. 9.7.3).

10.2.5 Linking Water Management, Ecosystem Services and Land Use/Spatial Planning

A great challenge is the **disconnect between both UWM and IWRM/IRBM and land use and spatial planning**. Despite some attempts at regional planning, mandates for land use and spatial planning are mainly local, while WRM responsibilities are at basin, state and national levels. However, the effectiveness of WRM instruments depends significantly on local factors. In particular, land use and building regulations and commercial or industrial developments—generally overseen at local levels—drive increases in water demand. Corruption, clientelism and weak human and financial capacity at local level facilitate unsustainable economic activities and irregular urbanization, including in protected areas. Harm to ecosystem services is directly linked to such land use changes, yet there is no regional planning that adequately addresses it. Strategic Environmental Assessments are not conducted at basin or aquifer scale, even when impacts on water resources are significant. Some respondents argued for the creation of a regional planning body with influence over land use and zoning.

Both the MVMC and MRSP have protected areas that are crucial for producing ecosystem services. For municipalities (or districts in the case of Mexico City) that have a significant part of their territory under protected status, this represents significant restrictions on their urbanization potential, economic development and tax revenue. **The metropolis (usually) relies on peri-urban or rural municipalities to monitor and prevent irregular occupations and preserve vital ecosystem services—even when these municipalities lack financial and administrative resources and economic incentives to do so**. Measures that acknowledge the reliance of the metropolis on ecosystem services from its rural hinterlands and align incentives (e.g. laws and zoning regulations, PES programmes) are limited, unenforced or absent.

Land ownership further complicates land use and spatial planning. In Mexico, most land is privately owned, making government interventions difficult. The land of indigenous communities and communal landowners protected by the National Agrarian Law, requiring municipalities to deal with federal institutions for any land and water-related matters in these areas. Moreover, many residents of metropolitan municipalities work in the urban core, where they consume water, produce wastewater, and are exposed to flood risks. Despite this, they have no influence on these issues and do not pay for the costs they create, as they live and vote in the neighbouring municipalities (and even neighbouring states in the case of the MVMC).

10.3 Implications for Metropolitan Systems in Federal States Around the World

Urbanization in Mexico and Brazil is expected to continue in the next decades but at a much slower rate than experienced during the second half of the twentieth century. Other countries, especially in Asia and Africa are still in earlier stages of this radical transformation of their societies, and many of their cities are expected to mushroom in the coming decades. For those in federal states, in particular, the cases of Mexico City and São Paulo can provide a number of lessons for addressing metropolitan water challenges. The particularities of metropolitan regions in federal regimes have implications for the provision of urban water services, for water resources management, the link between the two and the role of land use management.

10.3.1 Metropolitan Areas in Federal States

Metropolitan 'Government' Is Usually Not an Option

Federal regimes have more than one level of government, each with specific responsibilities defined within the Constitution. Arrangements at intermediary levels (e.g. metropolitan, basin) generally cannot have administrations with executive power. Unitary states can more easily redefine territorial boundaries for different purposes (e.g. creation of Grand Paris as a metropolitan area). Therefore, metropolitan regions in federal states usually cannot form a metropolitan government with powers similar to those of other levels of governments.

The Federal Constitution of both Mexico and Brazil define metropolitan regions and highlight the importance of regional action on issues of common interest. In 2013, the Supreme Court in Brazil stated that, in metropolitan regions, the state government and municipalities concerned must jointly address functions of common interest. In Mexico City, a Law for Metropolitan Coordination aims to set guidelines for a metropolitan planning strategy. In both cases, these efforts were met with resistance by those who see it as a threat to the federalist regime or to the decentralization of power to municipalities that followed democratization.

Metropolitan water challenges are better addressed at different spatial scales and through different levels of government or non-state actors depending on the nature of the challenge (e.g. bulk water at multi-basin scale). Metropolitan entities are therefore not necessarily appropriate, and metropolitan regions can better address water-related challenges through a range of arrangements across multiple scales.

Multi-scalar Water Governance in Metropolitan Regions

The Constitution in a federalist regime largely determines the balance of power between levels of government. In Mexico, more power is concentrated at central level, whereas states have significant power in Brazil and many responsibilities have been decentralized to the municipal level. This contributes to shaping water governance regimes in metropolitan regions and the levels at which policy instruments are designed. The types of jurisdictions included within a metropolitan region also play a role. Both metropolises include multiple states through inter-basin transfers and therefore also the federal government. In the MVMC, the national government controls major metropolitan water, wastewater and stormwater infrastructure that crosses state borders. The complexity of metropolitan water governance regimes is thus much greater than the jurisdictional composition of the metropolitan regions themselves.

To address metropolitan water challenges, it is essential that the Federal government and Constitution define the 'metropolitan region', its boundaries and expectations regarding who carries out which functions at that territorial scale. Nevertheless, the metropolis is not always the ideal spatial scale to address metropolitan water-related challenges. Depending on the nature of these challenges, different spatial scales may suit policies and policy instruments.

10.3.2 Implications for Urban Water Management

Water services are often considered of local interest and carried out by local actors. Research shows that while some challenges and solutions within metropolitan water governance are local, many are shared, as externalities spread beyond jurisdictions. Three factors suggest that (partially) managing UWM functions at a higher level in metropolitan regions can promote more inclusive and sustainable development:

- First, many metropolitan regions face severe inequalities, including in access to reliable drinking water. Peri-urban municipalities tend to have less capacity for providing these services independently, and informal urbanization makes it more costly (and less attractive) to utilities. Fragmented service provision is also linked to clientelism at local levels. If managed at a larger spatial scale, Wat&San services can benefit from cross-subsidies that make water tariffs more affordable and enable provision in marginalized areas.
- Second, metropolitan regions tend to be densely urbanized and share water supplies. Municipalities can benefit from combining financial and human resources and scaling up the provision of certain services (e.g. shared water and wastewater treatment plants). Coordinating water supply also enables municipalities with insufficient water resources within their borders to respond to local water demand. As water resources are often imported from beyond metropolitan boundaries, managing certain aspects of Wat&San services (e.g. regulation) at higher levels can better address potential externalities.
- Third, water runoff is not restricted by administrative boundaries within metropolitan regions. Actions in one municipality may aggravate flood-related

risks in a neighbouring jurisdiction. Implementing basin principles and harmonizing local stormwater plans involve a certain amount of coordination at higher levels.

The autonomy of local governments, as well as competition and tensions between them, can hinder the formation of this type of collaboration. For example, **party politics between Municipal governments and the State is a hurdle to integrate Wat&San policies** in the MRSP. Scaling up services is also often associated with centralized, top-down decision-making. Arrangements where metropolitan municipalities must share responsibilities with the State government, as was pushed by the Supreme Court in Brazil, can allow for a balance between local autonomy and regional interest. The State (through a regulatory agency, for instance) can define criteria for municipal Wat&San plans and verify their compatibility and overall coherence at regional level. Instruments such as cross-subsidies and redistributive policies benefit from economies of scale and can reduce inequalities across the metropolis (and beyond, depending on the reach of the provider). While many actions can be decentralized and may be more effective at local level, such as billing, pipe repairs, monitoring of small-scale stormwater and flood infrastructure, or flood evacuation strategies, in large metropolitan areas the adequate level for certain water governance functions may not be the local level.

10.3.3 Implications for Water Resources Management

In federal regimes, authority over water resources is generally shared (to different degrees) between central and state governments. Local governments may be responsible for streams, small rivers and other water bodies contained within their borders. WRM therefore requires vertical coordination between government agencies at different levels. It also involves horizontal coordination between states and between municipalities. Furthermore, where IWRM has been adopted, river basin organizations (RBOs) must also navigate this complex web of interactions. RBOs have been more successful in Brazil, but, in both cases they have struggled to compete with state institutions as they lack executive power. In metropolitan regions, WRM can therefore involve the three levels of government, the basin level and the metropolitan level. In fact, this research highlighted that metropolitan regions often involve multiple river basins and sometimes more than one state through shared water resources and infrastructure.

The state or central government often manages large-scale water infrastructure in metropolitan regions. If water bodies belong to the state or federal domain, the involvement of higher levels of government is inevitable. To include the interests of communities and protect ecosystems in donor basins, it is crucial to also involve jurisdictions concerned with the transfer of water and the discharge of wastewater and stormwater within and beyond the metropolitan area. Mechanisms that ensure accountability and transparency are key, as higher levels of government may have interests and favour certain jurisdictions, both within the metropolitan regions and in relation to its rural hinterlands.

For sustainable and inclusive water governance, actors in a metropolitan region must assess which water resources and respective ecosystems are used or affected one way or another. Metropolitan regions can then develop plans and instruments that consider the spatial scale(s) for their design and interdependencies. For instance, returning funds from water use fees to the (sub-)basins where they originated from strengthens the ties between the metropolis and basin(s) it relies on. A sense of a shared regional identity with awareness of the role of surrounding ecosystems can help increase public pressure for regionally coordinated action.

Due to the interconnections between metropolitan regions and water systems beyond their borders, utilities and other UWM actors should be involved in WRM if the aim is to achieve sustainable and inclusive development. Water for human consumption is a priority for WRM, but mechanisms can ensure that utilities also strive for efficient water use. Flat rates, the lack of water metres and water tariffs that are highly subsidized for most households encourage excessive use. As the case studies demonstrate, this ultimately hurts the poorest within the metropolis and in its rural hinterlands, as they remain excluded from formal services and pay much higher prices. Involving a regulator or third party that links services and water resources would facilitate relational and social inclusiveness.

10.4 Understanding and Overcoming Scalar Mismatches in Metropolitan Water Governance

Addressing scalar mismatches in metropolitan water governance is necessary, but not sufficient, to increase sustainable and inclusive development. This research concluded that five key elements in institutional and policy design can enhance policy effectiveness.

10.4.1 Defining Metropolitan Water Governance

Due to their concentration of people and economic activities, metropolitan regions have a water footprint and impact on people and ecosystems that expands far beyond their own borders. Identifying metropolitan demands and impacts locally and on the surrounding (or relatively distant) rural hinterlands may foster a shared conceptualization of the metropolis and its interlinkages with water systems. The concepts of *metropolitan water* and *metropolitan water challenges* are not static, but rather moving targets, as cities expand and increase their demands and impacts. Over time, institutional frameworks and instruments can become irrelevant or ineffective. This is the case of water allocation quotas for Mexico City and Mexico

State based on the MVMC's characteristics in the 1970s (see Sect. 7.4.1). In addition, *metropolitan water* can involve different spatial scales, and therefore different stakeholders, institutions and policy instruments, according to which challenges are addressed. Both IWRM/IRBM and UWM embrace decentralization, albeit to different spatial scales (i.e. basin and municipal), and bringing these actors together within a metropolitan water governance framework therefore enhances complexity. IWRM/IRBM and UWM actors have different approaches and policy responses to water challenges, different types of knowledge and unequal power. This complexity is further enhanced by party politics across the metropolis and its rural hinterlands and between levels of government. Reluctance or inability to act at different levels can lead to a mismatch between legal and policy frameworks for cities and basins. The experience of RBOs, for instance, demonstrated that municipal representatives typically maintain localist stances and require incentives to act in the interest of the river basin. In metropolitan regions, this is further complicated by the tensions between regional interest and political-administrative fragmentation. In this complex scenario, there are no one-size-fits-all solutions.

Metropolitan water governance actors are those within the metropolitan region or at higher levels of government that act on the metropolis. However, it also includes actors of neighbouring areas that are interlinked in water governance—for instance, because they influence water resources exported to the metropolis, or shape land use management in areas that produce these water resources. To address scalar mismatches, metropolitan water governance requires:

- Identifying the relevant spatial scale to address various metropolitan water challenges;
- A sense of a common interest and a shared commitment to its challenges;
- A recognition of the importance of knowledge- and data-sharing at the relevant scale;
- Clear mandates and guidelines for joint action, when this is beyond an actor's traditional attributes. Due to institutional inertia, actors do not easily change behaviour;
- Developing planning mechanisms at the relevant spatial scales (i.e. municipal, metropolitan, macro-metropolitan, (sub-)basin, multi-basin) with synergies between local concerns and interests and those at larger scales, such as ecosystem protection. These plans are not an amalgamation of local plans but are coherent with these while taking the regional interest as a basis. This can be facilitated by creating a set of common standards at local levels (Table 10.1).

10.4.2 Considering Different Types of Water

Large metropolitan areas have heavy footprints on different types of water, including blue, green and grey water. Sustainable and inclusive metropolitan water governance must aim to increase efficiency of use of different types of water and to diversify among them.

Table 10.1 Scalar mismatches within metropolitan water governance are reflected within four key elements

Scalar mismatches	Description
Different types of water	The water resources that are currently used or impacted in the metropolitan region, or that could be in the future (see Sect. 10.4.2). Besides blue water (i.e. surface and groundwater), this also involves green and grey water
Infrastructure	Infrastructure, from small-scale infrastructure to metropolitan-wide and beyond (see Sect. 10.4.3). Even small-scale infrastructure may be interconnected with larger-scale infrastructure and not operate in isolation. It is important to consider how infrastructure in one area may impact other areas of the metropolis
Ecosystems	Water resources (of different types) are connected to the ecosystems that sustain them (see Sect. 10.4.4). Often, little is understood about these ecosystems and the crucial role that they play in metropolitan water governance
Land use and urban sprawl	Metropolitan regions' boundaries are often ambiguous, especially as they are constantly evolving, and this has implications for how they manage land use and contain sprawl (see Sect. 10.4.5)

Blue: Towards Multi-basin Coordination and Conjunctive Use

Bulk water supply is based on surface and groundwater from within and beyond metropolitan boundaries. Regions exporting (surface or ground) water to the metropolitan region are often sidelined in decision-making processes, as state and national-level actors make top-down decisions. Inclusive and sustainable bulk water management therefore requires a framework with fair rules and shared water allocation responsibilities that incorporates the relevant basins and aquifers and local stakeholder interests. It is also necessary to address differences in cost structures and subsidies that lead to water uses inconsistent with overall water resources optimization (e.g. subsidies on inter-basin transfers that make water imports cheaper than more sustainable alternatives) and disincentivize water saving (e.g. utilities' revenue depends on maximizing water sales). Efforts that could reduce megacities' impact on other basins through demand management measures include:

(i) consistent use of water metres (for bulk use and for retail services) that must be installed within a set deadline to avoid a fine;
(ii) leakage detection and repair programmes;
(iii) incentives for installing or retrofitting water saving equipment and the use of more efficient irrigation systems in cities' rural hinterlands; and
(iv) the use of alternative sources such as recycled wastewater and rainwater harvesting (for non-drinking water purposes, including artificial groundwater infiltration). This is important to "close the loop" at (sub-)basin level and reduce the spatial scale of bulk water use. This implies strengthening coordination with Wat&San utilities, the industrial sector and rural water users.

Although IWRM/IRBM promotes the integration between surface and groundwater sources, many large metropolitan areas that rely on both do not address these together. Reliable knowledge on groundwater resources (i.e. recharge rates, actual extractions and contamination risks) is limited, preventing adequate planning of groundwater use. Groundwater extraction infrastructure is also often privately owned, as opposed to infrastructure for surface water abstraction, and monitoring groundwater extraction is resource-intensive, preventing it from being effectively integrated in a regional strategy. Consistent application of fines for water use infractions could support more widespread monitoring and enforcement.

Moreover, the boundaries of aquifers and river basins do not necessarily overlap. Aquifers may spread over more than one basin and vice versa. However, IWRM/IRBM promotes river basins as the unit for analysis and management, adding to the uncertainty regarding how to manage groundwater resources. The consequence can be unregulated groundwater use, with conflicts arising once the wells run dry.

Overcoming the mismatch between surface and groundwater management requires conjunctive use that integrates all river basins and aquifers used and affected by the metropolitan region. This can enhance climate change adaptation, as diversifying water resources will provide reserves during droughts or relieve pressure on over-exploited aquifers (Pincetl et al. 2019). With conjunctive management, an allocation framework can limit water use permits to sustainable abstraction levels. During heavy precipitation, measures such as green infrastructure and artificial aquifer recharge can store groundwater and complement water reserves in dams (which have drawbacks such as evaporation, sedimentation, and the use of scarce land) (Porse et al. 2015). The effectiveness of this measure depends on aquifers' storage capacity and their potential discharge. In periods of water scarcity, users could trade surface water permits for groundwater permits. Conjunctive management may require a single institutional framework for both types of resources, with structures at multiple spatial scales.

Grey/Black: Switching Blue Water for Grey Water

As metropolitan water demand rises and the accessibility of affordable and sustainable blue water decreases, another alternative is to recycle and reuse grey (and sometimes black) water for uses that do not require high water quality. Many industries that require water for industrial processes (e.g. cooling) can use grey water with no effect on the quality of their output. Win–win arrangements are possible if they transfer their (blue) water use permits to users requiring higher-quality water. Public water utilities, for instance, can sell treated wastewater to these industries at a much lower fees or for free in exchange for their water use permit. A challenge is that the cost of abstracting and distributing blue water is often subsidized or even not paid for by the user, which makes a switch to grey water significantly less attractive. To increase grey water recycling and reuse it is important to price blue water adequately and fairly.

Many grey water use measures are small-scale (e.g. grey water in toilets, rebates on water saving technology) and can be implemented in (new) houses, hotels, office buildings, schools and factories; retrofitting is more expensive and may need to be subsidized. They often rely on private initiative but are much more likely to spread if there are incentives or regulatory measures that promote this shift (see Sect. 9.7.2).

Green: Towards Understanding the Links Between Land Use, Green Water and Blue Water

As cities expand, agricultural land, forests and other ecosystems become urban land. This affects blue water (e.g. disappearance of springs) and green water (soil moisture). The role of green water, including its links to blue water, is still often misunderstood or ignored by water managers, and it is generally not integrated within IWRM/IRBM strategies in practice. A better understanding of green water could reduce the need for irrigation or increase its efficiency and free up blue water for other uses. It would also highlight the link to land use management and the protection of crucial ecosystems. Although IWRM/IRBM promotes the integration of water and land, water managers typically have no influence on land use management.

10.4.3 Metropolitan Water Infrastructure

Thinking Small (When Possible): Incremental Shift Towards a New Approach to Water Use

To a large extent, the modernist paradigm that promotes society's control of nature through hard engineering and a linear approach to water remains in place. The financial model of water management institutions also favours a linear approach to water supply. Their revenues are mainly based on water sales, creating a structural disincentive to reduce water supply. Larger, consolidated utilities are more capable of revising revenue strategies. Incentives and regulations could be designed at regional level with a medium to long-term perspective.

Shifting towards a more socially inclusive, ecologically sustainable approach that attempts to 'close the loop' of the urban water cycle is essential. Reducing pressure on donor basins can decrease regional inequalities and environmental stress. This requires the adoption of small-scale (sometimes disruptive) solutions that can be scaled up (e.g. water metres, artificial recharge, reuse of treated wastewater, and rainwater harvesting), that empower local actors and strengthen a multi-scalar water governance approach. However, this is generally thwarted by institutional inertia, path-dependency, transitioning costs and stranded assets, and it is therefore not always realistic in the short-term. An incremental shift that combines larger and smaller-scale infrastructure can help slowly phasing out or reducing mega-works that are neither sustainable nor inclusive. Both large and small-scale infrastructure must be integrated in regional planning, as small-scale measures may affect the effectiveness of large-scale measures once scaled up. Although some large infrastructure will remain indispensable in established metropolitan regions, there is an opportunity to avoid mistakes in rapidly growing cities, especially in the Global South. In Global

North cities, there is a window of opportunity with aging infrastructure to switch to a new system.

Avoiding White Elephants: Wastewater Treatment

The linear approach to wastewater management affects the replenishment of waterways and aquifers within urban areas. Large wastewater treatment plants allow for economies of scale and mean that fewer plants are needed, at least in theory. However, they can also become white elephants, operating far below capacity as sewage from distant neighbourhoods does not reach these plants due to topography, the extensive piping needed and the inability (for practical or legal reasons) to connect informal settlements. Localized sewage treatment can be a solution for certain housing complexes and industries (especially as these are privately financed), but not across the city as it occupies extensive space. However, large metropolitan areas offer sufficient scale for several medium to large-sized treatment plants. It can make sense to distribute these by sub-basin or sub-region, so that treated wastewater is discharged closer to the source, reducing the length of pipes needed and recharging local stream flows. With sewage treatment plants spread out across the metropolitan region, opportunities also arise for wastewater recycling and reuse as treatment facilities are located closer to the demand. In addition, it can lead to savings in the transportation of both untreated and treated wastewater (plants are closer to wastewater producers and to potential users of treated effluents). It also means that treatment involves smaller quantities and can be adjusted to the level of treatment necessary for different uses. In the city, it can be reused in industrial processes and large buildings (e.g. toilet water) and in peri-urban/rural areas it can be reused for irrigation.

Combining Local and Regional-Scale Approaches: Adaptation (in Particular to Flood Risks)

As the proportion of urban land in high-frequency flood zones is increasing and climate change is making extreme weather events more frequent and intense, inadequate urban stormwater and flood management will have severe consequences (Güneralp et al. 2015). Adopting measures that allow for redundancy and fast responses that increase climate adaptation is critical. This is the case of small-scale measures such as green infrastructure, warning systems, and community networks, which can be context-relevant and managed locally. Green infrastructure (e.g. parks, bioswales, green roofs) can complement conventional stormwater systems, and provide additional ecosystem services (e.g. leisure, heat regulation, water quality regulation).

However, in metropolitan regions, actions in one area can cause or exacerbate flood risks in other areas (e.g. dykes that redirect floodwaters to neighbours, erosion upstream causing siltation downstream, regional and global processes such as climate change). Local adaptation measures must therefore be supported by a metropolitan- or basin-scale plan that identifies and addresses linkages and shared risks across the urban space. Such a plan can tailor strategies to different areas of the metropolis. For instance, grey infrastructure is more cost-effective in densely urbanized centres, whereas infiltration is viable in peri-urban areas. Municipal governments or private actors can implement small-scale measures, while larger-scale measures may require involvement from higher levels of government.

Municipalities are reluctant to spend resources for results outside their borders but may be encouraged by funds made available for actions of larger-scale impact identified within a metropolitan or basin plan. Funding could be based on the user-pays or polluter-pays principles (e.g. a risk creator-pays mechanism), where developers that remove vegetation, increase soil-sealing or enhance flood risks in any other way must pay proportional compensation. The use of water fees for basin projects, including flood-related projects, in the case of São Paulo shows that this has potential to mobilize stakeholders around common-interest actions. Adding stormwater charges to water bills could also be an option, even if only during the rainy season, although this represents an additional financial burden for poorer households. Construction permit approvals can require developers to plan for on-site stormwater retention. In Washington D.C., stormwater retention credit trading enables developers to meet their on-site requirements or buy 'credits' for stormwater retention from other developers who voluntarily retrofit their properties through a stormwater credit-trading programme (DOEE, no date; Cardona 2019). This creates flexibility for developers according to available space and relative cost of foregoing development for stormwater retention, and this can provide incentives for preserving green areas in poorer parts of the metropolis. These credits can only be sold within the (sub-)basin where they are generated. This project is now expanding to other cities, such as Chicago, which is also dealing with stormwater flooding and a lack of financial resources (DOEE, no date; Cardona 2019).

Without a regional approach, municipalities have no incentives to unilaterally adopt stricter building and zoning regulations, as developers can simply move to the neighbouring municipality. Therefore, regulations pushing for adaptation in urban and housing development must be enforced at metropolitan or regional level.

10.4.4 Sharing and Compensating for Ecosystem Services/Nature's Contributions

Ecosystem services produced within and around metropolitan regions are highly valuable as they support large populations and major economic centres. Ecosystem approaches can be integrated in both water services and risk management and in WRM. This can foster a system of sharing and compensating the use of ecosystem services.

Upstream Effects

Surface and groundwater bodies that depend on healthy and productive ecosystems sustain water supply to metropolitan regions. The role of these ecosystem services is often misunderstood and undervalued. Even in contexts of (relative) water scarcity, bulk water supply is often wasteful (e.g. heavy water losses, inefficient irrigation practices, high industrial water use for non-essential products). This is linked to the linear approach to WRM that assumes water's inexhaustibility and ignores ecosystems' limitations. Even when water allocation systems recognize the discrepancy between supply and demand and impose restrictions, these are rarely associated with environmental preservation measures.

Sustainable bulk water management requires defining the relevant boundaries of the aquifers and basins that supply water to the metropolis. Identifying and mapping these areas allows for assigning legal protections and responsibilities to enforce these. Although local actors can enforce these measures, recipients of ecosystem services (metropolitan water users and others) can fund these through water use fees, compensation mechanisms, Payments for Ecosystem Services and more. Moreover, to ensure sustainable water supply, water allocation should not exceed what these systems can provide sustainably. Preserving minimum environmental flows and recharge rates within water allocation regimes is therefore crucial.

Downstream Effects

Ecosystems upstream of the metropolis are crucial for metropolitan water provision, but those downstream are negatively affected by untreated wastewater, contaminated stormwater and other diffuse pollution. The spatial scale of this contamination may extend beyond the immediate river basin and stakeholders downstream. Pollution reduction mechanisms such as wastewater discharge fees often fail as they are weakly enforced or because they are cheaper than reducing contamination. Wastewater discharge fees are often reinvested in wastewater treatment within the metropolis or even in other areas and downstream jurisdictions do not necessarily receive any type of compensation. In some cases, compensation is obtained through courts. Provisions can also be implemented to redirect (part of) the revenue from wastewater discharge fees to affected jurisdictions.

Water and Sanitation Services

Wat&San providers lack incentives to consider the sustainability management of the water resources they rely on or to address externalities from service provision onto other jurisdictions (e.g. lack of sewage treatment contaminating rivers). This disregard for the value of ecosystem services is reflected in the pricing of Wat&San services, such as highly subsidized tariffs for all consumers. Often subsidized tariffs are given to the poor, but the poorest remain entirely excluded from access. In rural areas, this is because their water is taken away, while in urban areas it is because they are in marginalized settlements. Measures that ensure access for all enhance social inclusion, but water tariffs should also relate to water availability and the cost of bulk water supply to consider environmental and economic sustainability. This will incentivize water savings by both consumers and utilities.

10.4.5 Addressing Urban Sprawl as a Regional Phenomenon

Land Use

Urban growth drives land use and ecosystem changes in the periphery and rural hinterlands. Through its economic and political power, the city obtains and monopolizes water resources and exports wastewater downstream. The challenge is that urban policies (i.e. land use, zoning, housing, infrastructure) are disconnected from environmental and water-related policies. As cities expand, farmland and green areas become urban land and farmland respectively. While urban demand for water and agricultural products rises, the land available for food production shrinks or shifts to more distant areas, expanding the territory on which the metropolis depends. This threatens both food and water security for metropolitan areas. In addition, urban expansion is a less conspicuous form of reallocating water from rural to urban areas, as water allocated to agricultural use is then (formally or informally) reconverted into water for urban uses.

Although IWRM/IRBM acknowledges the need to integrate water and land, it does not provide concrete tools for spatial planning. These are the mandates of municipalities and states, or even the federal government, and RBOs do not have any say on land use, zoning, or spatial planning. Land ownership also affects what measures can be implemented to control sprawl (e.g. difficulty of preventing private landowners from selling land to developers). Meanwhile, ineffective water allocation mechanisms facilitate urban expansion despite (relative) water scarcity if water use permits originally issued for agricultural use are transferred to urban use. At metropolitan or regional level, the policies of one municipality may be incoherent with or harmful to those of a neighbouring municipality. A regional land use policy is necessary to curb unregulated sprawl, push for densification and offset negative impacts. This is supported by the push for 'Compact Cities', promoted by the United Nations' Habitat III report, with the aim to limit suburban sprawl and preserve crucial ecosystem services (United Nations 2016).

Environmental Protection

Although certain areas valuable for their ecosystem services are protected on paper, it can be difficult to enforce near large urban areas. Private landowners face an opportunity cost by preserving rather than selling their land or developing it themselves. Expropriating this land for environmental protection involves significant compensation payments. Very restrictive legislation to protect areas can backfire without constant monitoring (which is also costly) and lead to informal occupations. Local governments are often responsible for land use, including preventing the occupation of protected areas, but municipalities in the periphery—where the pressure on land is highest—generally lack capacity to enforce these measures. As Wat&San utilities are often not allowed to provide services in informal settlements, residents install clandestine water connections and discharge their wastewater *in natura.*

Nature-based solutions need greater support, but basing these on a regional strategy rather than exclusively piece-meal initiatives by local governments is more effective. This can involve protecting green belts that provide ESS such as water recharge and water production, and linear parks that cross the metropolis. In addition, compensation mechanisms designed at regional level may reduce the disproportionate burden on municipalities in the periphery that must protect green areas on limited budgets. Although IWRM/IRBM actors have limited influence on land use and spatial planning, they may have indirect influence through other instruments, such as the classification of water bodies (see Annex G: Additional Instruments). By influencing the degree of protection granted to water bodies, RBOs can restrict the type of activities and land uses that can (legally) be developed near them.

Housing

As metropolitan regions expand, the demand for housing leads to the occupation of floodplains and areas of springs and aquifer recharge through both formal processes (i.e. real estate developments) and informal processes (i.e. informal settlements). Besides threatening crucial ecosystems, this often leads to inadequate housing (e.g. exposure to flood risks). Low-density sprawl causes environmental degradation and increases inequality between periphery and core, as low-income residents are increasingly far-removed from employment opportunities. It also affects water and wastewater infrastructure as pipes must cross greater distances.

In response to these challenges, a regional plan can identify areas adequate for development and aim for densification rather than sprawl. This may involve adjusting zoning regulations and building codes, addressing real estate speculation and regenerating or upgrading inner cities and informal settlements. Where legalization is unlikely to ever happen (e.g. settlements in areas at risk of floods or mudslides or crucial for ecosystem services), relocating residents through inclusive and participatory processes are key. Besides reducing the negative impacts from sprawl, densification reduces water use (e.g. smaller/no gardens needing watering, fewer appliances) and the unit costs for water utilities. Compact cities can host growing populations while minimizing impacts on the habitats that provide them with ecosystem services.

10.5 The Sustainable Development Goals: Including a Regional Approach into IWRM

The 2015 Sustainable Development Goals (SDGs) play an important role in shaping future metropolitan water governance regimes. Metropolitan water challenges relate directly to three of the 17 SDGs (see Sect. 1.2.2): SDG 6 on water and sanitation, SDG 11 on cities and human settlements and SDG 13 on climate change. Therefore, even if metropolitan water governance regimes do not explicitly adopt sustainable and inclusive development as a normative framework, shifts in related policy areas may influence how actors and institutions respond to metropolitan water challenges. This study relates to the interlinkages between how SDGs impact these challenges and how synergies can be harnessed for win–win strategies.

Metropolitan water governance requires a multi-level approach with coordination across levels of government, as well as or regional spatial planning. Many of the challenges concern land use changes (i.e. urbanization, agriculture) and involve multiple municipalities, as well as trade-offs and externalities between municipalities within the metropolitan region or between the urban and rural areas. It is essential that the spatial scale of implementation and effects of instruments are coherent with the scale of the issues they address. Otherwise, the costs (economic, social, environmental or relational) are likely exported elsewhere. *A regional approach for addressing metropolitan water challenges should involve the following steps*:

- ***Drawing boundaries***: Defining the urban area; Identifying at least the blue, grey and black water systems that are used or impacted by the urban area; Delimitating the areas that provide ecosystem services crucial to the city's water systems; Mapping the relevant macro-infrastructure for water management.
- ***Developing knowledge systems***: Collecting and sharing urban data across megacities and their rural hinterlands. Data is typically kept within organizations, sectoral silos and municipal or state governments.
- ***Elaborating a regional plan*** to 'close the loop' of the urban water cycle. Coordinating river basin plans (including groundwater management) with city plans and providing guidelines for local governments. Outlining measures to reduce sprawl, create buffer zones around water bodies to reduce contamination and protect ecosystems, and promote a shift towards decentralized grey and green infrastructure.
- ***Developing strategic water assessments*** for plans and programmes centred around water and related ecosystems. This could integrate metropolitan water and spatial planning through a focus on the changes necessary within the planning processes.

SDG 6 combines water services and IWRM, which is an important step forward in overcoming mismatches between urban and river basin scales. To support the integration of UWM and IWRM/IRBM, the creation of an additional indicator under Sect. 6.5 could promote the development of regional water plans for cities with more than one million inhabitants. Such a plan would be based on the four steps above. This regional planning framework for integrating urban and basin concerns in large cities can then be used to design, implement and evaluate policies and policy instruments related to different types of water, infrastructure, ecosystems and urbanization (see Fig. 10.1). Such an approach recognizes and addresses the multiple spatial scales involved in water-related challenges in metropolitan regions, thereby facilitating more inclusive and sustainable responses.

10.6 Final Reflections

This research explored the ways in which spatial scales in policy design play a role within metropolitan water governance and either hinder or support inclusive and sustainable responses to water-related challenges. Over the past decades,

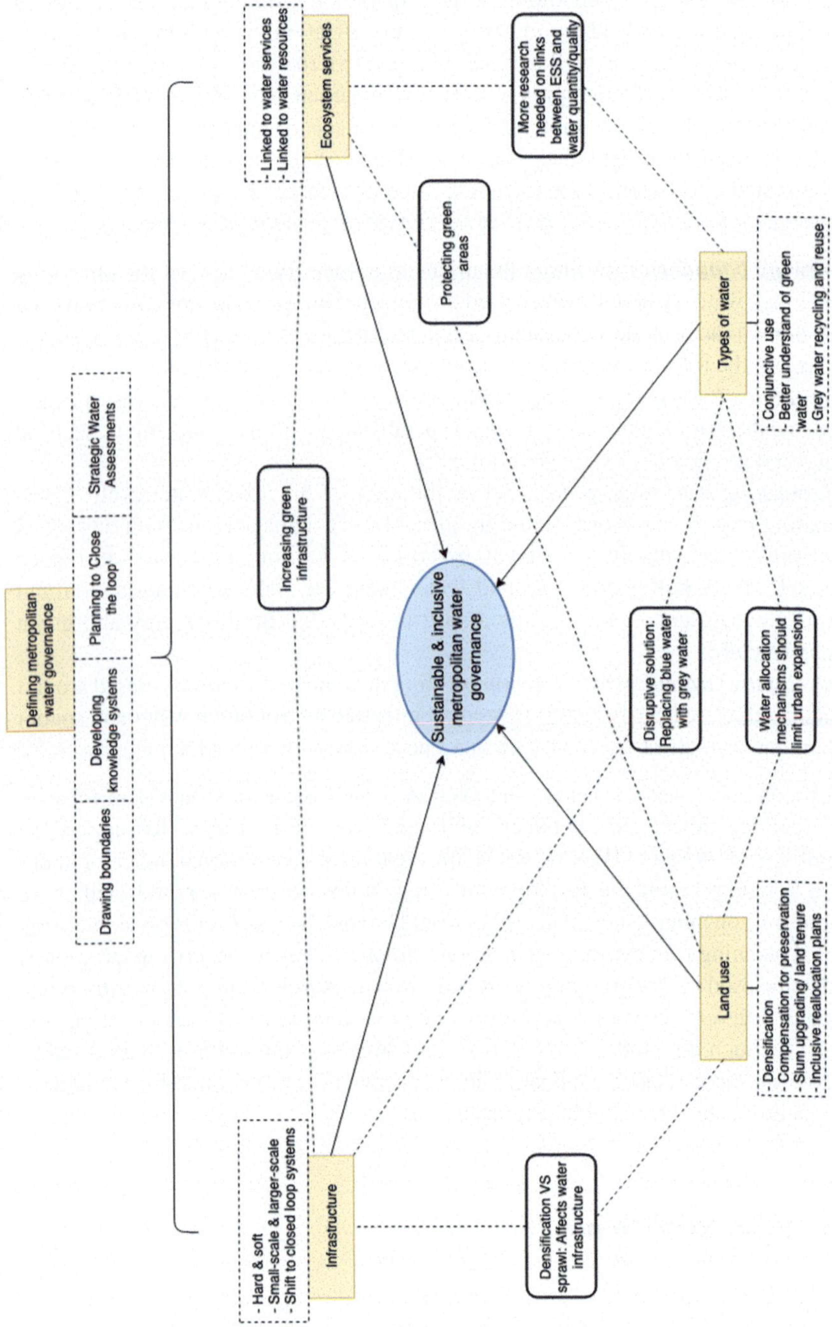

Fig. 10.1 Regional approach for integrating urban and basin concerns

the complexity of water resources in major metropolises was addressed through two main governance approaches—river basin governance and urban water governance. Although these approaches have promoted greater inclusiveness and sustainability, they focus on river basin and urban systems separately, which prevents each governance regime from addressing all relevant drivers and policy effects. This book proposes the concept of *scalar mismatches* and develops the framework of *metropolitan water governance* to bridge this gap.

Examining the case studies of Mexico City and São Paulo, two megacities that grew rapidly during the twentieth century, revealed what each could learn from the other, but also what lessons could serve cities in the Global South more broadly. As cities multiply and expand in the coming decades, and climate change puts additional stress on urban water systems, governance regimes must carefully consider where they draw boundaries and what, as well as who, these include. With adequate resources, political will and public support, policy (re)designs in metropolitan water governance can overcome potential *scalar mismatches* and support sustainable and inclusive development. These new concepts are open to further contributions from researchers and policymakers.

References

Cardona S (2019) Stormwater credit trading: lessons from Washington D.C. Metropolitan Planning Council, 23 Jan. Available at: https://www.metroplanning.org/news/8671/Stormwater-Credit-Trading-Lessons-from-Washington-D-C

DOEE (no date) Stormwater retention credit trading program. Available at: https://doee.dc.gov/src

Güneralp B, Güneralp İ, Liu Y (2015) Changing global patterns of urban exposure to flood and drought hazards. Glob Environ Change 31:217–225. https://doi.org/10.1016/J.GLOENVCHA.2015.01.002

Pincetl S et al (2019) Adapting urban water systems to manage scarcity in the 21st century: the case of Los Angeles. Environ Manage 63(3):293–308. https://doi.org/10.1007/s00267-018-1118-2

Porse E et al (2015) Pumping for the masses: evolution of groundwater management in metropolitan Los Angeles. GeoJournal 81:793–809. https://doi.org/10.1007/s10708-015-9664-0

United Nations (2016) HABITAT III—2016 ONE UNITED NATIONS FOR HABITAT III. Available at: http://habitat3.org/wp-content/uploads/One-UN-for-HabitatIII.pdf. Accessed 12 Sept 2019

World Bank (2013) Agua Urbana en el Valle de México ¿un camino verde para mañana? Washington, DC. Available at: www.bancomundial.org.mx. Accessed 8 May 2019

Xiao Y, Fang LP, Hipel KW (2018) Centralized and decentralized approaches to water demand management. Sustain 10:3466. Available at: https://www.mdpi.com/2071-1050/10/10/3466

Annex A
Literature Review

Search term(s)	Period	Fields	Numbers of results
Integrated water resources management	1970–2015	Title—Abstract—Keywords	1275 (271 with quotation marks)
Integrated river basin management	1970–2015	Title—Abstract—Keywords	487 (44 with quotation marks)
Integrated urban water management	1970–2015	Title—Abstract—Keywords	349 (23 with quotation marks)
Sustainable urban water management	1970–2015	Title—Abstract—Keywords	314 (17 with quotation marks)
Metropolitan water management	1970–2015	Title—Abstract—Keywords	113 (0 with quotation marks)

Source Generated from ScienceDirect

F. van den Brandeler, *Scalar Mismatches in Metropolitan Water Governance*, Water Governance - Concepts, Methods, and Practice,
https://doi.org/10.1007/978-3-031-08061-6

Annex B
Comprehensive List of Policy Instruments

	Regulatory	Economic	Infrastructure	Suasive
Water quantity	Water use permits	Water use fees (bulk use)	Rainwater harvesting systems	Collection and publication of data on water availability within a basin or aquifer (Mexico)
	Environmental flow standard (Mexico)	Water tariffs (consumers)	Water-saving technology	Registry of water use permits (Mexico)
	Water abstraction restriction zones	Electricity subsidies for irrigation pumping (Mexico)	Inter-connected regional water supply system	Water resources information system (Brazil)
	Transfer of water use permits between users	Sanctions for over-abstraction of bulk water	Water metres	
	Environmentally protected areas relevant for water resources	Cross-subsidies in water tariffs at state level (Brazil)	Inter-basin transfers	
	Classification of water bodies	Metropolitan funds (FUMEFI in Brazil and Fideicomiso 1928 in Mexico)	Artificial groundwater infiltration	

(continued)

F. van den Brandeler, *Scalar Mismatches in Metropolitan Water Governance*, Water Governance - Concepts, Methods, and Practice, https://doi.org/10.1007/978-3-031-08061-6

(continued)

	Regulatory	Economic	Infrastructure	Suasive
Water quality	Wastewater discharge permits	Pollution fees/fines	Water recycling	Awareness campaign of pollution in the Tietê River (Brazil)
	Quality standards for drinking water	Payment for ecosystem services linked to water resources	Treatment plants	
	Quality standards for wastewater treatment	Financing of protected areas through various funding sources	Water filters	
	Environmentally protected areas relevant for water resources	Environmental compensation mechanisms (Brazil)		
	Environmental licensing			
	Classification of water bodies			
Climate change adaptation	Environmentally protected areas relevant for water resources	Adaptation fund at city level (Mexico)	Basin-wide or regional cooperation on flood management	Flood alert systems
	Classification of water bodies	Adaptation fund at national level (Brazil)	Municipal, regional and national adaptation plans	National climate change assessments
	Environmental flow standard (Mexico)	Fund for natural disaster prevention (Mexico)	River basin organizations	National risk atlas (Mexico)
	Water abstraction restriction zones (Mexico)	Payment for ecosystem services linked to water resources	National water resources plan	
		Financing of protected areas through various funding sources		

Annex C
Policy Instruments Selected for Analysis

<table>
<tr><th>Instrument</th><th>Selection criteria</th><th>Brazil</th><th>Mexico</th></tr>
<tr><td rowspan="5">Water use permits</td><td>Scope</td><td colspan="2">Allocation of bulk water in and around the metropolis is a source of tension</td></tr>
<tr><td>Focus</td><td colspan="2">Main focus on water quantity, but may link to water quality and climate change adaptation</td></tr>
<tr><td>Instrument goal</td><td colspan="2">Inclusiveness and sustainability:
Ecological → Equitable access to/ownership of water resources and ecosystems' protection
Social → Guarantees water for small-scale farmers and preserves ecosystems for people's well-being
Economic → Crucial for industry, agriculture and energy generation
Relational → Criteria for granting permits that prevents power imbalances</td></tr>
<tr><td>Type of instrument</td><td colspan="2">Regulatory instrument</td></tr>
<tr><td>Levels of implementation</td><td colspan="2">State or federal level</td></tr>
<tr><td rowspan="3">Bulk water use fees</td><td>Scope</td><td colspan="2">(Differentiated) cost of water can lead to tensions between users</td></tr>
<tr><td>Focus</td><td colspan="2">Water quantity focus</td></tr>
<tr><td>Instrument goal</td><td colspan="2">Inclusiveness and sustainability:
Ecological → May incentivize rational use
Economic → Rational use and reinvestment of funds
Relational → A regulated system prevents the richest and most powerful actors to grab all water resources</td></tr>
</table>

(continued)

F. van den Brandeler, *Scalar Mismatches in Metropolitan Water Governance*, Water Governance - Concepts, Methods, and Practice,
https://doi.org/10.1007/978-3-031-08061-6

(continued)

Instrument	Selection criteria	Brazil	Mexico
	Type of instrument	Economic instrument	
	Levels of implementation	Basin level	Basin or aquifer level through regional organization
Water tariffs	Scope	Differentiated tariffs between and within municipalities and unpaid tariffs lead to tensions	Subsidized rates for public/urban use lead to problems with other users
	Focus	Water quantity focus	Water quantity focus with impacts on water quality
	Instrument goal	Inclusiveness and sustainability: *Social* → Subsidized tariffs for low-income households and cross-subsidies for state water company *Ecological* → Scaled rates aim for rational use *Economic* → Some tariffs are too low or not paid *Relational* → Low-income and informal settlement residents have lower tariffs, but often sub-par access	
	Type of instrument	Economic instrument	
	Levels of implementation	Municipal or state level	Municipal level
Wastewater discharge permits	Scope	Contamination of water bodies affects communities and ecosystems downstream, causing tensions	
	Focus	Water quality focus	
	Instrument goal	Inclusiveness and sustainability: *Ecological* → Reduces contamination by discouraging wastewater discharge without treatment *Social* → Low-income communities often most affected by contamination (e.g. waterborne diseases) *Economic* → Preventing contamination is cheaper than fixing it or having to import water from further *Relational* → All polluters are held to these standards	
	Type of instrument	Regulatory instrument	

(continued)

(continued)

Instrument	Selection criteria	Brazil	Mexico
	Levels of implementation	State or federal	Basin or aquifer level through regional organization
Payment for ecosystem services	Scope	Preserving ecosystems and water resources for the city has an opportunity cost for landowners and municipalities in the rural hinterlands upstream	
	Focus	Water quality focus, but also relevant for water quantity and climate change adaptation	
	Instrument goal	Inclusiveness and sustainability: *Ecological* → Incentivizes restoration/maintenance of crucial ecosystems. It also leads to climate regulation and protects habitats for biodiversity *Social* → Inclusion of rural communities and compensation for conservation efforts, possibly leading to poverty alleviation *Economic* → Loss of economic value in short-term but long-term benefits and cost savings *Relational* → Empowers rural landowners	
	Type of instrument	Economic instrument	
	Levels of implementation	Multiple possible levels, but mainly municipal	Federal or state level
Classification of water bodies	Scope	Defines restrictions on activities near water bodies and thus impacts economic development and environmental preservation	
	Focus	Water quality focus	
	Instrument goal	Inclusiveness and sustainability: *Ecological* → Determines where more restrictive policies can be implemented to preserve water resources *Economic* → Restrictions affect economic activities *Relational* → All actors must comply to same standards	
	Type of instrument	Regulatory instrument	
	Levels of implementation	State or federal	Federal
Metropolitan-scale water management system (water supply)	Scope	Tensions on water sharing across metropolitan region and between basins	
	Focus	Water quantity focus, but also relevant for climate change adaptation	

(continued)

(continued)

Instrument	Selection criteria	Brazil	Mexico
	Instrument goal	Inclusiveness and sustainability: *Ecological* → Rational use and conservation may be included *Social* → May help overcome relative scarcity that affects marginalized communities *Economic* → Guarantees supply for various users *Relational* → Less powerful actors must have equal decision-making power	
	Type of instrument	Coordination instrument	
	Levels of implementation	State level	Federal level
Macro-drainage	Scope	Floods and related issues affect the whole basin and municipalities can aggravate effects downstream	Large volumes of wastewater, low levels of treatment and topography create risks for population and ecosystems. Floods affect all lower lying areas of the metropolis, but especially lower-income neighbourhoods in periphery
	Focus	Climate change adaptation focus	Water quality focus
	Instrument goal	Inclusiveness and sustainability: *Ecological* → Ecosystem approaches, green infrastructure *Social* → Marginalized communities should be main target *Economic* → Cost must be compared to cost of inaction *Relational* → Less powerful actors must have equal decision-making power	
	Type of instrument	Coordination instrument	
	Levels of implementation	Basin level (but municipal and state levels may also be involved)	Federal and state levels

(continued)

(continued)

Instrument	Selection criteria	Brazil	Mexico
Climate change adaptation fund	Scope		Climate change risks affect the MVMC as a whole, but poorer areas are more vulnerable
	Focus		Climate change adaptation focus (but also relevant for water quantity and quality)
	Instrument goal		Inclusiveness and sustainability: *Ecological* → Green areas, water saving, resilience *Social* → Reduce citizens' vulnerability, increase adaptive capacity and risk awareness *Economic* → Economic benefits estimated to slightly outweigh costs in 6-year period *Relational* → Inclusive decision making
	Type of instrument		Economic instrument
	Levels of implementation		Federal District level

Annex D
Interview List

Code	Type	Level	Professional background	Country
Interview-M1	Government agency for environment and climate change	National	Climate change adaptation	Mexico
Interview-M2		National	Water management	Mexico
Interview-M3		National	Water management	Mexico
Interview-M4	National water agency	National	Water services policy	Mexico
Interview-M5		National	Water management/basin entities	Mexico
Interview-M6		National	Hydrologist	Mexico
Interview-M7		National	Hydrologist	Mexico
Interview-M8		National	Executive advisor	Mexico
Interview-M9		National	Former high-level official	Mexico
Interview-M10	National forestry agency	National	PES programmes	Mexico
Interview-M11	National housing agency	National	High level officials	Mexico
Interview-M12	National centre for disaster prevention	National	Extreme weather events expert	Mexico
Interview-M13	Basin management entity	Regional	Water supply infrastructure	Mexico
Interview-M14	Mexico City environmental department	Provincial	Climate change adaptation	Mexico
Interview-M15		Provincial	Environmental management, metropolitan planning	Mexico

(continued)

F. van den Brandeler, *Scalar Mismatches in Metropolitan Water Governance*, Water Governance - Concepts, Methods, and Practice,
https://doi.org/10.1007/978-3-031-08061-6

(continued)

Code	Type	Level	Professional background	Country
Interview-M16		Provincial	Climate change adaptation	Mexico
Interview-M17	Mexico City conservation agency	Provincial	Environmental conservation	Mexico
Interview-M18	Mexico City department of urban and land development	Provincial	Human right to water, urban water	Mexico
Interview-M19	Mexico City water and sanitation	Provincial	High level official	Mexico
Interview-M20	Mexico City government	Provincial	Metropolitan planning	Mexico
Interview-M21	Environmental regulator	Provincial	Environmental law	Mexico
Interview-M22	Environmental regulator	Provincial	Environmental law	Mexico
Interview-M23	CSO	Local	Political activist/local leader	Mexico
Interview-M24	District level hydraulic department	Local	Hydraulic engineer	Mexico
Interview-M25	District level hydraulic department	Local	Geophysical engineer	Mexico
Interview-M26	District level urban development department	Local	Urban planner	Mexico
Interview-M27	Local water and sanitation utility	Local	Hydraulic engineer	Mexico
Interview-M28	University/research institute		Water governance	Mexico
Interview-M29	University		Ecosystem management	Mexico
Interview-M30	University		Ecosystem management	Mexico
Interview-M31	University		Basin management, agronomist	Mexico
Interview-M32	University		Hydrogeologist	Mexico
Interview-M33	University		Political scientist/water policy	Mexico
Interview-M34	University		Climate change adaptation	Mexico
Interview-M35	University		Water justice/climate change adaptation	Mexico
Interview-M36	University		Human right to water	Mexico
Interview-M37	University		Urban and environmental management	Mexico
Interview-M38	University		Water governance/social justice	Mexico

(continued)

(continued)

Code	Type	Level	Professional background	Country
Interview-M39	University		Water management	Mexico
Interview-M40	University		Wastewater management	Mexico
Interview-M41	Private sector		Engineer	Mexico
Interview-M42	Private sector		Entrepreneur	Mexico
Interview-M43	Private sector		Commercial	Mexico
Interview-M44	Private sector		Commercial	Mexico
Interview-M45	International—NGO		Soil and water conservation	Mexico
Interview-M46	International—NGO		Environmental and urban policy	Mexico
Interview-M47	International—NGO		Environmental and social policy	Mexico
Interview-M48	International—NGO		Environmental conservation	Mexico
Interview-M49	National—NGO		Executive/hydraulic engineering	Mexico
Interview-M50	NGO		Conservation biologist	Mexico
Interview-M51	NGO		Economist	Mexico
Interview-M52	NGO		Basin management/activist	Mexico
Interview-M53	NGO		Water as a human right activist	Mexico
Interview-M54	NGO		Environmental policy	Mexico
Interview-M55	National—NGO		Basin conservation/forest conservation	Mexico
Interview-M56	NGO		Engineer/water saving technology	Mexico
Interview-M57	Independent		Climate change adaptation/urban water	Mexico
Interview-M58	Independent		Engineering/water policy	Mexico
Interview-M59	Independent		Drinking water access activist	Mexico
Interview-M60	Independent		Drinking water access activist	Mexico
Interview-B1	National water agency	National	Water resources management/financial instruments	Brazil
Interview-B2	SABESP	State	Urban water and ecosystem conservation	Brazil
Interview-B3		State	Former high-level official of Tietê River Project	Brazil
Interview-B4		State	High level official	Brazil

(continued)

(continued)

Code	Type	Level	Professional background	Country
Interview-B5		State	High level official	Brazil
Interview-B6	State Secretariat for Sanitation and Water Resources	State	Coordinator, bulk water supply/representative in basin committee	Brazil
Interview-B7	State Environmental Department	State	Environmental planning/involved in basin committee	Brazil
Interview-B8	CETESB	State	Climate change division, adaptation and vulnerability officer	Brazil
Interview-B9		State	Climate change division, mitigation officer	Brazil
Interview-B10	State Environmental Agency	State	Environmental management of metropolitan areas/representative in basin committee	Brazil
Interview-B11	State Regulatory Agency for Water and Sanitation	State	Regulation specialist, Water and Sanitation Directorate	Brazil
Interview-B12		State	Regulation specialist, Water and Sanitation Directorate	Brazil
Interview-B13		State	Regulation specialist, Economic-Financial Directorate	Brazil
Interview-B14	State Agency for forest conservation	State	Manager of conservation units in the MRSP	Brazil
Interview-B15	State Housing Agency	State	Architect/representative in basin committee	Brazil
Interview-B16	Research institute (state-affiliated)	State	Researcher, hydrogeologist	Brazil
Interview-B17		State	Researcher on erosion, flood risks	Brazil
Interview-B18		State	Technical agent of FEHIDRO	Brazil
Interview-B19	EMAE	State/metropolitan	Environmental department/representative in basin committee	Brazil
Interview-B20	Basin agency	Basin	High level official	Brazil
Interview-B21	Municipal environmental department	Municipal	Official of environmental department	Brazil
Interview-B22	Municipal water and sanitation company	Municipal	Director of the Department of Planning and Projects	Brazil

(continued)

(continued)

Code	Type	Level	Professional background	Country
Interview-B23		Municipal	Commercial/manager	Brazil
Interview-B24	Municipal department	Municipal	Drainage specialist	Brazil
Interview-B25	International—NGO		National Water Manager/water activist	Brazil
Interview-B26	International—NGO		Activist, climate and energy	Brazil
Interview-B27	International—NGO		Water activist	Brazil
Interview-B28	NGO		Water services/communications	Brazil
Interview-B29	NGO		Researcher focusing on water	Brazil
Interview-B30	NGO focused on river rehabilitation		Geographer, water and urban policy, public administration	Brazil
Interview-B31	NGO		Water quality and conservation, coordinator	Brazil
Interview-B32	NGO		Water resources and water services, urban environmental management	Brazil
Interview-B33	University		Water governance	Brazil
Interview-B34	Consultancy		Water services consultant	Brazil
Interview-B35	Lobby group		Industrial water policy	Brazil
Interview-B36	Water and sanitation workers union		Sanitation advisor	Brazil
Interview-B37	Low-income housing construction company		New real estate developments analyst	Brazil
Interview-B38	University		Spatial planning, environmental governance	Brazil

Annex E
Main Actors in São Paulo's Metropolitan Water Governance

Scale level	Actor	Responsibilities
Global	World Bank	Grants loans and provides for projects and infrastructure in the water and sanitation sector
	Inter-American Development Bank	Provides technical and financial assistance in the water and sanitation sector
National	National Water Agency (ANA)	Regulates and issues permits for the use of national water resources Carries out scientific studies and establishes an information system Acts as a conflict mediator [increased demand for water for diverse uses has increased conflicts (ANA n.d.)]
	National Council on Water Resources (CNRH)	Highest organ of the SINGRH, and by nature normative, deliberative and participatory Promotes the integration of the national and state government, users and civil society stakeholders in water management. However, it overwhelmingly represents government interests Elaborates the national water resources plan and deliberated on major issues and disputes Determines the creation of river basin committees for rivers pertaining to the Federal Union
	Ministry of the Environment (MMA)	Promotes the integration of sustainable development in public policies, in a participatory and democratic manner at all levels of government and society The Ministry of Environment coordinates policies related to fresh water, river basins (e.g. river revitalization programmes), aquatic biodiversity, water resources and coastal zones and oceans (OECD 2015: 64). It formulates policies in relation to climate change adaptation and the national policy on climate change (OECD 2015)

(continued)

F. van den Brandeler, *Scalar Mismatches in Metropolitan Water Governance*, Water Governance - Concepts, Methods, and Practice,
https://doi.org/10.1007/978-3-031-08061-6

(continued)

Scale level	Actor	Responsibilities
	Ministry of Cities	Develops urban development policy and coordinates with other government bodies on environmental sanitation (including water and sanitation)
	The river basin committees for rivers of federal domain (CBHs)	Approve the basin plan, arbitrate conflicts over water use, establish the values of bulk water use fees and more. They count on the participation of representatives of the States, municipalities, users, civil society and the federal government
Sub-national	DAEE	Manages water resources at state level. Grants water permits. Controls water use. Analyses proposed projects. Provides technical and administrative support to the CBHs
	CETESB	Manages environmental concerns at state level. Issues environmental permits. Monitors pollution. Analyses proposed projects. Provides technical and administrative support to the CBHs
	Basic Sanitation Company of São Paulo State (SABESP)	Responsible for water supply, and sewage collection and treatment in many municipalities of São Paulo State
	State Council on Water Resources (CRH)	Multi-stakeholder council that oversees and regulates IWRM in the state (i.e. discusses and approves laws related to the state water resources plan, mediates conflicts between CBHs, classifies water bodies, etc.). It is composed of 33 members, with equal representation from the state, municipalities and civil society (FABHAT 2016)
	Secretary of Energy, Water Resources and Sanitation (SERH)	Formulates and implements the state policies on water resources and sanitation and integrates these with the state policies on the environment, health, urban development and more DAEE and SABESP are subordinated to SERH
	State Environmental Secretariat (SMA)	Establishes the State Environment Policy. The CETESB is subordinated to the SMA

(continued)

(continued)

Scale level	Actor	Responsibilities
Basin	River basin committees	Approve and update Water Resource Plans. Deliberate on water-related decisions within the basin Reduce conflicts among stakeholders. Define water bulk prices The committee's creation is preceded by the elaboration of a basin Status Report that diagnoses its main challenges The Alto Tietê basin committee is composed of state, municipal and organized Civil Society representatives, each with 18 seats.[1] As there are 36 municipalities in the basin, there is a representative and a substitute of each sector for each municipality. Representatives are elected, and each municipality is either directly represented or through a substitute, except the municipality of São Paulo, which always has a seat (van den Brandeler 2013). The civil society bloc is composed of broad interests (e.g. industry and environmental preservation). The executive board is composed of a president, a vice-president and an executive secretary.[2] Meetings occur monthly and are open to the public, although the population is mostly unaware of the committees' existence (Interview-B31)
Metropolitan	Metropolitan regions	Regional units instituted by the States, formed by the grouping of neighbouring municipalities for integrating the organization, the planning and the implementation of "public functions of common interest" (Casa Civil 2015) São Paulo State's definition of 'metropolitan region' is "a grouping of neighbouring municipalities of national prominence, due to high population density, significant conurbation, highly diverse urban and regional functions, socio-economic specialization and integration, requiring permanent integrated planning and joint action by the involved public entities" (State 1994)
	EMAE (Metropolitan Company for Water and Hydropower)	Produces hydropower for the metropolitan region, through reservoirs and major engineering works that involved rectifying and reversing the Pinheiros River

(continued)

[1] Civil society representatives embody organizations such as NGOs, user associations, neighbourhood associations, and business or industry associations. While SABESP is 51% state-owned, it represents the State's interests in the committee and not that of consumers (i.e. civil society) (Interview-B35).

[2] The president is always a mayor, the vice-president is a Civil Society representative and the executive secretary is a State representative (from the DAEE or the CETESB). State representatives are technical experts and their leadership position in the execution of the committee's activities is based on the premise that they have more technical knowledge (Alvim 2006: 163).

(continued)

Scale level	Actor	Responsibilities
	State Secretariat for Regional Development	Formulates policies at the macro-metropolitan level on land occupation issues and compatibility with the region's sustainable development Aims to integrate sectoral, spatial and institutional projects and actions, and is focused on issues of mobility and logistics, environmental sanitation and housing
Municipal	Municipal governments	Responsible for land use and soil occupation, water services, urban drainage, civil defence, micro-basins and areas of springs in their territory, and local environmental issues The municipality of São Paulo's department of Works, which is responsible for drainage and flood control within the municipality, has the Emergency Management Centre that focuses on flood forecasts and triggers different alert levels. This centre is in close communication with the civil defence (Interview-B24)

Source Author

References

Alvim ATB (2006) A contribuição do Comitê do Alto Tietê à gestão da bacia metropolitanaentre 1994 e 2002. Pós Rev Programa Pós-Grad Arquit Urban FAUUSP 19:26–44. https://doi.org/10.11606/issn.2317-2762.v0i19p26-44

ANA (no date) Sistema Cantareira. Agência Nacional de Águas. Available at: http://www3.ana.gov.br/portal/ANA/sala-de-situacao/sistema-cantareira/sistema-cantareira-saiba-mais. Accessed 12 Feb 2019

Casa Civil (2015) Statute of the metropolis. Available at: http://www.planalto.gov.br/ccivil_03/_ato2015-2018/2015/lei/l13089.htm. Accessed 6 Feb 2019

FABHAT (2016) Relatório—I Plano de Bacia Hidrográfica do Alto Tietê. São Paulo, Brazil

OECD (2015) Water resources governance in Brazil. In: OECD Studi. OECD Publishing, Paris. https://doi.org/10.1787/9789264238121-en

State SP (1994) Lei Complementar nº 760, de 01 de agosto de 1994—Assembleia Legislativa do Estado de São Paulo. Assembleia Legislativa do Estado de São Paulo. Available at: https://www.al.sp.gov.br/repositorio/legislacao/lei.complementar/1994/lei.complementar-760-01.08.1994.html. Accessed 6 Feb 2019

van den Brandeler F (2013) Towards inclusive and resilient water governance? A case study on the implications of deliberation and participation in Guarulhos. University of Amsterdam, Brazil

Annex F
Main Actors in Mexico City's Metropolitan Governance

Scale levels	Actors	Responsibilities
Global level	United Nations	Promoting global norms, standards and targets (Human right to water, MDGs targets for water and sanitation)
	World Bank and other IFIs	Providing financing to the water sector
National level	Federal government	Regulating the use of water resources Contributing to the financing of investments
	SEMARNAT	Establishes official norms in relation to water management Supervises enforcement of norms
	CONAGUA	Responsible for water resources management in Mexico, including: Granting of water abstraction and wastewater discharge permits Supplying bulk water to the Federal District and to parts of Mexico state through the Cutzamala and Lerma systems Water policy, planning, irrigation and drainage development, water supply and sanitation, and emergency and disaster management (with an emphasis on flooding)
	Inter-ministerial Commission on Climate Change (CICC)	Supporting collaboration among federal and regional agencies, minimizing conflicts among sectors, and maximizing the benefits of synergies for the integration of a climate change policy
	SINAPROC (National Civil Protection Service System)	Informing authorities and society of imminent extreme events and risks, well as carrying out prevention signals, early warnings, evacuation procedures, providing provisional dwellings for affected people, damage control measures, thoroughly surveying damages and possible solutions for affected areas and helping the re-establishment of former living conditions (CONAGUA 2011)

(continued)

F. van den Brandeler, *Scalar Mismatches in Metropolitan Water Governance*, Water Governance - Concepts, Methods, and Practice,
https://doi.org/10.1007/978-3-031-08061-6

(continued)

Scale levels	Actors	Responsibilities
	CENAPRED (National Disaster Prevention Centre)	Develop risk reduction policies and coordinate information and warning systems
State level/Federal District level	SACMEX	Providing residents of the Federal District with drinking water, drainage and sewerage services, as well as the treatment and reuse of wastewater, in adequate quantity and quality Operating, maintaining and building water infrastructure
	Mexico State Government	Planning, regulating and developing infrastructure for water resources Providing bulk water Treating wastewater Assisting municipalities in providing water and sanitation services
	Mexico State Water Commission	Buying bulk water from CONAGUA, transmitting it through its own bulk water infrastructure and selling it on to 57 municipalities (4.1 million inhabitants) Monitoring water quality Providing technical assistance to municipalities in water disinfection and sewer cleaning Operating wastewater pumping stations and five wastewater treatment plants, emptying septic tanks Providing water in tankers in emergency situations Providing training and assisting municipalities in the establishment of municipal utilities
Basin level	Basin agencies (decentralized office of CONAGUA)	Formulating regional policy Designing programmes to implement such policies Conducting studies to estimate the value of the financial resources generated within their boundaries (water user fees and service fees) Recommending specific rates for water user fees and collecting them
	Basin councils	Guiding, together with CONAGUA, the Basin Agencies' work Coordinating government institutions Negotiating with water users and social organizations, with as main objectives the formulation and execution of programmes and actions to improve regional water management, support of hydraulic works development and related services, and preservation river basin resources
City level	Municipalities	59 municipal governments in Mexico State, one municipality in Hidalgo State and 16 districts in Mexico City are responsible for the provision of drinking water, drainage, wastewater collection, treatment and disposal for their constituents, as well as the management of solid waste and the creation of environmental protection zones

(continued)

(continued)

Scale levels	Actors	Responsibilities
	Municipal utilities	Municipalities can delegate the provision of water and sanitation services to a municipal utility or to the state water commission
Other	Irrigation districts	In Hidalgo State these oversee irrigation with wastewater from Greater Mexico City
	NGOs	Focusing on social and environmental issues linked to water (Guardianes de los volcanes; UN Habitat, Isla Urbana, Agua Para Todxs Agua Para la Vida and more)

Source Author

Reference

CONAGUA (2011) Lerma—Chapala Basin case study Mexico: a fruitful sustainable water management experience. Mexico City, D.F. Available at: http://www.conagua.gob.mx/CONAGUA07/Contenido/Documentos/LermaChapalaBasinCase.pdf

Annex G
Additional Instruments

Classification of Water Bodies in the MRSP

Description

According to the State Water Law (nº 7.633 of 1991), the State is responsible for the protection of aquatic flora and fauna and of the environment. This involves establishing priority uses for water and classifying water bodies in different categories according to their purpose for water users and the water quality levels required for these (Brazil 1997). These are then submitted as proposals to the basin committees for discussion and approval (CEDE 2015). The aim is to ensure water quality compatible with the more demanding uses and to lower the costs of combatting water pollution through permanent, preventive actions (Brazil 1997). There are five categories for surface freshwater resources and six categories for groundwater resources. Each category determines the level of protection by regulating the discharge of effluents and the licenses for activities with environmental impacts (PERH 2017). Water bodies that are deemed important for uses that require clean water or that are still well preserved, are typically classified within a category that involves more restrictions (CEDE 2015). This instrument links with the granting of water use permits, the charging for bulk water use and environmental licensing (CEDE 2015). This instrument can be considered a planning tool, as it not only looks at the current state of a water body but at the water quality it requires to respond to society's needs.

F. van den Brandeler, *Scalar Mismatches in Metropolitan Water Governance*, Water Governance - Concepts, Methods, and Practice,
https://doi.org/10.1007/978-3-031-08061-6

Effectiveness on Actors in Terms of Mandated Goals

All the waters of São Paulo State have been classified, which is not always the case in other states (CEDE 2015). The main challenges for the effectiveness of this instrument is the lack of adequate monitoring of water bodies and the need to calibrate cost estimates for the adoption of the measures to attend to the classifications (CEDE 2015). Nevertheless, in some basins, particularly those where the charging for the use of water is implemented, there are registration systems and monitoring networks (MMA 2010).

This classification significantly depends on the political will of a state or basin committee, as a more restrictive class means more restrictions on potential activities (CEDE 2015). The focus is on making water bodies meet the standards of their classification, which are sometimes very low, rather than promoting their preservation and restoration for purposes such as supporting aquatic life or improving the quality of life of riverside inhabitants.

Impact on Inclusiveness and Sustainability

Indirectly, besides a water quality instrument, this also represents a land use control mechanism as it restricts the installation of activities that would affect the quality of the water that a water body should maintain according to its classification (MMA 2010). As many sources of water contamination are related to land use factors, such as informal urbanization, the implementation of the regulations associated with this instrument has the potential to influence drivers of contamination.

The challenge is that contaminated water bodies often already have a lower classification, and therefore have weak regulations in place. Environmentalists criticize the category 4 class as so weak that it encourages the contamination of water bodies. In addition, the existence of the more permissive categories is at odds with the Law of Environmental Crimes that prohibits "pollution that results in damages to human health or that cause the mortality of animals or the destruction of significant flora" (Federal Law 9.605, 1998). This leads to legal ambiguity "where everyone does what they think is the right interpretation, or the most convenient". In the Alto-Tietê River basin, rivers in the urbanized areas fall under Class 4, the most permissive category, even when they are in still relatively good state, which reduces the incentive to reduce pollution.

Instruments	Effect on actors	Impact
Classification of water bodies	It has been implemented, although monitoring costs make it difficult to verify compliance	Ecol: 0 Class 4 leads to tacit acceptance of "dead rivers" and does not encourage improvement. However, classifying rivers allows for differentiated and more realistic standards

Redesign

This has the potential to influence land use management, as higher classifications restrict polluting activities near the water body in question. This instrument should be integrated with other planning instruments, so that discussions around it may include a broader range of actors, including the community to deliberate on what kind of river they want and what kind of river they can get. Informal settlements where sanitation cannot be installed remains a challenge and the solution must be negotiated with the housing and urban planning sectors, and with municipalities.

Some have argued that the most permissive class should not exist as it informally condemns the water body to remain in its contaminated state. However, eliminating it should be done within a longer-term planning horizon, so that current "polluters" have time to adjust.

Comparison in Terms of Design

In the case of São Paulo, the granting of a wastewater discharge permit depends on the classification of a certain water body. This classification ranks water bodies and waterways in terms of the level of restrictions and protections they require, and this depends on their condition and the type of use that is made from their water resources. Higher ranked water bodies, which may be well-preserved and used for water supply, therefore face more restrictions. While this protects these water bodies, it does not create incentives to reduce contamination in already polluted water bodies.

In the case of Mexico City, obtaining a wastewater discharge permit is also linked to the classification of the water body in question. The classification determines the water body's capacity to assimilate and dilute contaminants, the levels of contaminants that can be contained in discharged effluents and quality targets to be attained within specific timeframes. It is based on three quality parameters, which ignores other contaminants that can have harmful impacts and international recommendations on the number and types of parameters. Lack of compliance results (in theory) in a fee to be paid and that is (in theory) invested into the improvement of the appropriate infrastructure.

Comparison in Terms of Effect on Actors

The classification of water bodies has meant that clean water bodies, sometimes used for water supply, had a higher ranking and thus level of protection. However, the responsibility to maintain these water bodies at those levels falls largely on local governments, which must enforce zoning laws and control land use around these water bodies. They typically lack the human and financial capacity to fulfil these expectations. Moreover, water bodies that are given low rankings—due to their current state of contamination—are condemned to remain as such, as the classification does not create incentives for improving the ranking. In addition, critics have claimed that water bodies have been downgraded in their classification in order to allow more (polluting) activities near them. In the case of Mexico, although included in the National Water Law, the classification of water bodies had not been implemented everywhere. Overall, the responses were that regulatory measures had not remedied the problem of sewage contamination in the country.

A stricter application of wastewater discharge fees and more restrictive classification of waterways would incentivize polluters to treat their sewage. This requires a basin view, more difficult to achieve with highly fragmented sewage management, and a regional approach to land use and the challenge of informal settlements.

Wastewater Discharge Permits in São Paulo

Description

Wastewater discharge permits are the main instrument for water preservation (Brazil 1997). They allow water users to discharge effluents into water bodies for dilution, transport or final disposal, and hydropower use. Together with water use permits, discharge permits are a prerequisite for obtaining an environmental license and a key step to realizing any activity that can pollute or cause environmental degradation (Brazil 1981).[3] Although environmental licenses are granted by the CETESB, wastewater discharge permits are emitted by the DAEE or ANA (depending on whether the water body is of state or federal domain) (Brazil 1997, Art. 14). CETESB monitors the discharge of effluents at their source by collecting and analysing samples, and by verifying monitoring devices installed for this purpose (Interview-B7). The DAEE shares its database of water uses and permit-holders with CETESB, which then defines which use needs a license (Interview-B7).

Wastewater discharge permits are granted if the effluents meet qualitative requirements, or if the receiving water body has the necessary flow or volume to dilute the effluents so that it meets these requirements, thereby not compromising the water body's classification (CEDE 2015). This classification determines different standards for different water bodies according to their current condition and the quality requirements of users that rely on it. Polluting effluents must therefore undergo adequate treatment prior to discharge (Conama 2005). As with water use permits, a fee is charged for the discharge of effluents.

Effectiveness on Actors in Terms of Mandated Goals

Improvements in water quality may be attributed to water discharge permits, but the Tietê Project likely contributed by injecting large investments into SABESP's sanitation programmes. The difficulty of obtaining environmental licenses due to strict regulations may have curtailed the effect of discharge permits (Interview-B5/B37). The intended result (i.e. compliance with high standards and adjustment through stricter wastewater treatment processes) backfired and led to non-compliance

[3] Licenses are renewed periodically, and measurements are taken to verify compliance (Interview-B10).

(Interview-B5). The CETESB lacked the capacity to monitor potentially polluting activities, and although fines could be applied, in practice there was still a lot of impunity for polluters (Interview-B5). Meanwhile, the State Office of the Public Prosecutor operated in an isolated manner and presented polluters, such as SABESP, with billionaire fines rather than seeking comprehensive plans towards increasing sewage treatment (Interviews-B5/B22/B23).

Moreover, these permits are granted when pollutants can be absorbed without causing the downgrading of the water body's classification, but there are no requirements or incentives in place to improve the effluents' quality further (Interview-B5). This becomes a vicious cycle as polluted rivers are tacitly encouraged to remain so.

The effect of this instrument on industrial pollution is seen more positively than that on domestic sewage, which is partly explained by utilities' lack of financial capacity for their operations: In the largest wastewater treatment plant of the MRSP, "the pumping systems do not work and there are no interceptors, there is no maintenance" (Interview-B7). While the pressure is on sewage treatment plants and industries, diffused pollution from both urban and rural sources is left largely unaddressed.

Impact on Inclusiveness and Sustainability

Between 1990 and 2014, the length of the pollution slick in the Tietê River shrunk from 380 to 70 km, although it returned to 130 km after the water crisis, as investments in sanitation were diverted to increase water supply (SOS Mata Atlântica 2017). In addition, formal housing and commercial developments in areas that required higher levels of protection stopped, but irregular developments multiplied. This is particularly problematic in areas of springs, such as the Guarapiranga dam.[4] By focusing exclusively on quantitative and qualitative impacts on water bodies, the licensing process ignores the factors leading to the pollution (i.e. lack of affordable housing, informal urbanization) (Interview-B5).

There is a disconnect between wastewater discharge permits, which are emitted at state or federal level (depending on the domain of the water body), and sanitation planning, which is a municipal responsibility and does not (necessarily) integrate basin considerations. This leads to tensions between zoning, and the developments sponsored at local level, and the protection of water bodies by basin and state entities. State-level respondents particularly emphasized the municipalities' inability (or unwillingness) to prevent occupations by new informal settlements.

[4] Regulations in areas of springs aim to prevent housing densification by requiring a minimum surface area of 250 m^2 per housing unit (Interview-B37).

Instruments	Effect on actors	Impact
Wastewater discharge permit	The lack of enforcement encourages non-compliance. Fines have been applied sporadically and without attempts at remediation. Sewage treatment plants lack resources, and fines will not help	Ecol: – Discharge of untreated effluents remains high and cause water contamination Soc: – Marginalized populations are more likely to lack sanitation and live near contaminated water Econ: – – As local water resources are polluted, water imports become necessary. Costs are transferred from polluters to the State and donor basins Rel: – Lack of coordination between sectors and levels of government has contributed to informal growth, but permits do not apply in these areas

Redesign

First, compliance must be addressed. Environmental licenses should only be granted or renewed if it is verified that a wastewater treatment system is accounted for. Although constant monitoring is prohibitively expensive, sporadic inspections to a small number of permit-holders and the application of increasing fines (i.e. first-time offenders vs. recidivists) can encourage compliance. Part of the value collected by the permits should also be allocated to monitoring and compliance schemes, creating a possible snowball effect.

To address the main challenge (i.e. informal settlements), it is necessary to increase coordination between local and state actors. Approval of water and wastewater discharge permits should be connected to the municipal master plan and Wat&San plan, and if sewage treatment facilities cannot comply (e.g. large investments still needed, informal settlements that are not connected to the network), they must work with local authorities towards a solution (e.g. regularization and upgrading of areas or relocation, alternative treatment options).

Comparison in Terms of Design

In Mexico they are obtained together with water use permits. The classification of a water body determines the quality standards that the effluent must meet before discharge so that the water body concerned is deemed capable of assimilating and diluting the contaminants. In theory, a lack of compliance with these standards means that the polluter must pay a fee that is reinvested in the necessary sanitation infrastructure.

In Brazil, wastewater discharge permits are granted if the relevant authority (ANA or DAEE) estimate that the receiving water body has a volume or flow of water sufficient so that the discharge—together with the already existing discharges—do not affect water quality such that it would alter the water body's classification. Wastewater discharge permits are also necessary to obtain environmental licenses (for any type of potentially polluting activity). Similarly to Mexico, discharge permits require effluents to meet qualitative standards according to the classification of the water body.

In both cases, qualitative requirements tend to be stricter in areas further from the city, where water resources are better preserved. This instrument relies significantly on the classification of water bodies, and consequently so does its effectiveness.

Comparison in Terms of Effect on Actors

In both cases, wastewater is still often discharged without treatment, due to a lack of treatment capacity or—more often—a lack of piping infrastructure connecting to a treatment plant. This also concerns "polluters" with wastewater discharge permits and leads authorities to turn a blind eye in many cases. Even when "polluters" could be connected to the wastewater infrastructure, the lack of monitoring did not incentivize them to do so. In the MRSP, utilities that do not treat sewage could face significant fines, and SABESP faced legal woes for its slow progress in sewage management.

Overall, in both cases, industrial users are more likely to comply (as they are easier to monitor and fine). For some, this has contributed to an incentive to treat and recycle wastewater, although this is still a minority.

Wastewater Discharge Permits in Mexico City

Design

Wastewater discharge into national waters requires a CONAGUA permit (NWL 2004). This permit is obtained simultaneously with the water use permit and is also registered in the REPDA. In issuing this permit, CONAGUA considers the capacity of the concerned water body to assimilate and dilute the effluents' contaminants. Depending on the water body's classification, the effluents must meet more or less strict quality standards (CONAGUA 2015). More specifically, the effluents must comply with a series of indicators concerning water contaminating, the volume of wastewater and the pollutant load of the discharge (CNA-01-001). CONAGUA may charge fees for lack of compliance investing the money into infrastructure improvement projects.

Effectiveness on Actors in Terms of Mandated Goals

Wastewater discharge permits are ineffective due to insufficient enforcement. There was little monitoring, the few discovered infractions rarely led to fines and, when they did, polluters may ask to commute the fine into reinvestments into a poorly monitored water pollution reduction initiative (Interviews-M5/M9/M40/M52/M58).

Most importantly, effective water preservations policies require a shared, long-term vision of the river basin, with consideration of upstream and downstream linkages, communication between municipalities and other stakeholders of the river basin

(Interviews-M2/M9). This vision does not exist. For example, although a mega-sewage treatment plant has been built recently, there is no regional policy to prevent water contamination (see Sect. 8.4.3).

Impacts on Sustainability and Inclusiveness

Despite the emphasis on wastewater discharge on paper, in practice the large majority of wastewater effluents are discharged untreated and contaminated water bodies and agricultural fields beyond the VMB, affecting the health of locals and crops (Interview-M4/M9/M40/M50). Diffuse pollution and a lack of adequate solid waste management further degraded water quality and increased flood risks by clogging pipes and drains (Interviews-M5/M19/M50/M46). This was aggravated by informal urbanization as waste management services are not provided in these areas (Interviews-M5/M51).

The lack of domestic wastewater treatment and contamination from industrial activities, as well as soil subsidence—and the subsequent damage to piping—also threatened groundwater quality (Tortajada 2008; Pina 2011; Spring 2015).

Instruments	Effect on actors	Impact
Wastewater discharge permits	No enforcement and no regional view of water contamination challenges	Env: −− Untreated sewage discharges pose risks to public health and food crops Soc: − Informal settlements are most affected

Redesign See the redesign of water use permits, as similar recommendations are valid.

Climate Change Adaptation Plan and Fund of Mexico City

Design

The SEDEMA coordinates a climate change adaptation plan, the Climate Change Action Programme of Mexico City (PACCM). The plan, first introduced in 2008, had a timeframe for 2014–2020. Its main objective was to increase quality of life and sustainable development with low carbon intensity (SEDEMA 2014). The Environmental Fund for Climate Change (FACC), managed by the SEDEMA, was created in 2015 to implement the plan (Interview-M16). It received contributions from the national and local governments, and from external sources such as international foundations (Interview-M16). The PACCM had three main axes: Mitigation, adaptation and communication/education. The adaptation axis, with a budget of approximately

USD $160 million, consisted of 12 actions, including actions on sustainable housing, water saving systems, greywater reuse and energy saving systems (Interview-M14). Issues such as food security and water availability in adequate quantity and quality are emerging in discussions but not yet integrated into the plans (Interview-M14).

Effects on Actors

Mexico City detailed climate change programme reflected a recognition within the administration of the need to address the climate crisis and its local effects. This was supported by a broader legal framework, including the National Law on Climate Change. However, the plan put much greater emphasis on mitigation than adaptation (Interview-M57). This could be explained by the synergies of mitigation actions with addressing the city's severe air pollution. Nevertheless, the plan adopted ambitious targets for both mitigation and adaptation. Despite the FACC, funding remained limited in comparison to the plan's scope (INECC 2017).

Although all states and municipalities were supposed to prepare similar plans, many had limited or non-existent plans (OECD 2013) (Interview-M16).[5] Despite the shared challenges, adaptation measures were not coordinated with the MVMC's municipalities in Mexico State, beyond the coordination of macro-drainage (Interview-M14). Mexico City's districts were supposed to develop local plans, but differences in resources and political will led to important differences between them (Interview-M15). Some districts still lacked plans (SEDEMA 2017).

Moreover, there are no clear coordination mechanisms between levels of government to ensure a coherent adaptation policy (Interview-M15). The Federal Climate Change Fund could strengthen institutional capacity for adaptation at state and municipal levels, but funds often remained at federal level (Interview-M15). Ultimately, state and municipal governments designed and implemented separate climate change programmes. The result was fragmented policies due to different budgets and political will, with the further marginalization of peri-urban areas. Political will was low in most jurisdictions as there was little pressure from civil society and government administrations (Interview-M15).

Impacts on Sustainability and Inclusiveness

There are no clear evaluation measures in place to verify progress on the PACCM. In fact, it may be almost impossible to isolate the effect of the plan and its fund, as they supported many existing initiatives managed by other actors. This further limited the ability to effectively communicate results to the public (INECC 2017).

[5] As of early 2015, only 2.84% of municipal governments had a climate action plan (Delgado Ramos et al. 2015).

Instruments	Effect on actors	Impact
Climate change adaptation plan and fund	Ambitious targets despite focus on mitigation over adaptation. Disconnect between levels of government and across MVMC hindered coherence	Impacts are difficult to link to the plan, which mainly supported pre-existing initiatives. It raised awareness of the need for adaptation and strengthened capacity in sectoral departments

Redesign

It is necessary to design clearer evaluation methods, that can highlight which actions are effective. As many actions are implemented by sectoral departments or districts, this would require closer communication between entities.

All state had to develop such plans, but in the case of the MVMC this led to fragmented strategies and a loss of potential synergies and economies of scale. Mexico City, Mexico State and Hidalgo State should collaborate on a regional climate change plan (for mitigation and adaptation). This should be compatibilized with their individual plans and should not aim to centralize all efforts. Rather, this could help identify shared challenges and opportunities. This could include a regional approach to conservation and the preservation of ecosystem services that feed the basin's aquifers.

Comparison in Terms of Design

This plan represents a political recognition of the threat of climate change and a comprehensive strategy to address it in Mexico City. Multiple actions and a sizeable budget are dedicated specifically to adaptation, although mitigation is the plan's main focus. The actions are diverse, including actions related to housing, water saving systems, wastewater reuse, and more. The cost–benefit analysis suggests that water-related measures will bring more benefits than costs. The plan includes attempts to better integrate urban and environmental policies, in particular in the Conservation Land, which is crucial for climate regulation and aquifer recharge. In terms of extreme weather events, it only addressed risks from heavy precipitation (as opposed to droughts, forest fires, and other risks). In addition, while the plan recognizes that vulnerability aggravates risks, it does not specifically attempt to address or reduce drivers of vulnerability.

Another drawback is that, as the plan is developed by the Federal District, it does not include the rest of the metropolitan region. Many of the poorest areas, and those most vulnerable to the effects of climate change, are thereby excluded, and urbanization of the green belt is also an important factor aggravating risks for the entire MVMC. Neighbouring states, such as the State of Mexico, have climate action plans, as the General Law on Climate Change specifies that federal entities must develop their own actions. However, the State of Mexico's plan is focused on general guidelines rather than concrete actions, with timeframes and cost estimates.

	Brazil	Mexico
Wastewater discharge permits	0	0
Classification of water bodies	+	+
Climate change plan and fund		+

Comparison in Terms of Effect on Actors

While all federal entities must design a climate change action plan and fund, according to the Federal Climate Change Law, these have been slow to comply. In the case of the Federal District, there is a climate change mitigation and adaptation Law, which makes the implementation of the Plan and Fund mandatory. The higher level of human and financial capacity may further explain its greater progress. However, an effective climate change adaptation strategy was hindered by several elements. The plan lacked data and information to develop a baseline and specific indicators for adaptation, which led to mainly generic guidelines. The plan significantly depended on international funding, and resources are limited in comparison to the ambition of the plan. Many of the actions are the responsibility of different departments and the plan served more to put these existing actions under one umbrella, which did not give additional support to these existing actions. In addition, the plan only concerned the Federal District, and neighbouring municipalities of the MVMC did not have similar plans. Moreover, there was also no vertical coordination mechanism with the Federal government, to ensure coherence of measures. For some actions, isolated, local actions work well, as each district can focus on those that are most relevant. Addressing issues such as air pollution and water-related risks would benefit from a more regional approach as causes and effects spread across borders.

In the case of the MRSP, there is little official recognition of climate change risks and there is no specific strategy at city, metropolitan or basin levels. Climate change concerns could be incorporated into the basin plan and serve to reinforce existing measures that constitute win–win strategies (e.g. drought or flood prevention measures). Overall, climate change was still a taboo subject and many respondents argued that it is not clear whether the water crisis was linked to climate change or climate variability.

References

Brazil (1981) National environmental policy act. Brazil. Available at: http://www.planalto.gov.br/ccivil_03/leis/L6938.htm. Accessed 19 Apr 2018

Brazil (1997) National water law. Available at: http://www.planalto.gov.br/ccivil_03/leis/l9433.htm. Accessed 19 Apr 2018

CEDE (2015) Instrumentos de gestão das águas. Brasília

CONAGUA (2015) Estadísticas del Agua en México. México, D.F.

Conama (2005) Resolução n 357, 18 de março de 2005, Diário Oficial. doi: n° 053, de 18/03/2005

Delgado Ramos GC, de Luca Zuria A, Vázquez Zentella V (2015) Adaptación y Mitigación Urbana del Cambio Climático en México. Universidad Nacional Autónoma de México, México, D.F. Available at: http://computo.ceiich.unam.mx/webceiich/docs/libro/Adaptacion-web1.pdf. Accessed 1 May 2018

INECC (2017) Evaluación Estratégica del Programa Especial de Cambio Climático 2014–2018. Available at: https://www.gob.mx/cms/uploads/attachment/file/261388/Informe__evaluacion_PECC_final_limpio_1_.pdf. Accessed 22 May 2019

MMA (2010) Relatório técnico parcial 7. Brasília

NWL (2004) LEY DE AGUAS NACIONALES. Available at: http://www.diputados.gob.mx/LeyesBiblio/pdf/16_240316.pdf. Accessed 24 Sept 2017

OECD (2013) Making water reform happen in Mexico. In: OECD studies on water. Paris. https://doi.org/10.1787/9789264187894-en

Pina P (2011) Addressing water crisis in Mexico City: a political and institutional perspective

São Paulo (Estado) (2017) Plano estadual de recursos hídricos (PERH): 2016/2019. CRH/CORHI, São Paulo. https://doi.org/10.1902/jop.2010.104002

SEDEMA (2014) Programa de Acción Climática: Ciudad de México 2014–2020. Mexico City, D.F.

SEDEMA (2017) Publicados, seis Programas de Acción Climática Delegacionales. Available at: http://www.sedema.cdmx.gob.mx/comunicacion/nota/publicados-seis-programas-de-accion-climatica-delegacionales. Accessed 15 May 2018

SOS Mata Atlântica (2017) Poluir rios custa caro. Available at: https://www.sosma.org.br/artigo/poluir-rios-custa-caro/. Accessed 6 May 2019

Spring ÚO (2015) Managing water resources in Mexico in the context of climate change. In: Shrestha S et al (eds) Managing water resources under climate uncertainty. Springer. https://doi.org/10.1007/978-3-319-10467-6_18

Tortajada C (2008) Challenges and realities of water management of megacities: the case of Mexico City metropolitan area. J Int Aff 61(2):147–166

Annex H
Water Tariffs

MRSP: SABESP's Tariffs

		Water	Sewage
Normal residential tariff			
0–10	Reais/month	26.18	26.18
11–20	Reais/m^3	4.1	4.1
21–50	Reais/m^3	10.23	10.23
Above 50	Reais/m^3	11.27	11.27
Social tariff			
0–10	Reais/month	8.88	8.88
11–20	Reais/m^3	1.53	1.53
21–30	Reais/m^3	5.43	5.43
31–50	Reais/m^3	7.74	7.74
Above 50	Reais/m^3	8.55	8.55
Favela tariff			
0–10	Reais/month	6.77	6.77
11–20	Reais/m^3	0.77	0.77
21–30	Reais/m^3	2.56	2.56
31–50	Reais/m^3	7.74	7.74
Above 50	Reais/m^3	8.55	8.55

Source ARSESP (2019)

F. van den Brandeler, *Scalar Mismatches in Metropolitan Water Governance*, Water Governance - Concepts, Methods, and Practice,
https://doi.org/10.1007/978-3-031-08061-6

Mexico City: SACMEX's Tariffs

		Minimum rate	Additional fee per m^3			Minimum rate	Additional fee per m^3
Normal residential tariff				Middle residential tariff			
0–15	Pesos	534.07	0	0–15	Pesos	174.16	0
<15–20	Pesos	534.07	35.61	<15–20	Pesos	174.16	22.45
<20–30	Pesos	712.08	35.61	<20–30	Pesos	286.37	24.57
<30–40	Pesos	1068.11	35.61	<30–40	Pesos	532.05	29.09
<40–50	Pesos	1424.16	35.61	<40–50	Pesos	822.87	31.35
<50–70	Pesos	1780.19	43.33	<50–70	Pesos	1136.27	34.65
<70–90	Pesos	2647.07	47.22	<70–90	Pesos	1829.47	46.45
<90–120	Pesos	3591.36	62.70	<90–120	Pesos	2758.24	61.94
<120	Pesos	5472.14	97.53	<120	Pesos	4616.31	97.53
Popular residential tariff				High residential tariff			
0–15	Pesos	46.44	0	0–15	Pesos	208.98	0
<15–20	Pesos	46.44	4.02	<15–20	Pesos	208.98	23.52
<20–30	Pesos	66.56	6.38	<20–30	Pesos	326.57	26.30
<30–40	Pesos	130.30	13.02	<30–40	Pesos	598.51	31.21
<40–50	Pesos	260.44	18.99	<40–50	Pesos	901.58	33.34
<50–70	Pesos	450.34	28.28	<50–70	Pesos	1234.90	35.71
<70–90	Pesos	1016.05	35.86	<70–90	Pesos	1949.34	46.45
<90–120	Pesos	1733.15	61.94	<90–120	Pesos	2878.11	61.94
<120	Pesos	3591.22	97.53	<120	Pesos	4736.19	97.53
Low residential tariff							
0–15	Pesos	52.64	0				
<15–20	Pesos	52.64	8.98				
<20–30	Pesos	97.52	12.09				
<30–40	Pesos	218.37	17.13				
<40–50	Pesos	389.68	24.04				
<50–70	Pesos	630.04	30.40				
<70–90	Pesos	1238.25	37.19				
<90–120	Pesos	1981.90	61.94				
<120	Pesos	3839.98	97.53				

Source Mexico City Government (2020)

References

ARSESP (2019) Tarifas de Saneamento. ARSESP. Available at: http://www.arsesp.sp.gov.br/SitePages/informacoes-economico-financeiras/tarifas-saneamento.aspx. Accessed 19 July 2019

Mexico City Government (2020) ADMINISTRACIÓN PÚBLICA DE LA CIUDAD DE MÉXICO. Official Gazette of Mexico City, Mexico City. Available at: https://www.sacmex.cdmx.gob.mx/storage/app/media/tarifas/Tarifas_Agua_Art_172_2021_Pweb.pdf

The manufacturer's authorised representative in the EU is Springer Nature Customer Service Centre GmbH, Europaplatz 3, 69115 Heidelberg, Germany. If you have any concerns regarding our products, please contact ProductSafety@springernature.com

Printed and bound by CPI Group (UK) Ltd, Croydon, CR0 4YY
12/08/2026
02194522-0003